The Naked Neuron

Evolution and the Languages
of the Body and Brain

The Naked Neuron

Evolution and the Languages of the Body and Brain

Dr. R. Joseph

Plenum Press • New York and London

Library of Congress Cataloging-in-Publication Data

Joseph, Rhawn.
    The naked neuron : evolution and the languages of the body and
brain / R. Joseph.
        p.   cm.
    Includes bibliographical references and index.
    ISBN 0-306-44510-7
    1. Neuropsychology.   2. Psycholinguistics.   3. Neurolinguistics.
4. Brain--Evolution.   I. Title.
    [DNLM: 1. Nonverbal Communication.   2. Verbal Behavior.
3. Evolution.   4. Memory--physiology.   BF 637.N66 J83n 1993]
QP360.J664   1993
612.8'2--dc20
DNLM/DLC
for Library of Congress   93-9076
                                                              CIP

ISBN 0-306-44510-7

© 1993 R. Joseph
Plenum Press is a division of Plenum Publishing Corporation
233 Spring Street, New York, N.Y. 10013

Printed in the United States of America

Contents

v

NEURODYNAMICS

. . . for once ye eat of the fruit of the tree of knowledge . . .
thy eyes will be opened . . .
                 and ye shall see as gods . . .
                 and ye shall be as gods . . .
                 knowing good and evil . . .

In the Beginning . . .

. . . there was life.

Soon after the Earth was formed some 4.6 billion years ago, life began to proliferate within the crust of the Earth and throughout the nutrient-rich primeval ocean and salty seas. Sun-worshiping single-celled organisms quickly gained dominion over the planet and probably practiced a crude form of photosynthesis such that ultraviolet and other wavelengths of light were absorbed, which enabled them to survive.[1] Photosynthesis and sensitivity to the rays of the sun may well have provided the basis for the first forms of communion—a dialogue between planets and stars.

It is believed that these single-celled organisms contained DNA, probably a single strand washing about within the protoplasm of the cell. Hence, within each cell was a storehouse of information and a collection of ancient memories, as well as life plans for the future. The first cells, blessed with a struggling intelligence and the sequential language possessed by DNA,[2] were able to communicate within themselves. Moreover, their permeable cellular membranes enabled them to exchange material and information with the external environment as well as extract energy from the sun.

Scientists have referred to these first unicellular organisms as Prokaryotes.[3] It is supposedly from these creatures that multicelled organisms, including birds, bees, plants, dogs, cats, and even human beings, have evolved. Certainly the evidence in support of this notion regarding evolution per se is open to a number of explanations and interpretations. Nevertheless, if true, the ability to communicate and the languages of the body and the brain would then have similar origins.

Like their single-celled cousins, multicellular life forms may have been derived from the same cosmic soup that was the cosmos, prior to and after the "Big Bang" creation of *this* Universe some 18 or so billion years ago. That is, the "Big Bang"[4] may well be part of a repetitive cycle of expansion and contraction of all the matter and energy within this Universe. Instead of one "Big Bang," there instead may have been countless ones, a consequence of a continual pattern of waxing and waning which repeats itself endlessly and forever. When considered in regard to eternity and infinity, 18 billion years is not a very long time, and the likelihood that prior to the "Big Bang" there was nothing seems terribly unlikely. The origins of life may extend interminably into the long ago.

Certainly, given that this Universe had already been in existence for almost 12 billion years before the Earth was formed adds immeasurably to the likelihood that life has been around for quite a while, and predated the creation of our solar system. Thus, rather than having its source on this planet, instead the Earth may have been seeded by celestial clouds of living protoplasm or have been repeatedly struck by life-bearing meteors soon and long after its creation as it drifted through the ancient swirling Universe.[5] The seeds of life may have multiple sources.

Of course, many scientists hold to the rather outdated notion that the planet Earth is somehow isolated from the rest of the cosmos. However, the Earth is just one small island among many within the ocean of space with tons of debris, including blocks of ice as big as a house, washing to shore on a regular basis. That some of this debris may contain life does not seem all that unreasonable. If life were to appear on an Earth-based desert island, say in the Pacific or Atlantic Ocean, does it not seem equally reasonable to assume that it washed to shore as well?

Regardless of what theory one ascribes to, it also seems unlikely that all life is derived from one single Earth-based cell. Rather, even if life did have its origins on this planet, rather than one cell suddenly springing to life, there may have been billions, many of which may have independently given rise to a panoply of multicellular organisms.

If multicellular creatures are derived from their single-celled cousins, it is also likely that a variety of single-celled creatures underwent multicellular metamorphosis at different geological time periods and under various environmental conditions so as to create a multiplicity of life forms, that, again, have no identical single-cell origins. Hence, it is possible that some if not all five of the different kingdoms of life (i.e., bacteria, eukaryotes, plants, fungi, and animals) are in fact descendants of different multicellular and single-cellular organisms and are thus almost wholly unrelated except in regard to an origin which may predate the Big Bang. If this were the case, one might suspect that similarities in DNA are a consequence of this singular extraterrestrial origin, or to similarities in the adaptation of life to this planet. Of course, there are several reasonable explanations regarding the origins of life, none of which will be argued here.

Eukaryotes, Cooperation, and Communication

When the single-celled prokaryotes ruled the planet, there was no sex.[6] These cells merely divided and produced two identical copies of themselves and their DNA. It was probably not until about 1 billion years ago that sex first became a widespread form of reproduction, and this was first practiced by oxygen-breathing multicellular creatures who in turn became capable of indulging in more complex modes of communication.

In contrast to single-celled organisms, multicellular creatures contain eukaryotes,[7] that is, "the true kernel." Within each eukaryote can be found a nucleus which contains the double strand of chromosomal DNA. Thus, these cells were not only more complex but contained many times the memory, intelligence, planning skills, and capacity to communicate as compared to single-cell creatures.

The Evolution of Cooperative Communication

It has been postulated by some that the multicellular eukaryotes evolved from the combination of prokaryotic cells which had begun to live together in a cooperative fashion, thus forming a single organism. However, the ability to cooperate is dependent, at least in part, on the capacity to communicate not just within the cell and its external environment, but in a meaningful fashion between different cells. To strive for and achieve this degree of cooperation and complexity, that is, for two single cells to form a single multicellular creature, required that cellular language and intelligence be nurtured and further refined.

It is possible that this initial means of communication took place via electrical discharges[8] and that these cooperative efforts were not solely the result of chance or random occurrences. For example, it is well known that the surface membrane of a living cell is maintained by electrical forces and the interactions between macromolecules. Via their electrical charge, positively charged areas on one molecule are attracted to the negatively charged surface of a different molecule. Via alterations in these electromagnetic forces, molecular cellular structures are able to approach, position themselves, interact, and even exchange material and then separate. In that the most primitive means of true nerve cell communication also occurs via electrical and electrochemical interactions, it is possible that these first cooperative efforts were the harbingers of advances in communication yet to come; that is, the evolution of the neuron.

The First Single-Celled Nations

Certainly in addition to life, the first cells also maintained an ability to sense, in some fashion, their surroundings and were able to selectively attend to certain signals which were relevant to their continued existence. Indeed, in the beginning there was not only life, but the ability to communicate and eventually to cooperate to obtain ends that were highly beneficial first to the individual and then to the evolving social group. These capacities resulted in the formation of the first single-celled nations as well as the first pre–Babelic Towers billions of years ago.

For example, an examination of the fossil record reveals the traces of entire colonies of single-cell organisms in rocks more than 3 billion years old. Evidence for these same single-celled cooperative efforts can also be found in ancient layers of pillarlike billion-year-old rocks stacked in towering sheets called stromatolites. Indeed, living stromatolites can be found on the western coast of Australia.

Given these creative cooperative accomplishments, it is apparent that the ability of these cells to not only sense but to communicate is surprisingly complex given their unconscious origins, cellular simplicity, and great antiquity.

Multiple Languages of the Body and the Brain

Multicellular Similarities in Communication

As noted, it is certainly possible that all subsequent creatures may well have evolved from different multiple or single-celled organisms. However, it is possible that the array of multicelled creatures which eventually gave rise to the animal kingdom may have been more or less similar if not identical, thus providing all animals with a possible common heritage. Certainly if not by a similar cellular genesis, we are all linked by the manner of our creation. We are star dust, star light, and radiant energy, the residue of the romance between planets and stars. We still eat of the sun (by way of plants and direct exposure) in order to derive nourishment and energy, and in order to survive.

It is very likely that most or all members of the Animal Kingdom are linked not only by supposedly similar multicellular origins but also by the ability to exchange information and thus communicate, which is at least 3 billion years old. The ability to communicate did not evolve per se, however, but rather new language systems were developed in correspondence with the evolutionary metamorphosis of the body and the brain. The new systems did not replace the old but merely added a new luxuriant layer to the process as well as new body and brain tissue to mediate and control it.

Collective Limbic Origins

Possibly because humans emerged last and are the presumed beneficiaries of past evolutionary development, they also possess many of the same neural components which enabled other earlier-forming species and members of the Animal Kingdom of Life to proliferate and communicate. For example, within the belly of the human brain can be found the old cortical, olfactory-based motor systems and some of the same limbic system nuclei (which mediate sex, hunger, and emotion) and visually sensitive brain tissue (the optic lobe, or tectum/colliculi of the midbrain) as that possessed by sharks,[9] creatures who first swam the salty sea some 450 million years ago.

In this manner we and most other animals are all linked due to mutual possession of this collectively shared unconscious and emotionally charged limbic brain tissue.[10] We are linked by common bonds of sensing, feeling, and communicating at a variety of cellular levels and in a multiple manner.

The Multiplicity of Old and New Languages

There are multiple systems of language (e.g., olfaction, touch, gesture, sound, etc.), some of which have seemingly become hierarchically more complex as the host organisms evolved and increased in cellular complexity. However, the older language systems continue to function quite effectively as well and are often expressed simultaneously. For example, although we employ spoken language, a recent acquisition, we often accompany it by phylogenetically older systems such as gestures and other movements which not only emphasize various points but speak volumes on their own. These different language systems, can, in fact, act independently and divergently or may act in harmony. By attending to these different systems one can often determine, for example, if the speaker is sincere, a liar, or a sincere liar.

The expression of these older systems of communication enables our loved ones and our pets to recognize our moods and intentions, and to sometimes guess what we are going to say before we say it. This is why their own posture and movements, as well as the behaviors of a host of other creatures, are highly informative and speak volumes to a sensitive observer. Different animals possess certain similar cellular systems that perceive, process, express, and communicate certain forms of information in a similar manner.

The manner in which an insect and a human and many other members of the Animal Kingdom of Life communicate and gather information have many striking similarities. A dancer gyrating to a fast rhythmic and driving beat can communicate volumes of information as to his or her level of arousal, intentions, desires, feelings of confidence, and degree of alcohol intoxication, and so, too, can a dancing bee.[11] Moreover, both can do so in the absence of the spoken word.

A dog in heat will attract numerous sexually excited male admirers, via her scent alone, whereas a woman may apply perfumes derived from the anal glands of various animals and will hormonally secrete a pheromonal–olfactory molecule which when sniffed by her male companions can also induce arousal and attraction. Olfaction is a form of communication and has a language all its own.

Human beings as well as other members of the Animal Kingdom are capable of communicating a veritable treasure house of information, including details best kept secret for self-preservation. This is accomplished by smell, smile, gestures, posture, movement, and touch, all in the absence of

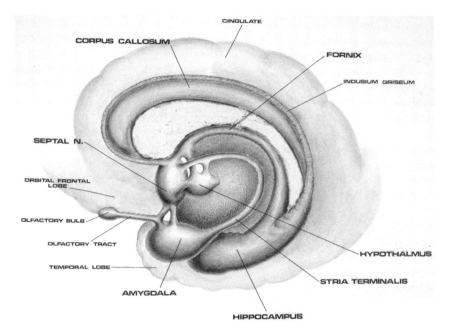

Figure 1. The limbic system, depicting the amygdala, hippocampus, septal nuclei, hypothalamus, and the more recently evolved transitional limbic-neocortical, four-layered cingulate gyrus.

spoken language. This is because each system of communication is in fact a *language* of signs, symbols, or molecular codes that predate and do not rely on spoken words for comprehension. All were well established long before human beings and our *Homo* and Australopithecine cousins first lurched upon the scene some 5 million years ago.

Miscommunication between the Old and New Brains

Although we are able to communicate, often unconsciously, with other creatures as we possess much of the same old brain tissue, we also utilize new brain tissue and new language systems they do not possess and cannot comprehend. Most creatures are lacking the most recently acquired six- to seven-layered neocortex (new brain) which covers the old limbic brain like a

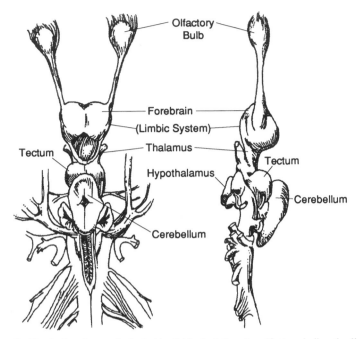

Figure 2. The brain of a small shark (dogfish), depicting the olfactory bulbs, the limbic forebrain which houses the two cortical-layered amygdala, hippocampus, and septal nuclei. The thalamus, visual and primordial auditory tectum (or colliculi), cerebellum, and the brain stem and spinal nerves are depicted.

cloak. Even those with considerable neocortex, such as our fellow primates, do not possess a brain area unique to humans; that is, the angular gyrus, which is essential in the production of complex spoken language.[12]

Although the new brain (the neocortex) has in part developed to serve the needs of the old limbic brain, neocortical tissues often have difficulties understanding the impulses or languages originating in these more ancient regions, although both are located within the same brain. However, the converse is also true. The ancient two-layered limbic cortical tissue has difficulty communicating with and comprehending data processed with the new brain as they speak different languages.[13] What occurs in one brain area is sometimes unintelligible to another as each has its own specializations and limitations.[14] Some forms of information which are being processed or expressed by old limbic brain tissue may not be recognized or compre-

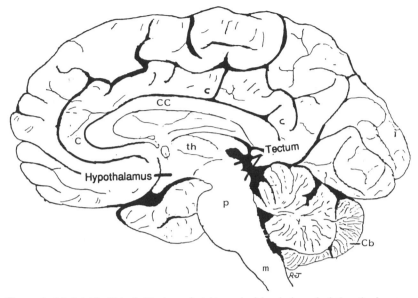

Figure 3. Medial ("split-brain") view of right cerebral hemisphere depicting the human midbrain including the visual and auditory colliculi (or tectum) and thalamus (th). The brain stem (p = pons, m ⁻ medulla), cerebellum (Cb), corpus callosum (CC), cingulate gyrus (C), and hypothalamus are also indicated.

hended by the new brain, and vice versa, even when such messages originate and are processed within a single individual's head.[15]

Right and Left Cerebral Miscommunication

Due to the tremendous complexity and specialization within the new brain, however, similar problems in comprehension occur even within the neocortex; difficulties which are exacerbated due to the limitations in information transfer between the right and left half of the brain.[16] That is, the right and left half of the brain utilize different means of communication and sometimes rely on different language systems, as they are concerned with different aspects of experience.

For example, the left half of the brain (in the majority of humans) is very efficient at processing spoken language but has great difficulty processing social or emotional sounds, which are usually under the guidance of the

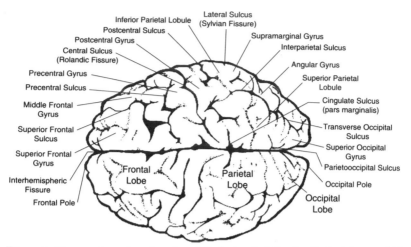

Figure 4. The superior surface, top view of the brain, depicting longitudinal fissure which divides the brain into a right and left cerebral hemisphere.

right half of the brain, which is deficient in language skills. Hence, some forms of information are processed only by one versus the opposite half of the brain. And when they attempt to exchange impressions and transfer occurs, there sometimes result errors in translation.

In some instances, there is no transfer whatsoever such that one half of the brain may well be completely in the dark as to what is occurring in the other.[17] However, both halves of the brain are also quite capable of sharing a considerable degree of information as well. It is due to this acquisition of neocortical tissue and the subsequent differential organization and specialization of the two cerebral hemispheres that linguistic consciousness, the unconscious, creativity, the arts, and the experience of neurotic obsessions, conflicts, and fears have come into being.

The Neuron

DNA

Admittedly, it is possible within the Animal Kingdom that some species may have different multicellular origins. Nevertheless, it is apparent that despite the diversity within the Animal Kingdom, which ranges from sponges to insects to humans, that all are linked in multiple ways, including the fact that all possess DNA.

DNA has long served as the intellectual center of all cells as well as the memory center which, when combined and interchanged with the DNA of a second cell, makes possible the creation of yet a third organism. This third cell, being in possession of recombined DNA, is thus also the recipient of some of the genetic plans and memories of its parents.[18]

Be it worm or human, the tissues of the body, perhaps like our ancestral multicellular cousins, are derived from a single cell—the sexually fertilized ovum of the female. This primal cell quickly undergoes a transformation called meiosis and divides; this is a process which is repeated by all its daughter cells thus creating a complex multicellular organism. Each daughter cell, based on the DNA instructions possessed within each of its nuclei, in turn quickly develops an identity and is sent to a specific location within the developing body to form certain interconnections and to perform specific functions. This is made possible through the highly complex yet ancient languages spoken by the genes.

The Nerve

About 700 million years ago, long after the invention of sex, a cellular metamorphosis of paramount importance resulted in the creation of a completely unique type of cell, a nerve cell—the neuron. Nerve cells, or rather, neurons, are the result of this process of division and differentiation. However, rather than divide in order to pass on complex memories and life plans, neurons are able to think and in fact create memories and plans as well as communicate this information to a second neuron. Where before cells might emit an electrical discharge or secrete or detect chemically coded messages in a global and diffuse manner, nerve cells eventually developed highly specialized nerve endings and nerve cables (called dendrites and axons) for reception versus transmission of electrical and chemical messages.

When the first neuron was born, the creature who possessed it now possessed the first cellular *brain*. This brain, however, was in the form of a single cell that was fashioned so as to process, store, and express a billion times more information than any former sensory or motor cell could muster. Over the course of evolution, as the number of secreting and transmitting nerve cells that a creature possessed increased, a network of interlinked neurons, called the "nerve net," was soon fashioned.[19] With the establishment of the nerve net, the organism could now behave as a complete unit, in a controlled, highly coordinated, and directed manner as different regions of the body could now communicate together almost simultaneously.

Although multicellular organisms without neurons show intelligence, learning, memory, and the ability to communicate,[20] the development of the first neuron and then the nerve net gave all subsequent life forms that possessed them tremendously advanced powers of mental and sensory acuity. Eventually, as the size and complexity of the nerve net increased, the first true brains came to be fashioned.

The Neuronal Kingdom of Life

Only members of the Animal Kingdom of Life possess neurons. Although intelligent and capable of communication and interaction, the members of the remaining kingdoms of life (e.g., bacteria, algae, mosses, fungi, and plants) are completely without such neural components. They have no brains.

However, not all members of the Animal Kingdom are neuronally blessed. Indeed, the most primitive and ancient animals, such as sponges (which are a step below coelenterates; e.g., jellyfish, sea anemones, and Hydrozoa), have only a very primitive organization of nervelike tissue, much of which is concentrated within and around their external orifices and pore sphincters through which sea water freely circulates.[21] The outer sensory surface of their body is very poorly differentiated and is without specialized receptors, and no nerves or neurons are found. Moreover, in that they spend their whole life fastened to the mud, sponges are not capable of locomotion, other than very slow, generalized, and isolated reflexive movements which are made possible via motor cells which respond directly to stimulation.

Nevertheless, it appears that sponges are capable of a very slow protoplasmic form of information transmission. For example, if a finger of a sponge is severed, after a few minutes, tissue in distant regions will begin to contract.[22] Hence, sponges show a very primitive type of neuroid transmission from which true nervous conduction may have evolved. Sponges, in fact, contain a number of neurosecretory substances including recognized neurotransmitters such as serotonin, noradrenalin, adrenalin, and acetylcholinesterase.[23] These chemical transmitters are found in high concentrations within the human brain and are highly involved in memory, emotion, and movement.

Planaria, Flatworms, and the Memories of the Dead

As noted, it was not until around 700 million years ago that the first true neurons appeared. This coincided with the appearance of worms and other soft forms of complex life which began to wiggle and swim the seas; that is, the coelenterates. The more advanced coelenterates, such as planaria and flatworms are carnivores and will eat any creature that comes to be grasped by their anterior head region or which can be held by adhesive organs which they use for attaching to various substrates. These creatures are very sensitive to the chemical presence of food and will locate it and proceed directly to it in a straight line once it is sensed via the chemoreceptor sensory organs located in ciliated pits within the head area. Hence, they are much more complex and intelligent than sponges, which, as noted, are a step below the coelenterates.

The coelenterates (e.g., planaria and flatworms), however, are the lowest of the metazoa and, in fact, are the most primitive members of the Animal Kingdom to possess a "nervous system." This first nervous system is composed of distinct sensory versus motor neurons as well as a third type of cell called a ganglion cell (or protoneuron).[24] However, these creatures, too, were without true brains. Instead, they possessed tiny neural ganglia, which are essentially a colony or cluster of similarly functioning nerve cells located in the anterior head region.

Planaria are very sensitive to light and in fact can quickly learn to avoid it. They also possess cephalic eyes which contain retinal cells that have long axons that terminate in a ganglion of visually responsive neurons.[25] Moreover, they are also responsive to changes in temperature and are capable of not only learning in response to electric shock and light,[26] but are capable of passing on their memories and learning experiences to others. In fact, untrained flatworms who eat the ground-up brains of former cousins who had been trained to perform a specific task, quickly demonstrate an equal mastery.[27] One might assume that 700 million years ago these or like-minded creatures were similarly endowed and likewise capable of passing on ancestral memories not just through sex and by their DNA but via the ingestion of the brains and neurons of their friends, siblings, and fathers or mothers.

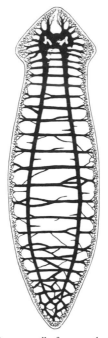

Figure 5. The nervous system, "nerve net" of worm, the triclad turbellarian. The nerve net forms two large pairs of ganglia in the anterior head region. A bundle of axons forms a commisure connecting the two ganglionic lobes and the dual nerve cords extending the length of the body. Eye cups and optic nerve are depicted in the anterior ganglia. From T. L. Lentz. *Primitive Nervous Systems*. New Haven: Yale University Press, 1968.

Naked Neurons

The first neurons, however, were not well developed. Indeed, it is likely that these first neural cells were without axons and were probably also without dendrites.[28] Instead, these first primitive neurons simply secreted electrical and chemical substances which acted on other cells in a generalized manner. Later, neurons developed the ability to grow a single, long, thin axonal transmission fiber through which these same electrochemicals could be selectively secreted to a second neuron. The dendrite of this second neuron in turn would absorb these chemicals molecule by molecule via a selective receptor surface located along its terminal junctions.

It was only long after the development of axon transmission fibers that

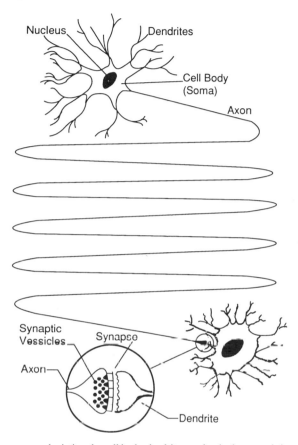

Figure 6. The neuron, depicting the cell body, dendrites, and a single transmitting axon which travels sometimes considerable distances to make contact with the dendrites of other neurons at a gap junction, the synapse. The transmitting axons contain tiny transmitter-filled bubbles called vesicles. When sufficiently activated, the axon will release neurotransmitter substances into the synaptic space, where they make contact with the dendritic receptor surface.

insulation in the form of myelin came to be invented. Myelin is a fatty cell which wraps itself around the axon and through which it derives its sustenance. It is the fatty myelin sheath which prevents information leakage as data are transmitted from one neuron to the next. These first neurons, that is those without myelin, were thus essentially "naked," and were thus not very efficient and transmission resulted in considerable loss of information and even miscommunication.

"Naked neurons," that is, cells with axonal cables that are without an insulating myelin coat, are potential causes of communication failure and information leakage even within the human brain; at least as compared to the efficiency of the more evolutionarily advanced myelin-coated neuron. Just as a stripped electric wire is less efficient as compared to one with insulation. Even so, the naked neuron was a monumental and literally world-altering development as neurons are the building blocks of what would become the brain.

In modern-day worms and, one might presume, among ancient worms, the cell bodies of these original neurons are essentially externally located outside the body and were quite generalized although especially responsive to light as well as olfactory and pheromonal messages. It was only over the course of evolutionary metamorphosis and the ensuing eons of time that these cell bodies began to migrate inward and to eventually collect together within the body so as to form collections of nuclei and the various lobes of the brain. By the time that the first vertebrates and fish begin to swim the oceans, around 500 million years ago, the first primitive lobes of the brain had also become fashioned through the collectivization of these neural ganglia.

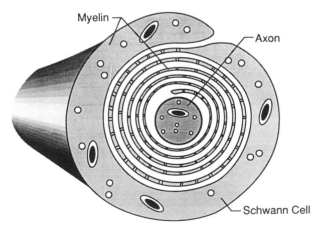

Figure 7. The myelin cell forms a fatty sheath of insulation as it coils around the axon cylinder.

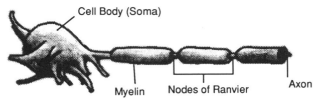

Figure 8. The myelin cell coils around the length of the axon. The space between each myelin cell is called a node of Ranvier. The myelin is a form of insulation and serves to prevent information leakage as it travels down the axon.

The Olfactory Forebrain

Be it worm, fish, or human, externally located neuronal cell bodies are similarly still present, although in humans they have become concentrated within the nose. Like the sensory neurons of more ancient creatures, these nasal-located neurons perform the same vital functions. They act to analyze the pheromonal, olfactory, and chemical nature of the environment for clues concerning sex, food, the weather, or whatever.

Over the course of evolution, one group of these primal sensory cells formed an internally located ganglia of like-minded cells which is called the olfactory lobe; that is, a rounded projection of what would become part of the forebrain. Yet another group of sensory cells which act to analyze visual input formed an optic lobe (the tectum or colliculi) in what is called the midbrain.

By time true mammals appeared, about 100 million years ago, the forebrain had continued to expand and evolved into the right and left cerebral hemispheres. Indeed, it is via the expansion and axonal–dendritic interconnections of these first two ancient lobes (the visual and olfactory) that the modern brain came to be fashioned.[29]

However, it is from the olfactory system that the ancient limbic lobe of the brain evolved, a series of structures that encircle the brain stem (which controls heart rate and breathing) and midbrain (which receives visual and auditory information). The limbic system is concerned almost exclusively with feeding, fornicating, fighting, or fleeing.[30] Indeed, by 450,000 years ago, the first sharks had acquired a limbic system, which they, like modern humans, still possess today.

Cortex

In fact, the olfactory–limbic system eventually gave rise to what would become the first layers of cortex.[31] That is, as more and more nerve cells accumulated for the purposes of analyzing olfactory–pheromonal messages, they began to ball up and form layers. Brains that consist of only single thin sheets of neurons are thus without cortex. It is only when similarly functioning cells that serve a similar purpose begin to form layers that cortex appears.

This first limbic cortical tissue essentially consisted of one layer of receiving cells, above which was superimposed a second layer of outgoing (or effector) cells. Moreover, this same olfactory–limbic cortical tissue, by time the first sharks and fish appeared, also gave rise to the first true motor cortex (the basal ganglia which controls movement), as well as what would become the two cerebral hemispheres which in humans completely encase the rest of the brain. Hence, much of the brain evolved from the olfactory–limbic system.[32]

The Diversity of Life

It was about half a billion years ago that many different types of vertebrates, including jawless fish, began to swim the ocean and plants began to invade the land. There then followed an explosion of life with plants

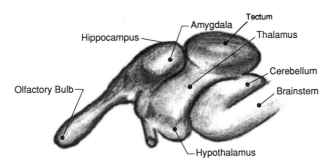

Figure 9. The left lateral, side view of the lizard brain. The olfactory bulb differentiates into the amygdala and hippocampus and the rest of the forebrain. The midbrain which consists of the tectum is connected to the forebrain by the thalamus and the hypothalamus.

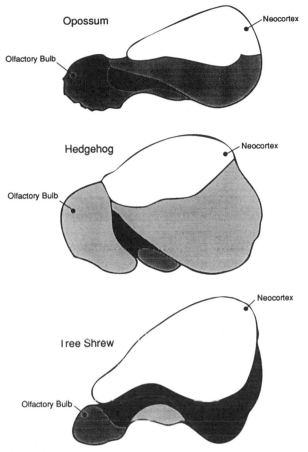

Figure 10. The brains of two ancient mammals and that of a primitive protoprimate, the tree shrew. The neocortex expands and increasingly enshrouds the forebrain and limbic system in the course of the transition from primitive to more advanced mammals. The shroud of neocortex progressively expands to cover the forebrain and the olfactory bulbs.

rapidly invading and devouring the Earth. Seeds were invented and great forests grew (some 350 million years ago during a time period referred to as Carboniferous) only to die and later become the fields of coal that sustain much of our energy needs today.

There was a tremendous proliferation of not just plants but various tiny

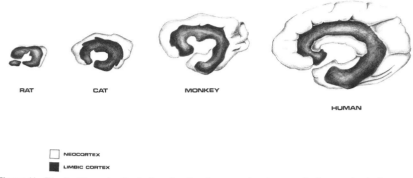

RAT   CAT   MONKEY

HUMAN

☐ NEOCORTEX
■ LIMBIC CORTEX

Figure 11. The limbic cortex (including the cingulate gyrus) and neocortical expansion in four species. Note similarities in size, location, and extent and progressive expansion of the neocortical mantle. From R. Joseph. *Neuropsychology, Neuropsychiatry, and Behavioral Neurology*. New York: Plenum Press, 1990.

aquatic insects who followed, feeding upon the luxuriant varieties of blossoming and blooming life that led the way. By 350 million years ago, not only insects but a variety of amphibians, some up to 15 feet long, were swarming over the Earth. Over the next 50 million years, the first reptiles appeared which were followed by the reptomammals (the therapsids) and then the first tiny dinosaurs around 250 million years ago.[33] Like so many first creatures, many of these same types of insects, amphibians, and reptiles live even today, although the great majority died out and became extinct long ago.

When animals first ventured forth from the sea, since there were so few competitors, these creatures proliferated rapidly, and a tremendous diversity of forms was unleashed upon the planet and multiplied. Presumably such diversity was a consequence not only of adaptation to selective and environmental pressures, but of molecular DNA competition and metamorphosis, a genetic process referred to by many as "mutation."

Cortical Competition between Mammal and Dinosaur

Over the course of evolution some creatures became more adept at gathering and consuming plants, others at preying upon the herbivore herds and even eating their own cousins. Some took to the sky, and others returned

Figure 12. One of the first large land-based predators, the fin-backed Dimetrodon, a very primitive and archaic mammallike reptile. The fins may have acted as a solar panel which aided in thermal regulation. The three potential prey are amphibians, Erypos. Reproduced with permission from G. S. Paul. *Predatory Dinosaurs of the World*. New York: Touchstone, 1988.

Figure 13. An early dinosaur, *Herrerasaurus ischigualastensis*, and its young being harassed by a Theocondont, a protodinosaur which was still half lizard. Reproduced with permission from G. S. Paul. *Predatory Dinosaurs of the World*. New York: Touchstone, 1988.

to the sea. However, all increased their capacity to gather information from their environment and to share it with similar creatures. The development of the first complex languages and the capacity to communicate and comprehend was greatly enhanced.

The brain also evolved and underwent continual environmentally assisted metamorphosis. A shark, fish, amphibian, reptile, reptomammal (i.e., the therapsids), dinosaur, and bird possess only very ancient two-layered cortical motor centers and limbic-system tissue. Once the true mammals appeared (about 100 million years ago) new layers of cortex began to spread and enshroud the old brain as it expanded layer by layer until finally forming the cerebral hemispheres and becoming seven layers thick. This new six- to seven-layered cortex is called neocortex.

Indeed, it was probably the development of this new brain and mammalian neocortex (which has six to seven layers, in contrast to the old cortex, which has only two) which gave mammals a sufficient competitive intellectual edge so as to take advantage of the meteor-induced catastrophe which presumably wiped out most of the large land-based creatures, the dinosaurs, some 65 million years ago.[34]

Figure 14. Several Chasmosaurus confront a Tyrannosaurus torosus. Reproduced with permission from G. S. Paul. *Predatory Dinosaurs of the World.* New York: Touchstone, 1988.

If not for the development of mammalian neocortex and the tremendous intellectual powers that it provided, the dinosaurs might well have been able to recover from this cataclysm, just as they had bounced back following previous catastrophes induced by meteors striking the planet. For example, it appears that a similar meteor-induced catastrophe occurred 225 million

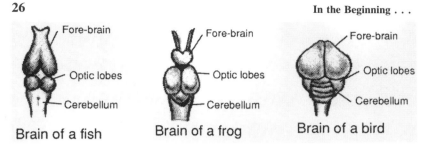

Figure 15. The brain of a fish, frog, and bird, with the limbic forebrain, the optic lobes (tectum/colliculi) of the midbrain, and cerebellum depicted.

years ago and resulted in the death of over 90% of life on Earth due to the tremendous climatic and environmental changes it wrought.[35] It was this event which may have killed off many of the therapsids (reptomammals), thus giving dinosaurs a competitive advantage.

Following this most recent catastrophe 65 million years ago, with the exception of those raptors who took to the sky, that is, modern-day birds, most of the small terrestrial dinosaurs who remained were probably simply no match for the equally small but much more intelligent mammals who killed them off.

On the other hand, adding to this confluence of events that gave rise to the fall of the dinosaurs and the ensuing dominance of the mammals is the possibility that just as plants, animals, humans, and even stars have life spans, at the end of which they die, that species, too, have life spans. Perhaps the life span of the dinosaurs as a species may have been 165 million years, at the end of which they simply died out regardless of external and other contributing circumstances.

The Human Brain

Freed of their oppressors and killing off and eating up the rest, mammals explored and conquered the many diverse niches that had been formerly occupied. They rapidly evolved, proliferated, and multiplied and soon dominated the planet, just as had their ancestors, the reptomammals, the therapsids, some 250 million years ago.

The complexity of their mammalian brains also exponentially expanded, and layer upon layer of neocortex began to spread and proliferate

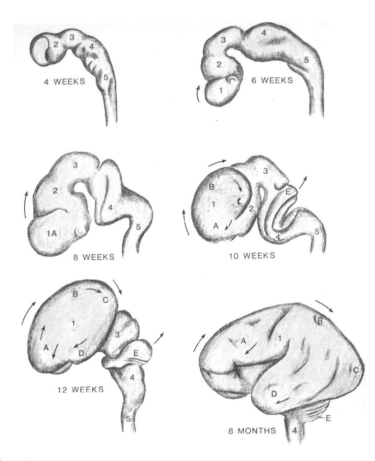

Figure 16. Sequence of embryonic and fetal development of the human brain. Not drawn to scale. Arrows suggest patterns and direction of growth. At approximately 4 weeks, a primitive neural tube has formed and grown several large swellings. This includes (2) the olfactory forebrain and limbic system, (3) the visual and auditory midbrain, (4) the brain stem, and (5) the spinal column. The (1) cerebral hemisphere (the telencephalon) differentiates out of the limbic forebrain and forms the (A) frontal, (B) parietal, (C) occipital, and (D) temporal lobes, which grow up and over and encase the midbrain and brain stem. The (E) cerebellum differentiates out of the brain stem. From R. Joseph. The neuropsychology of development. *Journal of Clinical Psychology*, 38 (1982), 3–33.

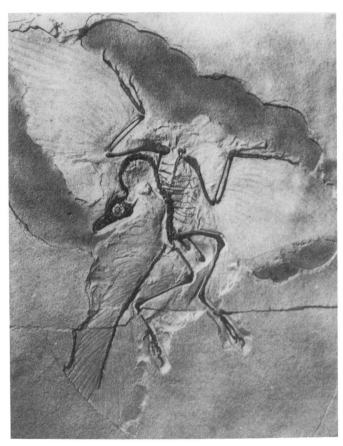

Figure 17. The fossil of possibly the first bird, Archaeopteryx. It is possible that it was also a tree climber.

and undergo environmentally assisted, via selective pressures, metamorphosis. The old brain was now covered by a shroud of neocortex, six to seven layers thick. A giant step from lizard to human had taken place.

Due to the continued accumulation of neocortex as one ascends from primitive mammals to primates and then humans, gyri were soon formed, due to the curling up of neocortex up and over itself as it expands and strives

Figure 18. Archaeopteryx, the first "bird." Reproduced with permission from G. S. Paul. *Predatory Dinosaurs of the World.* New York: Touchstone, 1988.

for space within the cramped confines of the hard, unyielding skull. It is these expansions in neocortex which gave rise to tremendous developments in the frontal and remaining lobes of the human cerebrum and which greatly contribute to making our humanness somewhat unique.

By 5 million years ago, the first protohumans began to emerge from the mists of time and soon diversified and spread across the face of the planet.[36] By 130,000 years ago, our modern human ancestor, the Cro-Magnons, appeared in North and South Africa and by 50,000 to 40,000 years ago they had spread upward into the Middle and Far East, then into Europe, China, the Americas, and Australia.[37] It was the Cro-Magnon who first possessed the necessary neural tissue to issue forth temporal sequential grammatically complex spoken language, as well as a throat and mouth which would enable them to articulate consonants and vowels.[38] Unlike their Neanderthal cousins, they were also blessed with massively developed frontal lobes, which conferred upon them powers of foresight, cunning, and planning that completely dwarfed all who came before. Indeed, the frontal lobes and the human brain reached its greatest size among the Cro-Magnon people, for it even exceeded that of modern man and woman.

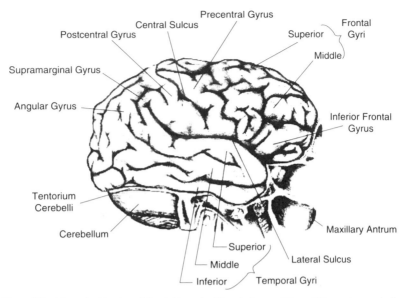

Figure 19. A lateral, side view of the right cerebral hemisphere, sitting within a cutaway skull.

The Multiplicity of Language

Born of multiple cellular components and comprised of neurons, ganglia, and layers of brain, the human organism is a veritable living museum, containing a multiplicity of sensory receiving centers, each with tremendous powers of memory and intelligence. It is this complex neural cellular arrangement which makes possible and can sometimes give rise to tremendous creative leaps of the imagination and intellect when the different parts of the brain independently struggle over and produce solutions. Of course this can also result in the experience of anger, unhappiness, and confusion, when these different brain regions act at cross-purposes each according to their own agendas and perceptions.[39]

For example, the old limbic system responding to a pheromone olfactory message indicating sexual availability in a member of the opposite sex may initiate sexual arousal or aggressiveness which overwhelms the neocortex. The person may then act on his or her limbic lusts and sexual desires and then later, when the equilibrium between the limbic and neocortical brain has been reestablished, the so-called "higher" neocortical centers may then become confused or guilt ridden by their behavior or the consequences. Sometimes within a single human head there is much confusion due to failures to transmit, receive, or recognize certain forms of information as well as errors that occur in translation.

When we consider that other humans are also utilizing multiple means of communicating not only within their brains but with us, such as through olfaction, pheromones, gestures, facial expression, melodic vocal intonation, and posture, the possibilities for creativity, intimacy, understanding, as well as misunderstanding become nearly infinite.

Fortunately, it is these multiple senses and multiple modes of communicating, the languages of the body and the brain, which make possible intuitive understanding and the search for knowledge and the creation of information for its own sake. When one considers the multiplicity of cells contained within the human body and brain, the possibilities are boundless. It is these cellular multiplicities, the residues of the romance between planets and stars, which give rise to life, love, art, desire, intimacy, science, literature, technology, medicine, and that endless ceaseless desire to search for and acquire knowledge—the languages of the body and the brain.

. . . and the spiraling Universe coiled back and swirled round on planets knees and shooting stars to ponder its own depths in the Temple of Human Consciousness.

. . . in the mirror of the sea of human consciousness.

. . . to peer and reflect upon its own soul as mirrored in the rising tides of human consciousness. . . .

I

LANGUAGES OF THE BODY AND BRAIN

1

Emotional, Musical, and Grammatical Speech

Ralph had worked for over 20 years building up his tile company into a multimillion-dollar-a-year business, and yet he forever remained a difficult person to work with. Unfortunately for family, employees, and business associates, as he aged he did not mellow but only became more irascible and ornery and was quite willing to give anyone a good verbal lambasting for the most trifling of transgressions. Even his pastor would not be spared his verbal slings and arrows, if they happened to disagree on some noncelestial matter, such as which hymns were to be sung the following Sunday. Ralph loved to sing in the choir and had a very pleasant baritone.

Although he had been warned about keeping his cholesterol levels down, Ralph always felt proud of his robust health and ate whatever he liked. Hence, he was mighty surprised while in the midst of an argument with a supplier, his right arm and hand suddenly went completely numb. He then became so weak that he dropped the phone.

"Goddamn it!" he loudly proclaimed and was then at a complete loss for words. In fact, he was speechless.

When his secretary came into the office, she found Ralph slumped in his chair and drooling out of the right side of his mouth. Glancing up, he immediately began uttering a bewildering array of profanities, cursing at the top of his voice. Nevertheless, although he could swear, he still could not talk. Nor could he walk or get out of his chair as he was paralyzed on the right side of his body.

35

Ralph had suffered a stroke. A major artery supplying the left frontal region of his brain had become clogged with thrombotic debris thrown from his heart. In consequence, the area of the brain that subserved speech, that is, **Broca's expressive speech area**, and the adjacent neocortical "motor" region which controls movement of the right half of the body had been destroyed. This is why he was paralyzed and had lost the ability to talk, a condition referred to as "**expressive aphasia**."[1] Nevertheless, he could still swear a blue streak and could still read and understand what others were saying.

Comprehension of spoken and written language was preserved in Ralph's case, because that region of the brain which subserves the ability to comprehend and understand speech, located in the left temporal lobe, was undamaged. This expanse of neocortex has been called **Wernicke's area**, and if destroyed, the comprehension of even individual words becomes severely compromised.

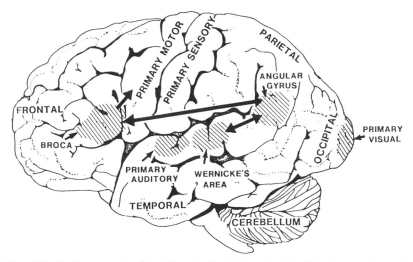

Figure 20. The language axis of the left cerebral hemisphere, depicting Broca's expressive and Wernicke's receptive speech areas, and the multimodal association area, the angular gyrus of the inferior parietal lobule. Arrows and heavy lines indicate probably interactive pathways involved in the expression and comprehension of spoken language. From R. Joseph, The neuropsychology of development. *Journal of Clinical Psychology*, 38 (1982), 33–34.

If, in Ralph's case, the artery supplying the left temporal lobe had become clogged instead, he would have retained the ability to talk but would not understand anything anyone said to him. This condition is referred to as **Wernicke's receptive aphasia**.[2]

Among the majority of right-handed individuals, and approximately 75% of those who are left-handed, the left half of the brain mediates the ability to talk and to understand written and spoken language. Broca's area controls the ability to speak, whereas Wernicke's area mediates the ability to understand what has been said. If a person were to repeat what he or she had heard, this information would be transferred from Wernicke's to Broca's area via a bundle of axonal nerve fibers called the **arcuate fasciculus**.

It was soon discovered that in addition to Ralph's preserved ability to curse, he could still sing and in fact would sing along to the hymns he heard over the radio. Later, he was able to accompany the choir when he attended church on Sundays. Paradoxically, Ralph was able to sing words that he could not say.

Why is that? Swearing and singing are controlled by the right half of the brain, which is dominant for most aspects of emotional expression. This includes the ability to perceive and express emotional nuances via the melody of one's voice.[3] The right half of Ralph's brain had not been injured by the stroke.

What Do You Mean?

Soon Ralph was released from the hospital, and his family became adjusted to the new routine of running the business, having him home, and bringing him to therapy. After a few weeks of this, Tony, Ralph's son, decided to go out on the town to unwind from the stress. Tony gravitated to his favorite country music nightclub, and after quaffing several "brewskis," became engaged in a conversation with a rather attractive young woman whose exotic perfume only added to her allure. By her smile and facial expression, the way she leaned close, gently stroked his hand, tilted her head, and looked into his eyes, it seemed to Tony that she was enjoying their discussion. In fact, anyone who was watching would realize the same even without ever hearing a word she might say.

Suddenly someone poked him on the shoulder. Turning, Tony looked into the eyes of another man who said, "Why don't you and me step outside?" Tony, displaying absolutely no emotion, simply stared intently at the man so as to determine what he meant by that question.

What did he mean? To correctly determine what people mean by what they say requires that we attend to their facial expression, gestures, posture, encroachments on our personal space, the tone and melody of their voice, and if they are touching us, then the amount of pressure being exerted.

In this particular instance, the spoken words clearly indicated a question referring to the outdoors. However, what exactly was meant could involve any number of actions and alternatives. Was Tony being asked to go out onto the terrace where they could enjoy a smoke or talk in private, or was he being threatened with a punch in the nose?

Even if he had not turned to look and the man had not touched him, Tony would still have been able to discern the fellow's attitude and intentions based on his melody and tone of voice. This would have been true even if Tony did not quite hear what was articulated.

This is because *meaning*, as conveyed through speech, although structured by the grammatical relations and denotative features of what has been uttered, is also determined by the loudness, intensity, cadence, melody, and emotional qualities of the voice.[4] If speech were stripped of these paralinguistic variables, including body language and vocal melody, it would sound like a monotone and quite boring, and *meaning* would become ambiguous. You would not be able to determine if a person were sincere or sarcastic, insulting or sympathetic, frightened or sad, humorous or humorless, and so on. Vocabulary and syntax alone are not a sufficient means for communicating or transmitting this type of information. Much more is required so as to transmit feeling, motivation, intent, and context.

Discerning and Intuiting Meaning with the Right Half of the Brain

Among most people, it is the left half of the brain which is responsible for the grammatical and denotative aspects of speech, including the ability to read, write, and to solve math problems. By contrast, the right half of the cerebrum has great difficulty producing or understanding words or sentences.

On the other hand, the musical, emotional, and melodic qualities of the voice are produced and comprehended by the right half of the brain.[5] Indeed, the entire right half of the cerebrum is dominant for and controls the ability to sing, curse, pray, and even coo words of love and sorrow, whereas the left hemisphere experiences considerable difficulty with these nuances.[6] The right hemisphere is also responsible for expressing and analyzing information conveyed by the face and through touch.[7] It is via the right half of the brain that we are able to determine what a person means or implies by what he or she is saying, and how he or she feels about it.[8]

When an individual loses the ability to understand speech, such as following a massive stroke involving the left temporal lobe, they are said to be suffering from Wernicke's receptive aphasia. However, since the right half of their brain remained healthy, they would still be able to make a number of correct inferences as to what was being said or asked. This is because the ability to extract meaning would be intact via the right half of the brain's preserved ability to perceive tonal, melodic, and emotional vocal cues.[9] Because of this preserved ability to perceive these nuances, sometimes family members and even physicians and nurses may fail to realize that someone has suffered a severe brain injury and no longer comprehends what is being said to or asked of them until accumulated evidence makes the condition undeniable.

Once, after I had diagnosed an elderly woman, "Lilly," as suffering from receptive aphasia, her nurse disagreed. According to the nurse, Lilly was able to understand and respond correctly to many questions. For example, when Lilly was asked, "How are you today?" she responded, "Fine," and smiled and shook her head appropriately. Because of this, the nurse assumed that comprehension was intact. However, when later I again examined this patient, I said to her, "It's raining outside?" and used the same *tone* and *melody* of voice *as if* I had said, "How are you today?" Lilly, responding to the melody of my voice smiled and shook her head and replied, "fine, fine." This is because she correctly interpreted the tone and melody of my voice but was unable to discern my words.

Her right hemisphere was listening to the melody, prosody, and emotional features of what was being said and was deducing that concerned questions as to her health were being asked. Wanting to be pleasant and sociable, she replied accordingly. Of course, her undamaged right hemisphere was also attending to my body language and facial expression which even in the absence of words can communicate volumes of information.

When Ralph suffered the stroke involving Broca's speech area, in the left frontal portion of his brain, thus disrupting his ability to talk, he was able to continue swearing and singing because the right half of his brain remained undamaged. If instead the stroke had injured the right frontal region of his brain, the capacity to sing and swear, as well as the musical, emotional, and tonal qualities of his voice, would have become disrupted. He would no longer have been able to sing or carry a tune and most certainly would have been kicked out of the choir.

However, in Ralph's and Lilly's cases the ability to perceive music as well as other nonverbal sounds, such as the emotional qualities in another person's voice, was preserved. This is because the ability to perceive these emotional and melodic nuances is mediated by the temporal lobe of the right hemisphere, as well as the limbic system, which in these instances remained undamaged.[10] If the right temporal lobe is destroyed, then vocally expressed emotion would no longer be perceived and instead it might sound abnormal, or bland and empty like a monotone,[11] whereas music would sound like noise or maybe even what has been referred to as "hip hop" and "rap

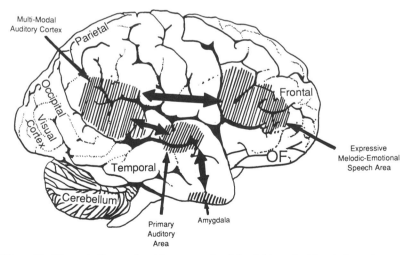

Figure 21. The melodic–emotional language axis of the right cerebral hemisphere. Arrows and heavy lines indicate probably interactive pathways involved in the expression and comprehension of emotional language. The amygdala is buried within the depths of the temporal lobe. From R. Joseph. *Neuropsychology, Neuropsychiatry, and Behavioral Neurology.* New York: Plenum Press, 1990.

music," where rapid speech and rhythm are emphasized at the expense of melody.

Two Brains, Two Minds: Right- and Left-Brain Cooperation

If we were to remove the top of someone's skull, we would discover that the brain is divided into two halves by a large interhemispheric fissure. If we squeezed our fingers down into this fissure, our progress would soon be interrupted by a large rope of nerve fibers which interconnect and allow information to be transferred between the two brain halves. This axonal fiber bundle is called the **corpus callosum.**

The two halves of the brain, however, are concerned with different aspects of experience and have their own superiorities and weaknesses. For example, the left half of the brain has trouble perceiving emotional nuances as conveyed by the voice, face, or body, and is socially quite dense and

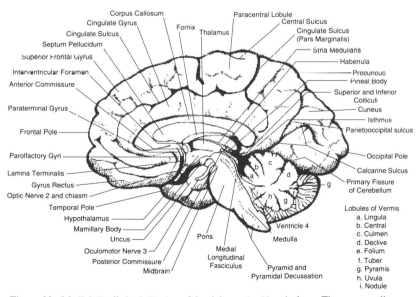

Figure 22. Medial ("split-brain") view of the right cerebral hemisphere. The corpus callosum is composed of a thick bundle of axons which originate in cells within the neocortex of one hemisphere and cross the interhemispheric fissure to make contact with neurons on the opposite side of the brain.

concrete. The right half of the brain can't read or write but is the depository of our emotional memories and mediates our ability to sing, dance, or leap about in space, and to recognize the faces of friends and loved ones.[12]

Since they are concerned with different aspects of our mental life, each half of the brain in fact enjoys a mental existence that is somewhat different from that possessed by the opposite half of the cerebrum. We not only have two brains but two minds.[13]

Each half of the brain has its own memories, attitudes, and goals for the future, which are influenced by the different aspects of our experience they are concerned with—music, emotion, math, speech. Because of this, sometimes we feel in mental conflict. What we rationally know and what we emotionally feel may not coincide, a condition that may be exacerbated by the limbic system, which may have its own agenda. We may feel torn over various issues or when making certain decisions, or, at a minimum, on occasion we may have trouble making up our mind. The two halves of the brain and the limbic system do not always agree.[14]

Although they sometimes act in opposition, the two halves of the brain often act cooperatively as so much of our daily life requires their harmonious interaction. For example, it takes both halves of the brain to make music, the right providing melody, the left the rhythm. In order to read we must not lose our place in visual space (a right-brain function) and must attend to the words and their order (left-brain function).

Similarly, whenever we speak or listen, both halves of our brain participate and offer their own unique contributions. The left frontal region (Broca's expressive speech area) programs and produces the words we say, and the melodic–emotional speech area located in the right frontal lobe produces and modulates the melody and emotional, intonational nuances. These two areas thus act in concert, like two wheels joined together by an axle.

When we listen to others as they speak, their words and the grammatical structure of their sentences are perceived and comprehended by Wernicke's receptive language area, located in the left temporal lobe. This region is joined, via the corpus callosum, to the temporal lobe of the right cerebrum, which perceives the melody and emotional features of what has been uttered.

When Tony was asked if he would like to "go outside," his left hemisphere analyzed the words, and it was his right hemisphere which processed the prosodic intonational qualities so as to extract the meaning and any emotional messages being conveyed. That is, his right cerebral hemi-

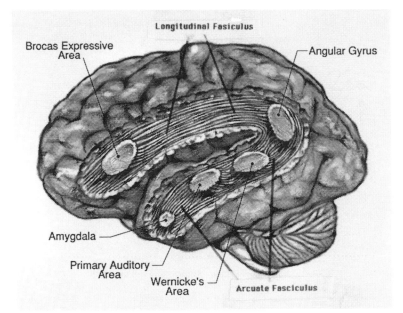

Figure 23a. The language axis of the left half of the brain, the amygdala, primary auditory area, Wernicke's area, the angular gyrus, and Broca's area are interlinked by an axonal fiber bundle which runs beneath the neocortex, the arcuate, and longitudinal fasciculus.

sphere determined if he was perhaps being threatened with a punch in the nose, or if the communication was friendly.

If the temporal lobe of the right half of the brain were damaged, an individual would lose the ability to hear and appreciate music and would no longer be able to deduce the emotional messages being conveyed. If someone were to say, "I love you," the message would only be heard by the left half, which has great difficulty discerning context, mood, humor, or emotion.[15] In consequence, *feelings* of love would not be perceived. The words might be flat and unconvincing, or sound strangely altered and incongruent.

However, if the temporal lobe of the left half of the brain were severely damaged and the same words were uttered, the person would realize that they were being spoken to in an emotional and caring manner.[16] This would be accomplished by their intact right hemisphere. They just wouldn't be sure as to what was being said.

Hence, just as the right and left frontal regions of the two brain halves

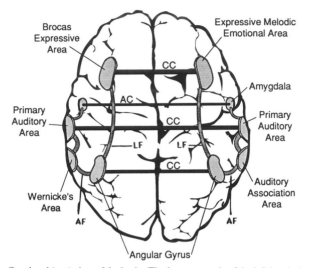

Figure 23b. Overhead (top) view of the brain. The language axis of the left hemisphere and the emotional–melodic speech areas of the right hemisphere are linked together by axons which travel via the corpus callosum (CC) and the anterior commissure (AC). The axonal fiber pathways, the arcuate (AF) and longitudinal (LF) fasciculus, interlink Broca's area, the angular gyrus, Wernicke's area, the primary auditory area, and the amygdala, and a similar axonal circuit performs the same within the right hemisphere.

cooperate in producing speech and melody, the right and left temporal lobes act together so as to perceive not only what is being said, but how and in what manner it is being said so that meaning and emotional messages can be discerned.

"It's Not What You Said, It's the Way You Said It"

It sometimes happens that the right and left half of the brain do not work in harmony and have different concerns, attitudes, and even agendas. This is because we have two cerebral hemispheres, the right and left halves of the brain, which have different specialties, their own unique memories, goals, and plans, and are concerned with different aspects of what appears to be the same experience. Due to this unique relationship, sometimes the left half of the brain does or says one thing, while the right hemisphere has something a little different in mind. This is why when speaking or behaving, different and conflicting messages are sometimes transmitted or acted on. For instance, the left hemisphere may rationalize "kissing up" to a potential employer, whereas the right half may resent the indignity implied. Suddenly the job seeker astounds himself not by his poor choice of words, but the sullen, insulting, sarcastic tone of voice he used in delivering them in response to certain questions. Out on the street he admonishes and yells at himself and wonders why he keeps doing this to himself. But, what half of the brain admonishes which?

A few days before Ralph had suffered his stroke, he called for his secretary and asked her where a certain file was to be found. His secretary, having handed it to him that very morning, and feeling angry that he was again behaving in this accusatory manner, told him that the file was sitting right in front of him, but used a tone of voice (right brain) which communicated that she thought he was an old fool and a jerk. Ralph's right brain, having fully discerned her attitude, became activated, and he began swearing at her. It wasn't what she said, it was the way she said it.

Coworkers, friends, and lovers often have similar difficulties when communicating and are often hurt or surprised by what they or others have said. Again, it is not just what is said, but it is how it is said, and this is sometimes a function of each half of the brain independently acting on different agendas, goals, and even different memories.

For example, Lisa and John made up after an intense argument. Later when attempting to rationally and logically work out a scheduling problem, Lisa (right brain) used a very sarcastic tone of voice because she (her right brain) is still feeling irritable about their earlier argument. John (his right brain) surprised himself by swearing at her. They began fighting again. In this case, although their left hemispheres were attempting to work on a problem that had nothing to do with their earlier uproar, the right half of their cerebrums were still arguing and had not yet made peace.

Multiple and Emotional Means of Communication

Languages of the Body and the Brain

There is so much to communication, all of which can also disrupt or enhance the effectiveness and outcome of what is being transmitted. Indeed, there are so many facets that have in fact nothing to do with the production of speech, that multiple messages are often being expressed and discerned simultaneously and sometimes in a contradictory manner. This is because communication can be through facial expression, posture, movement, dress, and even hairstyle, and it may be transmitted through hesitations, pauses, and the melody of speech, gesture, touch, and even smell as well as via the limbic system and the right and left half of the brain.

In some cases, these multimessages all reinforce one another and enhance what has been said, whereas in others, contradictory signals are being conveyed which may confuse the receiver or which may reflect ambivalence on the part of the sender. Sometimes the messages are meant to be confusing and ambiguous such as when a prey is attempting to elude a predator, or when a man and a woman cautiously flirt after meeting for the first time. What eventually transpires depends on their success at interweaving and separating out these competing messages. Even animals and insects which live a solitary life are dependent on the ability to correctly perceive and respond to visual, auditory, tactual, and olfactory signals. For if they fail to do so, they die.

Although humans are able to walk in space, the mind is still adapted to a life spent living among the elements and as such remains profoundly aware and subject to environmental influences be they transmitted through smell, touch, movement, gesture, sights, sounds, dreams, images, voices, or visions. The human body and brain remain geared to sense and respond to the known and the unknown and to emergencies which threatened existence for millions of years. The brain and its many constituent elements also retains and is sensitive to all those means of communicating, perceiving, knowing, and remembering which enabled first primitive animals and now human beings to evolve, flourish, and survive.

2

The Nose Knows
The Language of Smell

Olfactory and Pheromonal Communication in Insects and Humans

Peter stepped into the tiny restaurant and was immediately delighted by the aroma of freshly baked bread. It permeated the air and made his mouth water. However, the faint presence of another familiar but completely unexpected fragrance stopped him in his tracks and killed his ravishing appetite. It was that special scent of perfume which he always associated with Jessica.

Overwhelmed by a flood of memories, Peter felt both depression and longing. He had not thought of her in years and had been long convinced he was completely over her. Even so, he remembered her funeral, which he had tried so hard to forget, and for a brief moment he could even recall the sickly sweet smell of the orchids and heavy fragrance of the freshly turned earth which sat in two big piles on either side of the open grave.

Feeling queasy, Peter turned and slowly walked back out into the street, tears welling in the corners of his eyes. How could a simple odor trigger so many memories and cause so much confusion and anguish?

47

The Nose Knows

Odors, be they the stink of rotting flesh, the aroma of a freshly brewed cup of coffee, or the fragrance of an expensive perfume, are often comprised of a number of divergent chemical agents which can often exert profound influences on memory, motivation, sexual behavior, and emotional arousal.[1] Indeed, this is true of humans as well as the most lowly of organisms and is a function of the chemical composition of the cerebrum. That is, the brain communicates within itself via various chemicals, called **neurotransmitters**.

Nerve cells secrete neurotransmitters, via nerve fibers called **axons**, in order to transmit information and issue commands to other nerve cells located in various regions of the brain. Conversely, a neuron is able to receive these chemicals via specialized receptors located at various points on short branching fibers called **dendrites**. Every neuron has an axon (via which transmission takes place) and multiple dendrites (which receive messages). The axons and dendrites of separate neurons make contact at synaptic junctions. The brain is literally awash in chemical agents, and any alteration in its chemical composition in turn alters brain functioning.

As noted, single-cellular creatures and other ancient and primitive animals receive chemical and pheromonal messages via receptors located on the body surface. Over the course of evolution, although the receptors have remained externally located, the actual cell bodies have migrated internally, some of which eventually formed ganglia and then lobes, and finally the brain.[2] Hence, even when the dendrite surface is still externally placed, the cell body which is attached to the long dendritic cable is found deep within the body or even within the brain. It is in this manner that external sensory stimuli are transmitted to receiving stations within the cerebrum.

Many of these dendritic receptors are also still sensitive to the presence of external chemicals such as those we consciously recognize in the form of various odors. It is in this manner that chemicals arising in the external environment, be they secreted by plant or animal, are able to act on the brain so as to trigger sometimes reflexive and involuntary responses among individuals who chance to come in contact with them.

The nose is one such olfactory receiving site. In this regard, the nose, like the retina of the eye, is a literal extension of the brain. As is apparent, based on the size of the human nose, chemical–olfactory signals, including what we consciously recognize as odors, play an extremely important role in our lives. Indeed, fragrances in the form of incense, perfume, and edible

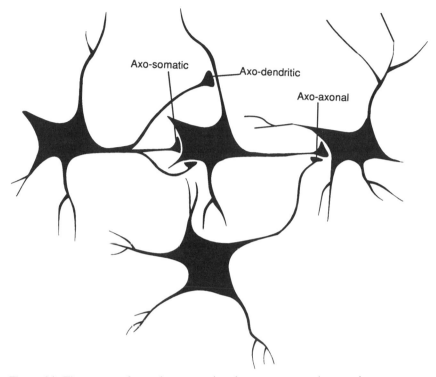

Figure 24. Three types of axon interconnections between neurons. Axosomatic synapses consist of axons which synapse on the cell body, axodendritic synapses are those where the axon synapses with the dendrite, and axoaxonal synapses are those where two axons form a synapse. Axoaxonal synapses are often inhibitory such that the axon from one cell fires so as to prevent the firing of a second axon.

delicacies can exert subtle or profound effects, depending on their chemical composition. They form a rich textured backdrop to religious ceremonies, business activities, and social functions, and contribute tremendously to our eating pleasure, and even our sex life.[3] It is thus not at all unusual for people who suffer a loss of smell to also lose their sex drive. Commonly, their ability to derive pleasure from food is markedly reduced as well.

Bodies as well as homes are perfumed or sprayed with scents, as are brands of toilet paper, furniture polish, disinfectant, douches, shampoos,

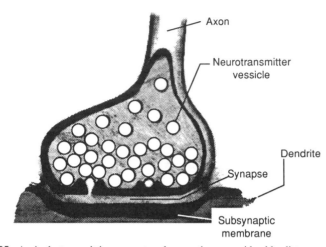

Figure 25. A single transmitting axon travels sometimes considerable distances to make contact with the dendrites of other neurons at a gap junction, the synapse. When sufficiently activated, the axon will release neurotransmitter substances which are stored in synaptic vesicles.

laundry detergents, and the list goes on. Cars have artificial "new-car" scents applied to them, and an entire industry has developed so as to create artificial scents that make us think of apple pie, pizza, fresh bread, a newly mowed lawn, expensive leather, and so on. Moreover, olfactory cues are employed for indicating sexuality (e.g., perfume). Odor also exerts a powerful influence on what is considered socially acceptable—hence, the abundance of artificial chemicals designed to eliminate various body odors. This includes the elimination of those body fragrances which act naturally to increase feelings of sexual arousal and attraction in a partner.[4]

Olfactory chemicals, be they transmitted via cars, homes, clothes, the unwashed, perfumed, or menstruating body, are an important form of communication. These chemicals serve as a nonverbal, silent language that is used as a means of information exchange between lovers and even total strangers. Olfactory cues can inform us as to the economic, social, as well as sexual status of our fellow humans as well as their possible sexual availability. Helen Keller indicated that she could quite easily and accurately determine a person's occupation, or even what room of the house (e.g., bathroom, kitchen) they may have just left, based on smell alone.[5] Similarly, via a single sniff, adults and children are able to tell if a piece of clothing has been worn by a man or a woman, and women can pick out articles worn by their children, husbands, or lovers based merely on scent.[6]

The presence of familiar odors can also exert considerable feelings of emotional comfort and security. Nancy Reagan, for example, after her husband President Reagan had been shot, is reported to have slept with his shirt that night and each day he was hospitalized as this gave her a sense of comfort. Some women have responded similarly when mates are away for long time periods or have died. However, it is not the just the article of clothing but the fragrance of their lover that can serve not only to soothe but to sexually excite.

Odors can in fact exert a variety of influences on our behavior ranging from feelings of revulsion, disgust, and gastric upheaval, to sensations of gustatory ecstasy and the sublime. These chemicals, in fact, inform us as to the existence of dangerous or undesirable conditions such as the presence of rotting food, a lurking skunk, a poisonous plant, and even the possibility of contagion by disease. Experienced nurses and observant physicians, for example, are often able to detect and diagnose certain illnesses based on the smell of the body; patients with diabetes smell like sugar, and those with measles smell of feathers. Some people are even able to predict the weather, such as an approaching storm, based on smell. This is not entirely surprising, for when atmospheric pressure changes from high to low numerous odors rise into the air from rocks, plants, and various nooks and crannies. Noses have been employed in this regard by human beings and their ancestors for many millions of years.

Most of these influences, however, occur unconsciously at the level of the limbic system, as our ability to consciously appreciate even their presence or what they imply has been greatly reduced. The ability to chemically and nonverbally signal socially among humans was probably much more pronounced early and until recently in our evolution. Before we lost our hairy coats, men and women (like our hairy, modern-day primate cousins chimpanzees and gorillas) probably gave off a variety of very strong odors, all of which served different communication functions such as status and group or clan identity.

Just as homes and hives have their own characteristic odor, it is likely that bands and tribes of human did so too. It has often been said that blacks, whites, American Indians, and Japanese all give off different odors. It has also been suggested that this is due to differences in diet. In any case, group odors probably served then, as they do now, to promote cohesion as much as social signaling and identity. These are just some of the functions olfaction retains even today.

Smelling, Tasting, and Eating

The olfactory system is of course related to eating and thus to the gustatory and limbic system. One need only suffer a severe cold in order to appreciate the dominant function of smell in our ability to fully taste and savor the flavor of food and thus experience pleasure when dining. Much of what we experience as flavor is really an olfactory experience. This is why some connoisseurs prefer to sniff their wines, which, without their fragrance, would be fairly tasteless.[7]

However, the olfactory system is completely independent of the taste buds, which are located along the tongue. Although these systems interact, they are in fact separate sensory systems and take different paths to the brain.

The olfactory and gustatory systems may have started out as one generalized chemoreceptive system and only over the course of evolution became distinct. Thus, among humans and other primates, taste buds became confined to the mouth, and the olfactory nerves to the nasal mucous membrane. In reptiles and many other animals, however, an auxiliary (vomeronasal) olfactory organ is located within the roof of the mouth, and such creatures may in fact use their tongue to taste various substances and then dab the tip against this structure. However, they, too, have olfactory structures that are completely separate from the taste buds, as do humans.

This was a very important adaption, for some things, although smelling quite good, taste terrible and have no nutritive value. Consider for example, what many children and some adults consider to be the wonderful smell of "play dough." Play dough, however, does not taste good at all. Some pleasant-smelling chemicals can in fact poison and kill the body.

Thus, originally and early in the course of evolution, via the analysis of odor alone, a potential edible substance might be deemed as good to eat and was consumed. With the separation of these two systems, before it could be eaten it had to be tasted in the mouth in order to be deemed palatable (the taste test), thus giving one the immediate option of swallowing or spitting it out. Of course, some things smell and taste great, or have no taste at all, but can still be poisonous.

The Conscious Nose

Human beings are often not consciously aware of the presence of many olfactory chemicals, much less the sometimes reflexive reactions they elicit. Certainly we are quite capable of consciously appreciating the odor of freshly baked bread or apple pie and are more than able to distance ourselves from unpleasant smells such as rotting fish or garbage. However, many other externally arising chemicals which affect memory, emotional, and even sexual functioning often exert their influences unconsciously at the level of the limbic system.[8] Odors can alter brain-wave electrophysiological activity even when the subject is not aware that any particular scent is even present. This is because the structures involved in the secretion and detection of various odors important in the survival of the individual and the species evolved and took control over behavior hundreds of millions of years before the appearance of human beings, the conscious mind, or the development of the cerebral **neocortex** (or new brain).

Take for example, a recent, widely reported experiment which was conducted by Dr. Alan Hirsch, in the gambling casinos of Las Vegas. Gamblers who were secretly exposed to a special scent that he concocted spent 33% to 53% more money and time gambling than those who were not. Individuals who were exposed to yet other special odors in certain regions of a department store were more likely to linger, make purchases, and even spend more than nonexposed shoppers.

Olfactory chemicals, for the most part, exert their influences on the most ancient and primitive regions of the cerebrum, that is, the seat of the emotions, the **limbic system**.[8] As noted, the limbic system began to evolve well over half a billion years ago.

Communication via chemicals, some of which we recognize today in the form of specific odors, is the earliest and most ancient means of affecting motivational and emotional functioning. In fact, chemical communication not only preceded the development of refined visual, tactual, and auditory modes of interacting, but acted as an impetus which promoted the formation of these senses as well as the complex behaviors which were made possible following their creation.[9]

Being chemical creatures whose cellular surface also functioned at the behest of chemical messengers, ancient primitive life forms evolved and soon developed a sensitivity and the capacity to respond to external chemical agents and to secrete them as well. Eventually, the ability to detect and search out food or mates based on the presence of chemicals excreted by other life forms came into being as specialized receptors especially adapted for this purpose evolved.

Although light-sensitive, but basically blind and deaf, these ancient and primitive creatures (such as sponges and primitive worms) were still able to richly sample their world and detect and analyze other forms of life even when separated by enormous distances.[10] In fact, even among higher forms of marine life, such as fish, the olfactory system remains extremely sensitive to a wide range of amino acids, which indicates that it is specialized for detecting that which is alive, either so as to serve as food, or to be avoided, or mated.[11]

As noted, chemical communication with the external environment also wrought enormous changes in behavior and adaption. That is, the ability to physically approach and manipulate the object of their olfactory desires also evolved so as to accommodate this sensory system and vice versa. There thus resulted expansion in the olfactory–limbic brain.

Over the course of evolution, part of the olfactory–limbic system became modified into what has been referred to as an olfactory striatum and then the corpus striatum (of the basal ganglia), which in part serves as a motor center that enables the organism to act on its emotional and motivational needs.[10] Parkinson's disease (which Muhammad Ali purportedly suffers from) involves destruction of the basal ganglia, and typically the body becomes stiff and the face completely emotionless and blank.

Hence, the olfactory-limbic system promoted motor development, and this is because the ability to move, crawl, coil, whirl, and swim became important when mates and food sources could be detected at a distance. Indeed, early in the course of evolution, creatures did not travel for the pleasure of traveling, but for the reward that awaited one at the end of their journey—food or a mate. As noted, however, the olfactory system first gave rise to the limbic system, which mediates all aspects of emotion including the desire for and feelings of pleasure when one eats or has sex.[12]

Since it is not very adaptive merely to absorb, or eat whatever comes into one's mouth, the ability to make distinctions, to recognize, to make comparisons with memory was also necessary, and this too was accomplished through smell and the limbic system. Otherwise, one might consume or approach an object or animal that could make one ill or even lead to death. Hence, learning and memory came into being.

Indeed, the ability to recognize, learn, and remember are properties that smell and olfactory communication retain today, even among humans.

In fact, the limbic system brain structures involved in memory, the **hippo-campus** and **amygdala**, are part of what is called the **rhinencephalon**, or rather, the *"nose brain."* It is referred to as the nose brain because the nerve fiber pathway which leads from the olfactory mucosa of the nose to the olfactory bulbs terminates in these brain regions.

The olfactory system in fact gave rise to the evolution of a very primitive amygdala and rudimentary hippocampus probably some 500 hundred million years ago, when the ancestors of the eel-like cyclostomes and the wormlike, burrowing, limbless Gymnophiona first slithered upon the scene; creatures which possess a well-developed hypothalamus, thalamus, and rudimentary hippocampus and a well-developed, almost mammalian amygdala.[13]

Be it cyclostomes, sharks, fish, amphibians, reptiles, birds, mammals,

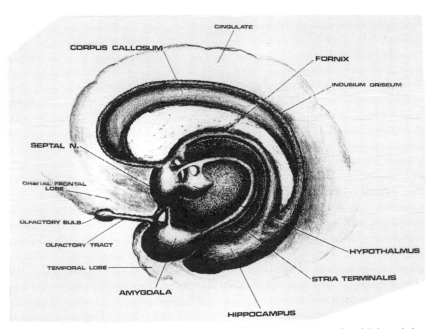

Figure 26. The limbic system, depicting the amygdala, hippocampus, septal nuclei, hypothalamus, and the more recently evolved, transitional limbic–neocortical, four-layered cingulate gyrus.

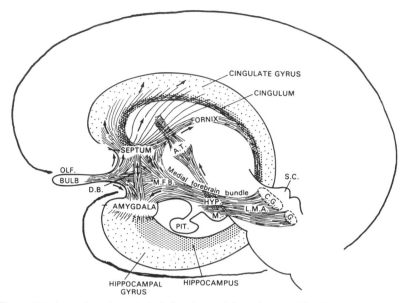

Figure 27. Axons from the optic bulb flow into and form the amygdala, hippocampus, and septum. Unlike the two-layered old limbic cortex, the cingulate gyrus is transitional, being formed out of the hippocampus and consisting of four layers. From P. Maclean. *Evolution of the Triune Brain.* New York: Plenum Press, 1990.

or humans, these regions of the limbic brain (i.e., the hypothalamus, amygdala, hippocampus) are collectively referred to as the limbic system, a term coined by Dr. Paul Maclean. These limbic-system nuclei are extremely important not only in olfaction, but in feeding, fighting, fleeing, and sexual functioning as well as in learning, memory, and motivation. It is the limbic system that controls the ability to experience orgasmic pleasure, suicidal depression, or a blind, raging, out-of-control desire to kill or fornicate.[14]

Long before the limbic system, the nose brain, and even the nose itself evolved, the ability to detect externally originating chemicals (hereafter referred to as **pheromones**) was made possible by specialized receptors that were located on various parts of the organism's body, such as within the abdomen or along the legs.[15] These receptors enabled many ancient and still enable modern-day species to continue to communicate and sense pheromonal alterations in their environment.

In fact, although humans and other mammals presumably no longer appear to have the ability to perceive pheromones or odors via the body surface other than through the nose, the capacity to excrete these chemicals through the skin has been retained. Humans possess, for example, scent-producing glands, some of which are referred to as axillary organs.

Axillary organs are located under the arms and are maximally developed among females. Their purpose is presumed to be sexual and social communication via the secretion of various scents, like a flower. Other mammalian scent glands are located in the anal and genital region and beside the teats.

Via their secretion of pheromones, males and females of many species are enabled to find mates, and those secreting organs located near the breasts enable a baby to recognize its mother and the source of its sustenance so as to orient in order to feed. Be it adult or infant, these sometimes subtle pheromonal messengers remain a highly important means of communicating for amoeba, insect, reptile, mammal, and man and woman alike.

Pheromones and Insects, Sex and Humans

The ability to communicate via pheromonal chemical agents serves first and foremost to promote the survival of the individual, be it insect or human. As noted, the manner in which pheromones are perceived and sensed remains very similar regardless of species. Of course, the manner in which these chemical agents are secreted and responded to can also differ drastically, depending, for example, if one is an insect or a human being. The similarities are nonetheless striking.

Male and female insects contain glandular cells on many different regions of their body which synthesize and release pheromones. When these chemical messengers are released from one of these glands, they may be applied directly to some surface that the insect wishes to mark, including, for example, its mate, nesting area, or food source.[15] Or they may be released directly into the air as a puff of compact pheromonal filaments and molecules. Once airborne, the puff breaks up and the tiny filaments are spread far and wide by air turbulence and the competing eddies and currents of the wind.

The secretion of these chemicals, however, be it by man, woman, or insect, is not a continual ongoing activity but is subject to the control of the

Figure 28. An ant lifts its abdomen and sprays the air with pheromones. From D. H. Patent. *Looking at Ants.* Photo by W. F. Wood. New York: Holiday House, 1989.

nervous system and the hormonal biological block which ticks away within us all. Numerous other factors which also play a significant role in their release, include the amount of ambient light, temperature, humidity, age of the host, hormonal fluctuations, and even previous sexual activity all of which can greatly affect pheromonal production.[16]

Lacking a nose, the perception of pheromones in insects is made possible via specialized receptors which may be located on the legs, antennae, or other regions of the body. The receptor surface is in turn part of a nerve fiber, a dendrite. When one of these pheromone molecules makes contact with a dendritic receptor, there results an electrical–chemical reaction which causes the receptor and its dendrite to become excited and to **depolarize**, creating what is called (once it reaches the **axon hillock**) an electrical **action potential** that shoots from the cell body down its axon. This enables the messages contained in the chemical molecule to be transmitted from one cell to another. It is in this manner that an external chemical comes to exert influence on the brain and subsequent behavior.

Consider, for example, a light bulb. When the light switch is flicked on (when the dendritic receptor is depolarized by the transmitter substance), electricity flows up the wire (the dendrite) to the light bulb (the cell body). If there is sufficient electricity the light bulb will go on (is depolarized). Once it is depolarized, an action potential is initiated down its axon which then releases its transmitter substance at the threshold of the next dendrite.

Among humans and other animals this initial olfactory-limbic receiving area is in fact shaped like and is referred to as a bulb, that is, the **olfactory bulb** which is made up of million upon millions of neurons. Here the message undergoes considerable analysis (the light bulb goes "on"). This information is then transmitted from the olfactory bulb to yet other regions of the brain. This is essentially what occurs in the case of an insect, mammal, or human. Insects, however, do not possess an olfactory bulb per se.

Among insects, whose cerebral organization is quite different from humans and mammals, these impulses are transmitted to a brain structure referred to as the **"mushroom body,"** so called because it resembles a mushroom. They do not possess an amygdala or a "nose brain." Conversely,

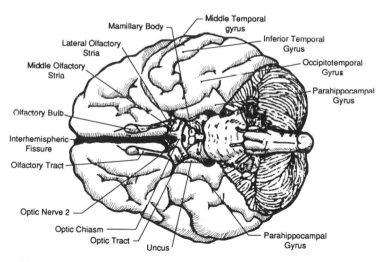

Figure 29. An inferior, bottom view of the brain depicting the olfactory bulbs, olfactory tract, temporal lobes, cerebellum, and brain stem.

humans and mammals do not possess a mushroom body. Instead, vertebrates possess an amygdala, so named because it looks like an almond.

In any case, it is in these brain structures, the mushroom body in insects and the amygdala in humans, where further sensory analysis and integration occurs such that the olfactory input is assimilated with tactile, visual, gustatory, sexual, and other information. It is due to this multiple convergence of sensations in the amygdala that one can form multiple classifications. In response to certain orders we can determine what or who produced it, what they may look and sound like, as well as how good they may taste or feel to make contact with.[17] Among insects and even mammals, much sensory processing and integration occurs at the initial olfactory receptor site (be it located within the mammalian nose or along an insect's leg) before this information ever makes it to the brain.

Specifically, among humans and mammals, these tiny airborne chemical molecules are drawn into the nasal cavity where they make contact with microscopic hairs called cilia, which are located throughout the nasal mucosa. It is here where the initial dendritic receptor is located. The dendrite then leads to cell bodies which are located in the mucous-covered olfactory epithelium (i.e., skin) of the inner nose.

As noted, in addition to a dendrite, each cell body contains an outgoing nerve fiber called an axon. Axons transmit and release neurotransmitters at junctions called **synapses**, directly opposite which are the dendritic receptor surfaces of the next cell. Axons arising from these nasal nerve cells collectively form the **olfactory nerve**, which travels from the nose and terminates in the olfactory bulb. Within the olfactory bulb there is a considerable degree of convergence, with axons from many different cells forming synapses on the dendrite of a single neuron. This receiving neuron in turn acts to analyze and integrate the many messages received.

In some animals, particularly those in which smell is essential to sex, survival, and social relations, the olfactory bulb is quite large and is considered a lobe of the brain. Thus we find that it is very large among dogs and even larger among those whose survival depends on smell, such as anteaters, sharks, and bottom-feeding fish. By contrast, birds have a very poorly developed olfactory bulb, and some species of whales and porpoises have no olfactory bulb, although they still possess an olfactory system, which leads to the brain.

If for some reason the olfactory nerve is severed (which sometimes occurs following a head injury), the individual will completely lose the

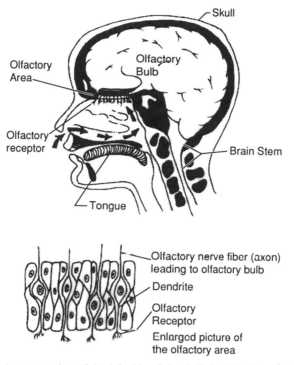

Figure 30. A cutaway view of the left side of the skull depicting the olfactory area, the olfactory receptive dendrites, and the olfactory bulb.

ability to detect odors or to taste most flavors. The same can occur if the olfactory bulb is destroyed or if the **olfactory tract** (the collection of axons which exit the bulb and terminate on other limbic brain structures and the temporal lobe) is compromised such as due to infection or invading tumor.

However, if the temporal lobe (within which are buried the amygdala and hippocampus) and limbic system are damaged, rather than a loss of the sense of smell, olfactory functioning may instead become quite abnormal and altered. Some patients in fact often complain of the presence of foul odors emanating from their body or some unknown location, when in fact there is none present, or of smelling the wrong odor when sniffing food. Gastrointestinal functioning may also become altered, as well as their sex drive.[18]

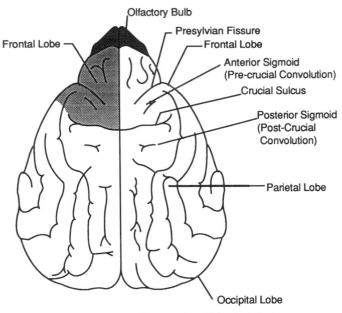

Figure 31. Superior, top view of the dog's brain.

Olfactory Blends: A Symphony of Smell

As noted, various odors are comprised of a multitude of chemical agents. Indeed, some so-called connoisseurs have likened smell, particularly that of perfumes, as a little bit like a symphony, for to make a perfume requires multiple blends that must harmonize together so as to make a particular pleasing impression.[19]

Similarly, when an insect or mammal releases a pheromone, it is not composed of a single olfactory chemical. Rather, like high-grade gasolines and premium whiskies, they are composed of blends of many different chemical entities. Thus, the release and perception of a pheromone can exert a number of different influences because a variety of chemicals have been transmitted and received.

In perceiving these pheromone blends, a different pattern and number of dendrites may fire depending on which chemicals were received. That is, if a blend of four different pheromone molecules is received on one occasion, and a blend of three different agents on another, and yet a blend of

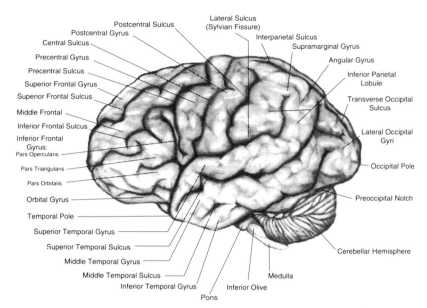

Figure 32. The lateral neocortical surface, side view, of the left cerebral hemisphere.

all seven agents on yet another occasion, at a minimum three different patterns of excitation will result depending on the composition. It is as if these molecules were punching in a code. It is via these different patterns and codes that the identity of certain individual odors is indicated and which determines what type of behavioral response might result.[20]

Since there are many different receptors which are sensitive to different pheromonal chemicals, once contact is made, different behaviors can then be triggered. Hence, a considerable degree of complex communication is possible as a number of different messages can be transmitted. However, only a fraction of whatever has been released is ever perceived, if at all.

The Grammar and Syntax of Smell

Sometimes all components of the pheromone must be perceived for a complete behavioral response to occur. Just as gasoline requires certain chemical additives in order to fire up an automobile engine, complex behaviors often require a complex combination of stimulants in order to be initiated. All the olfactory molecules, like the grammars comprising spoken language, must be perceived, sometimes in sequence, or the message will be incomplete or confused and not understood.[21] This is true for woman, man, and insect. However, like humans, many insects also use certain postures, gestures, and even dance movements to communicate their intentions to their companions prior to, during, or following the release of their pheromones.

Sex and Pheromonal Sequences

Consider a sex pheromone. Once he has perceived it, the male animal may cease to engage in whatever activity is occupying his time, orient toward the source, approach it, locate the female, mount, and so on. Hence, a complicated behavioral sequence is initiated which in turn may be reinforced and strengthened as more of the chemical is sensed. If only certain components of the blend are perceived, the actual sex act may never occur as the behavior sequence may never reach that stage. In fact, sometimes several different pheromones are released in sequence so that appropriate actions are triggered in the proper order. For example, the insect female secretes an attractant pheromone and when the male has approached, she then secretes a pheromone which stimulates the male to mount her, and so on.[22]

It is obvious not only among insects, but dogs, primates, and even humans that smell serves a sexual purpose. This is why so many animals spend time sniffing and licking each other's genitals or thoroughly investigating urine or feces. Moreover, just as some insects will manufacture a chemical which acts as a sexual attractant to other species (so that they can eat them when they arrive drunk with desire), the pheromones produced by many animals exert similar, albeit much less-pronounced effects on humans as well as on unrelated species of animals, acting to promote sexual arousal and related physiological responses.[23] Certain animal odors can affect the human menstrual cycle and in some instances can cause females to ovulate and to even conceive more easily.[24] However, among certain groups of mammals, such as mice, the odor of a strange male can act to terminate the pregnancy of the female.[25]

Similarly, a woman's scent can exert profound effects on the hormonal and sexual functioning of other women as well as men. It has sometimes been noted that female roommates who live together for long time periods begin to menstruate at the same time, and similar phenomena have been reported in all-women dorms.[26] Women who are simply exposed to the sweat of other females on a daily basis for five minutes within a few months begin to cycle at the same time. This effect is not limited to just humans but occurs among dogs, cats, and other social varmints who live in close proximity.

Although not always consciously realized, the sexual nature of these chemicals is why many women inundate themselves with a variety of perfumes, the purpose of which is not merely to smell pleasant, but to exude almost narcotic sexual allure so that men are overwhelmed by an "obsession," and so on. Of course, such behavior and its motive are not completely unconscious. Some of the most popular and expensive perfumes have such obvious names: "My Sin," "Tabu," "Decadence," "Opium," "Indiscretion."

One need not cover the body with artificial scents in order to elicit sexual arousal. Natural pheromones exuded by men and women often contain subtle sexual messages including when a woman is most likely to become pregnant. For example, in one experiment it was reported that men find the vaginal smells of women most pleasant when they are at that point of their cycle when they are most likely to conceive.

However, men are supposedly not as sensitive to smell as are women. These findings, however, may be misleading. That is, women are better able to name smells than men, which may be more an indication of superior language capability and less a more discerning nose.

Pheromones are produced by males and females and do much more than signal sexual status. In a variety of species, pheromones indicate social status, aggressiveness, fear, threat, alarm, and the need to aggregate or disperse.[27] Different pheromones are also employed to stimulate grooming, the disposal of the dead, regurgitative feeding, and emigration. Others promote individual recognition, determine caste assignment, inhibit ovarian functioning, promote the construction of enclosures where insect queens might be reared, and are used to mark food sources.

There are also those which can stimulate mating, and others which inhibit it. Those which attract males to a willing female, and others which are released by the first lucky consort, which has the effect of driving away other suitors. Among mammals, a stallion will urinate on the feces of his mare, which essentially says to other males, "she belongs to me."

Some pheromones are released by insect larva which acts on the mother so as to promote her presence and care and which enable her to recognize her own versus that of a stranger. Indeed, be it insect, bird, mammal, or woman, mothers are often able to recognize their newborn infants by smell alone and later by smell can pick an article of clothing that their own child may have worn. Among insects, however, some pheromones are released by mother or infant which have the effect of driving away other pregnant mothers so that they will lay their eggs at some other site.[28]

Aggregation and Alarm Pheromones

Many pheromones also promote the seeking of close physical contact as well as causing insects to aggregate together so that a collective response can be made. For example, when a colony of bees is planning to swarm to relocate to a new home base, they often release chemical messengers which impel the entire colony to take flight.[29]

Sometimes the same pheromone will affect different members of the same species in a diametrically opposed manner. Among wasps, if an alarm chemical is released deep in the interior of the nest, queens and males will disperse and retreat, whereas wasp workers will congregate around the emissive site so that they can stage a group attack on whatever is threatening their home. They then form a ring around the offending intruder, and attack pheromones are released. During the ensuing battle more attack and aggregation pheromones are released so as to stimulate further attack behavior and to recruit others to their cause. This of course requires that pheromones be released in their proper temporal sequential order so that the right response occurs at the right time and the right place. If, for example, insects were to immediately fly into an attacking frenzy before they had a chance to aggregate and face the enemy, they would most likely lose the battle.

Among termites, once these alarm, aggregation, and then attack pheromones are sequentially released, soldiers will launch into a biting and snapping frenzy, attacking and inflicting often fatal injuries on predatory infiltrators, such as ants. As they fight, they secrete additional attack and alarm chemicals so as to keep the battle at a high pitch. Moreover, intruders will often be marked by these chemicals so that they can be hunted down and dispatched even as they try to flee.[30]

This also occurs among honeybees. The sting shaft is saturated with alarm pheromone which then attracts other bees to follow and attack the intruder even after it has run or flown quite a distance away. Some species of ants will, via their mandibles, spray attack chemicals directly into the wounds of the intruder, which not only elicits further attack, but which has a venomous effect on the nervous system of the wounded interloper.

Conversely, in response to an intruder, dispersal pheromones can be

Figure 33. Soldier ant. From D. H. Patent. *Looking at Ants*. Photo by E. S. Ross. New York: Holiday House, 1989.

released, which enables the colony to escape quickly. Once the threat is gone, aggregation chemicals will be released so the colony can reform. On the other hand, some predators are able to spray a disabling pheromone into the nests of those they wish to plunder. This has the dual effect of disorienting their victims and attracting additional raiders.

Hence, sometimes aggregation messengers are released so that once the group collects together, other forms of communication and behavior are made possible. Aggregation stimulants can act to promote feeding, fighting, reproduction, as well as a multitude of complex behaviors in sequence, such as those involving sex, but in particular, defense.

Pheromonal communication can thus be broken down into four basic major components: feeding, fighting, fleeing, and sexual activity. Be it insect or human, these functions are essentially completely limbic in orientation, and it is from these four components that more complex olfactory–pheromonal messages and responses evolved.

Pheromonal Communication between Different Species

Just as a shark can detect chemicals indicating that a potential prey is wounded and bleeding, or a dog or grizzly bear can determine that a human female is menstruating, a pheromone released by one species can be perceived and responded to by a different species, such as a predator or an insect involved in the slave trade.[31]

In fact, some species will release chemicals which mimic the aggregation or the sex pheromones of a completely different species and then attack and eat them when they arrive. The bola spider, for example, releases a chemical sex attractant which mimics the sex messengers of her favorite delicacy, the moth. When the eager male arrives in search of his potential consort, the spider jumps out and swings a sticky ball of webbing directly at him, in which he becomes hopelessly enmeshed.

Dogs are especially adapted for making numerous determinations regarding the activities of any number of species, including humans. Indeed, via a few careful sniffs your dog can determine if and what you have recently eaten, if you visited a friend and if they have a dog, where you might have spent your afternoon, if you had sex and possibly with whom, as well as if you recently urinated or defecated.

Territorial, Trail-Making, and Scenic Scents

In addition to sex, aggregation, and alarm pheromones, among insects and mammals there are territorial pheromones and those which are used for marking trails. Terrestrial trail markers are deposited along the soil, on plants, or upon objects so that others who follow will not stray or become lost. Many species of termites and ants use this not only to lead other workers to food sources but so as to maintain the social cohesion of the colony, particularly when migrating to a new home.[32]

Even human beings inadvertently leave pheromone trails, as any sheriff with a bloodhound could easily attest. Dogs, as most of us know, employ pheromone markings, via their urine, as a complex social–territorial signature. However, they avoid urinating on or marking themselves. Perhaps in this manner they are a step above many primates, such as the Galago, who purposefully urinate on their hands and not only mark everything they touch, but will rub their urine on their bodies and on those of their mates.

Nevertheless, dogs, females in particular, often love to roll in and cover themselves with the feces or rotten flesh odor of some dead animal. Dogs are not alone in engaging in these practices; some human societies apply fecal matter to their hair. Moreover, in addition to the natural chemical they secrete, modern-day humans anoint their bodies and homes with a number of colognes and perfumes, many of which are derived from the anal sacs of deer, cats, boars, and other little creatures.[33]

Perhaps the anal origin of so many pleasant odors accounts for the willingness of some animals to frequently sniff and lick one another's "perfumed" bottoms and genital regions, an activity among humans that sometimes takes the form of fellatio and cunnilingus. Similarly, not just dogs and humans, but monkeys and apes will sniff, lick, and suck on the genitals of their prospective mates and will lick and smell fingers which have been inserted into vaginas.[34] Of course, it is not just for the nice smell that humans and other animals engage in such practices, but due, in part, to the almost aphrodisiac sexual effects that such odors and pheromones induce.

Sex, aggregation, and trail-making chemicals are often dispersed directly into the air. For flying insects, this makes sense since they generally don't spend much time traveling on the ground. Hence they need only follow the direction of stable chemical traces floating in the air. If concentrations diminish rather than remain stable or increase, they know they are headed in the wrong direction.[35] Moreover, what is airborne is then able to travel great distances, certainly much farther than one could see or hear. This is why dogs and insects are often able to detect a sexually available female many miles away.

Neocortical Control

Most higher-order mammals have an olfactory system which in fact is quite similar to those of lower-order animals and even the most primitive vertebrates. However, some vertebrates, such as reptiles and some mammals, have an additional accessory olfactory system including a supersensitive vomeronasal olfactory organ which is located in the roof of the mouth. As noted, certain reptiles will sample the air or various items on the group with their tongue and then rub the tip against the roof of their mouth so as to increase their olfactory acuity. Indeed, this ancient olfactory organ is even present in human embryos and in adults, indicating it has not yet become completely vestigial in humans.

Even with a minimally developed vomeronasal olfactory organ, the overarching influence that olfaction and pheromones have on certain aspects of primate, and thus human, behavior remains very powerful. Humans, in fact, are capable of detecting and remembering as many as 10,000 odors. About 1% of the human genetic blueprint remains devoted to olfactory functioning, and there are almost 1,000 different genes which are responsible for the creation of odor-detecting nerve cells.

Nevertheless, in comparison to other species, the importance of smell has certainly diminished in regard to the tremendous control it exerts over behavior. In part this is a function of the development and evolution of the neocortex, the layers and expanse of which have reached their greatest dimensions among humans. Creatures such as bottom-feeding fish, sharks, amphibians, reptiles, and insects are completely lacking the 6- to 7-layered neocortex, and their brains are consequently more under the control of olfactory–pheromonal and limbic influences.

Many mammals are also highly dependent on olfactory input in regard to social, sexual, and feeding behavior, and are more or less ruled by their limbic systems. In contrast, humans are the beneficiaries of the tremendous development which has occurred in the neocortical regions of the frontal lobes and are often able to resist these olfactory and limbic influences, which instead may be countered or postponed until one is no longer in the mood, or hungry, or whatever.

In most other species the effects of many pheromones on the limbic system and behavior are almost reflexive, and at the minimum, sometimes overpowering. For example, in response to a sex pheromone being actively secreted by a nearly receptive female, the vast majority of insects, cats, dogs, and many other species are compelled to react and will persist even when actively rebuffed or there is a threat of being killed or eaten.

In general, humans, and but much less so, other apes and monkeys, are able to resist these urges (such as in the presence of a dominant male), or employ neocortically derived strategies that would enable them to meet the woman in a socially acceptable manner and offer pickup lines, ask her for a date, and so forth, so that what is desired might be acquired. Humans, in fact, can make a decision to give up sex altogether.

Nevertheless, among humans, despite the tremendous expansion in neocortex, particularly within the frontal lobe, emotional upheavals frequently occur, often with murderous or sexually inappropriate consequences. Among humans, these old limbic and rhinencephalic influences remain quite powerful and exert almost continual streams of influence which affect many different aspects of our lives including not just sex and the consumption of food, but the manner in which we interact with strangers and loved ones alike. Although largely unconscious, we still employ and rely upon olfactory and pheromonal communication, the essence of which not only tells us about the world, but through which the world is told about us.

3

Touch Me—Feel Me—Feed Me— Kiss Me!

Mary was not sure what was more surprising, that her handsome, swarthy boss had asked her out, or that she was sitting next to him on these soft, velvety cushions. Gently, softly, he took hold of her fingers with his strong, coarse hands, hands that were calloused and roughened from working on his ranch every weekend. He felt so strong and powerful; it made her heart pound against her chest. Entwining his toughened fingers about her own, he pulled her close. She could feel the indentation of his ring from the pressure of it against her skin as well as the hard, rippling muscles of his thick, strong chest beneath the silky smooth fabric of his shirt. His body felt hard as a rock. Running her soft hand across his firm, unyielding chest, she could feel each individual sinewy muscle, each coil of which made her pulsate with excitement. As she stared into his deep brown eyes, she sensed the presence of two of his cool fingers upon her hot knee. And then he drew her soft, moist lips next to his. The coarse, tiny hairs of his beard tickled her with pleasure. It made her cool skin tingle and her heart thump against her chest more rapidly as those strong, hard fingers slowly and softly inched up her velvety thigh. . . .

Lisa stiffly and reluctantly picked up her baby and held her uncomfortably next to her tense body. Her body trembled with irritation as she tried to avoid those deep brown, curious eyes with those thick, dark lashes, which all her friends described as so beautiful. Dave's eyes! Gritting her teeth, she felt

Figure 34. Michelangelo. "The Creation of Adam."

Figure 35. Brancusi. "The Kiss." 1908.

her whole body grow even more rigid and then she started to shake angrily. Slowly, rigidly, she held her little daughter at arm's length and then rapidly set her back down, releasing her just inches above the bed. She began tapping her foot irritably against the floor and then kicked one of the legs of the crib. It was hard to believe she was almost a year old. A whole year since he abandoned them. Suddenly she stomped her foot so hard on the floor that the crib shook from the vibration. Then she went away.

It has sometimes been reported anecdotally that some infants, sleeping in another room with the doors closed, often seem able to discern when mommy and daddy are about to make love, even when they are very quiet. Presumably, this crying response is triggered by olfactory cues.

Although infants respond strongly to a variety of aromas, humans appear to first experience or express emotionality in relation to the body and in response to touch or tactile sensations, sudden movement, or rapid changes in position or posture.[1]

Among human infants, the earliest smiles are induced through touch (e.g., light stroking or even blowing on the skin), whereas loss of physical support or the sensation of falling is the most powerful stimulus for triggering an alarmed emotional reaction in the newborn.

However, the earliest and most consistent manifestation of emotion in the newborn infant consists of screaming and crying whereas positive emotions are limited to an attitude of acceptance and quiescence[2]— emotions which are first mediated by the hypothalamus of the limbic system.[3] It is only over the course of the next several weeks and months that the remainder of the limbic system begins to exert its influence, and several months more before nerve fiber interconnections within and between neocortical tissues begin to mature and are more fully established.[4]

Hence, initially, the world of the newborn infant is a somewhat confused matrix of physical, tactual, and olfactory sensations, coupled with sounds and sights which mean little or nothing to its neocortex except in regard to the reflexive reactions elicited within the limbic system and other ancient cortical nuclei. This includes the old nuclei within the **brain stem** and the **midbrain**, which mature before the neocortex and which enable the infant to respond to a variety of visual, auditory, and particularly physical sensations.

It is via the body and through the tactile stimulation provided by the primary caretaker that the infant first makes emotional contact with the outside world. It is through this contact and physical interaction that the infant first begins to establish a sense of self and begins the long road to emotional self-sufficiency or instability.[5]

Determining Emotional Significance

In their journey from the external to the internal environment, tactile (as well as olfactory, auditory, and visual) inputs are transmitted to various limbic nuclei such as the amygdala. These structures in turn scan external signals so as to detect those that have some emotional or motivational significance. They then compare them with memories of similar experiences, and if necessary motivate the organism to make an appropriate response.[6]

As we've noted, the amygdala as well as other limbic structures are essential in discerning social–emotional cues from all sources of sensation, and all sensory systems transmit to it. When someone touches us in a stiff manner this information is analyzed for its emotional significance and is then compared and perhaps integrated with cues from the other sources of sensory input, for example, facial expression, tone of voice, memory, and so on.

For infants and adult human beings, touch is an extremely important form of communication, and it shapes our cognitive, intellectual, emotional, and social development, and mental health. Individuals express fondness and acceptance by holding hands, hugging, slapping one another on the back or the butt (especially professional football players), and both children and adults seek comfort in the arms of loved ones. One of the earliest as well as the most pervasive aspects of emotion—fear—often triggers extreme hugging and huddling not only among infants, children, and even chimpanzees, but hardened soldiers on the front lines of battle.[7]

Touch serves to soothe and also acts to strengthen the bond between members of the same social group or between mates or mothers and their young. Not only humans but many species of birds and mammals engage in mutual touching in the form of grooming and even kissing. They will comb through feathers, or will lick, touch, rub, or sift through the fur of their young or their mates for hours on end. This mutual touching serves not only to reduce tension and increase feelings of closeness but can greatly lessen the likelihood of being attacked by a dominant member of one's social group who just happens to be feeling ornery.[8]

Touching also serves as a means of achieving intimacy and physical closeness and can serve as an exuberant prelude and accompaniment to an exciting evening of sexual acrobatics. Even male monkeys and apes greatly increase the amount they groom their lover, once she goes into heat.[9]

Touching, in fact, can exert not only soothing and sexual feelings, but even healing influences. This has been exploited by some religious groups who practice a laying-on of the hands so as to dispel evil, sickness, and devils from the body. However, rather than dispelling demons, some health practitioners utilize touch as a means of dispelling muscle tension and toxins from the skin, as well as to stimulate circulation. As formalized systems of touching, these include Swedish massage, which uses long strokes toward the heart; Reichian massage, which uses long strokes away from the heart; reflexology, which focuses on the feet; and Japanese shiatsu, which is like finger acupuncture.

Touch is important because the body is often used as a conduit of emotion. This is a function of emotional expression being rooted in the body and manifested through muscular reaction. This is also why most people *feel* their emotions rather than hear or see them. Of course, one can express their emotions in a variety of ways. Nevertheless, emotions are first and foremost experienced in relation to the body, and it is via the body that they are expressed.

Since the body and emotions are linked, sometimes people who are unable to articulate or act on their feelings instead express them physically in the form of bent posture, facial grimaces, and tense movements. Males are particularly likely to express their feelings in this manner, or in the form of actual physical violence, as their ability to verbalize feelings is much more poorly developed as compared to women.[10]

When emotions are channeled and expressed through the body, one unfortunate side effect of all this emotional tension and turmoil is body aches and pains. When individuals cannot face or even recognize whatever might be bothering them and instead keep their feelings locked up, they may instead develop headaches, backaches, stomachaches, ulcers, hives, shingles, various allergies and asthmas, as well as tumors and cancers, a consequence of the body's literally taking an emotional pounding as their feelings strive for expression. Of course, not all such symptoms are due to the bottling-up of emotions.

It is well known that many of the muscles relax and contract as we move about, breathe, and interact with the environment. Sometimes the relaxation phase is incomplete, however, because the cycle associated with its activation/relaxation is incomplete. If an emotion is held in, then the muscles may fail to relax and are held in as well, and the musculature becomes tense and tight. This failure to relax the muscles is due to unreleased or suppressed emotional tension and reflects a high degree of arousal in the amygdala. As such, emotionally, intellectually, and physically the ability to function effectively can become impaired. In consequence, those who suppress their feelings, as well as those who are subject to emotional upheavals and trauma, often experience sometimes debilitating muscle, back, and head-aches due to prolonged defensive or suppressive musculature contraction and limbic system arousal.

Often these suppressive influences are most keenly felt at night, paradoxically right at the time they are most trying to relax so that they can sleep, which they cannot. My advice has been as follows: When you go to bed, tell your shoulders, arms, legs, and so on to relax and then deflate them as if they were dead, like you are letting air out of a balloon. Once relaxation is achieved, the body melts into the soft contours of the bed and the angels of sleep descend. This is why massage sometimes is an excellent way of not only relaxing the musculature but also of releasing associated suppressed feelings, as the limbic system mediates not only emotional arousal but also the emotional characteristics of touch.

Many people, however, have lost the ability to see the interwoven connection between their mind, their emotions, and their body. Many are so out of touch with who and what they are, as well as their own need to touch and be touched, that the very unity of their psyche always remains perilously close to fracture. Males are particularly subject to these terribly disruptive influences as they are much less willing to engage in tension-reducing nonsexual hugging, touching, and caressing, and are generally reluctant or unable to verbalize their feelings and emotions, except when angry or sexually aroused.[11,12]

Limbic Linkage of Touch and Emotion

How you are touched, how someone touches you, even your desire for someone to touch, fondle, stroke, and kiss various regions of your anatomy are experienced and mediated by nuclei of the ancient limbic lobe, as is one's capacity to experience an orgasm.[13] The limbic system in fact motivates and urges an individual to physically approach and make contact with the object or person of his or her desire, via its extensive interconnections with the ancient old cortical motor centers as well as the neocortex.

The limbic system is also structurally and functionally organized differently in males versus females, a consequence of the presence or absence of masculinizing hormones (testosterone) soon after birth or, in humans, early in pregnancy.[14] It is due to this difference that males and females think, feel, and react quite differently about what appears to be ostensibly the very same experience.

Among modern humans as well as ancient and more primitive life forms possessing only a limbic system and little or no neocortex, the amygdala maintains rich interconnections with a series of ancient two-layered cortical **motor nuclei**, collectively referred to as the **basal ganglia**. The basal ganglia, in fact, evolved out of the olfactory limbic system. Hence, the basal ganglia are involved in the expression of many aspects of emotional as well as routine behaviors.

Among modern humans these same old cortical motor centers control facial expression and gross body movements, including those of the arms, hands, legs, and feet. It is due to the limbic interconnections with the basal ganglia that individuals may act on their emotions without thinking by running away, striking out, or via facial expression. Conversely, this is why, for example, when the basal ganglia are injured or damaged, such as occurs with Parkinson's disease, movement becomes quite difficult or spasmodic, and the face becomes masklike and devoid of emotion—of course, these are just some of the functions of the basal ganglia.

Be it man or beast, given a suitable emotional stimulus, the amygdala in conjunction with other brain regions can activate these ancient motor centers so as to run away or toward what has arrested its attention. In this manner the limbic system is able to take control over behavior even when the neocortical regions of the brain may be opposed or still trying to come to a decision. When this occurs, an individual may be overcome with emotion, act heroically, irrationally, courageously, ferociously, sexually, or destructively, and then later proclaim in all innocence that he doesn't know what came over him.[15]

The capacity to respond spontaneously and quickly to certain situations is also very adaptive, and the amygdala has evolved accordingly. The limbic system is concerned first and foremost with survival, escaping pain, obtaining pleasure, and establishing intimate social, emotional, and physical contact with others. The limbic system also contains opiate receptors and

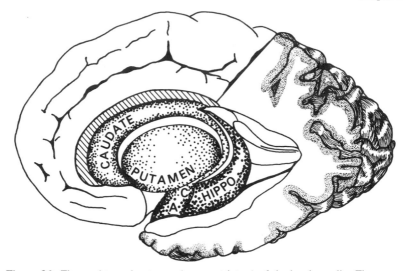

Figure 36. The caudate and putamen (corpus striatum) of the basal ganglia. The corpus striatum evolved from the olfactory striatum and thus from the limbic system and the amygdala with which it maintains rich interconnections. From P. Maclean. *The Evolution of the Triune Brain*. New York: Plenum Press, 1990.

narcoticlike neurotransmitters referred to as enkephalins and endorphins and is capable of generating profound, even mystical states of orgasmic pleasure, or profound misery or rage. In this manner it is able to exert immediate and profound influence if not control over behavior, and for good reason. If a dangerous animal is about to leap upon you, the body must be able to respond immediately and either flee or fight. In such a situation, the generation of thought, which is under the domain of the new brain (neocortex), may instead result in the generation of a meal for some predator. However, it is the accumulation of this same neocortex which makes for a successful hunter, and mammalian predators are well endowed although they too are subject to the whims of their limbic brains.

The limbic system enables us to survive and mediates, promotes, and rewards a whole host of behaviors designed to maximize one's ability to thrive whether one is an infant or adult. Morals, politics, strategy, and long-term consequences are not as important to it as this is the domain of more recently evolved neocortical tissues. Among human infants and so-called

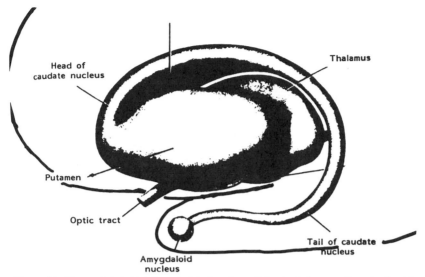

Figure 37. The caudate (which is part of the basal ganglia) extends in an arc from the frontal into the temporal lobe where it merges with the amygdala. From *Human Neuroanatomy*, by R. C. Truex and M. B. Carpenter, 1969. Courtesy of the Williams & Wilkins Co., Baltimore, MD.

lesser animals, these more intellectually related concerns are not only irrelevant but could be dangerous. If for some reason limbic needs were overruled, neglected, or denied, the result is likely to be death.

Contact Comfort

During the late 1930s, the Nazi Germans conducted an experiment designed to raise "supermen." Couples were selected according to strict physical, health, and racial criteria and were allowed to live in special secret camps until children were born. Soon after birth, the children were removed and reared in a special home designed to maximize their superiorities with the single exception that mothering was not provided as it was thought undesirable to expose these impressionable youngsters to those who might instill in them feelings of nurturance, compassion, and the like. Within 2 years, 19 of the 20 children raised in this "superior" environment became severely developmentally and emotionally abnormal and would "lie like dead fish," and in all respects behaved like "idiots."

R. A. Spitz[16] studied over 200 children, half of whom were raised in a prison nursery and the rest in a foundling home where mothering and social stimulation was quite minimal due to the high ratio of children to staff. Within 1 year those reared in the foundling home became unresponsive to social stimulation and would lie passively on their beds. They would scream and try to avoid strangers or novel objects or toys, and spend hours on end engaged in repetitive, stereotyped, and bizarre movements designed to maximize self-stimulation such as rocking, head banging, or pinching precisely the same piece of skin until sores developed. Most of these children became permanently emotionally and socially scarred.

As is well known, for the first several years of life, physical and social interaction with others is critically important to psychological, neurological, and physical development. The nervous system of an infant, in fact, has been long adapted to receiving vigorous physical stimulation. For much of human evolution, infants were often carried about by their mothers who in turn would bend, twist, walk, run, climb, gather, and what not with a baby strapped or clutched tightly to their bodies. This is also why infants retain the ability to clutch back, the grasp reflex; they are designed for mobility and holding on to something, such as mom's hair or breasts, because they cannot walk. In fact, although the female breasts are in part designed to indicate continual sexual availability, since humans have long lost their hair coat, they, too, serve as convenient handles upon which infants may hang and grasp for support.

In any case, if diverse forms of stimulation are not provided during early development, the growth of nerve cells, dendrites, axons, and the establishment of certain neural circuits throughout the brain becomes retarded due to cell atrophy or death, and all subsequent behavior is severely affected.[17] Babies need to be held, touched, caressed, rocked, and carried about for a good part of every day, for these are the adaptive requirements of our 65 million-year-old mammalian nervous system. If denied this stimulation, infants have no way to provide it on their own and in consequence will either die or fail to thrive.[18]

Not surprisingly, so intense is this need for stimulation that during much of the first year of life children will indiscriminately seek contact with and will smile at the approach of anyone, even complete strangers.[19] During the first 8 months, they will also appear to become upset or will protest when someone attempts to leave the room as their desire is to receive as much social stimulation as possible. These behaviors in turn maximize opportunities for physical interaction and, in part, are a function of the maturation of the limbic system and in particular the amygdala, which mediates and requires social–emotional stimulation, particularly in the form of touch.

Septal, Cingulate, and Amygdala Interactions: Sex and Attachment

Like the neocortex, the limbic system has also evolved, although retaining its essential concerns regarding survival. Soon after the evolution of the amygdala, about 500 million years ago, an additional limbic structure began to appear, and this is the septal nucleus, which is an outgrowth of the hypothalamus. The septal nucleus, like other limbic brain tissue, is very concerned with sex and the functioning of the genitalia. In conjunction with the amygdala and the more recently evolved, four-layered transitional limbic cortex, the cingulate gyrus, the septal nucleus also promotes feelings of pleasure, aversion, and selective attachment to loved ones such as the mother or one's mate.[20]

The septal nucleus, cingulate gyrus, and amygdala have reached their greatest degree of development in humans and have expanded in size in correspondence to increases in the mass and density of the neocortex. It is the appearance of the six-layered neocortical mantle, in conjunction with these limbic developments, that have made possible the establishment of long-term maternal child care and the establishment of the family.[21] This is why creatures who are devoid of neocortex, such as reptiles, amphibians, fish, sharks, worms, and so on, are devoid of maternal feelings and why their young often must hide immediately after they hatch, to avoid being cannibalized, often by their own parents.

However, it is the differential rates and time of maturation of these three limbic nuclei and the neocortex which make possible feelings of selective attachment to one's mother, and later, to a spouse, lover, or infant.[22] For example, at about 7 to 9 months of age, the infant becomes less likely to put everything she sees or touches in her mouth, and she is more discriminate in her interactions and less likely to seek contact from just anyone. Prior to that, contact seeking is rather indiscriminate, and the child is extremely oral, putting everything into her mouth, and this is due to the immaturity of the amygdala. It is around 8 months of age that a very intense and specific attachment is formed and strengthened; for example, to one's mother—an attachment which becomes progressively more intense and stable over the ensuing months and years. This is a function of the counterbalancing influences of the septal nuclei and cingulate gyrus as they begin to mature.

Again, this first stage of indiscriminate orality and contact seeking is due to the slow maturation of the amygdala and the lack of input from the septal nuclei and cingulate, which have yet to appear. Hence, the oral stage of development is also mediated by the amygdala.

The process of selective contact seeking is due to the later appearing and also slow maturation of the septal nucleus and cingulate.[23] For example, once the septal and cingulate nuclei begin to mature, they begin to provide oppositional equilibrium within the limbic system so that the amygdala is not allowed to function unchecked and indiscriminately. In consequence, the seeking of contact comfort becomes increasingly narrowed and restricted to the mother or primary caretaker. The child also becomes less oral in orientation and progressively more inclined to experiment with the genitals as it moves from the amygdala to the septal stage of development.

If both amygdala were for some reason surgically removed or destroyed, this narrowing would in fact become impenetrable, and the animal or human so affected would refuse contact and completely avoid others. They would also become hyperoral and again begin to place things in their mouth indiscriminately.

Social–Emotional Agnosia

Among primates and mammals, bilateral destruction of the amygdala significantly disturbs the ability to determine and identify the motivational and emotional significance of externally occurring events, to discern social–emotional nuances conveyed by others, or to select what behavior is appropriate given a specific social context. Bilateral lesions lower responsiveness to aversive and social stimuli, reduce aggressiveness, fearfulness, competitiveness, dominance, and social interest. Indeed, this condition is so pervasive that subjects seem to have tremendous difficulty discerning the meaning or recognizing the significance of even common objects. They respond in an emotionally blunted manner and seem unable to recognize what they see, feel, and experience. Things and people seem stripped of meaning.

Like an infant (who similarly is without a fully functional amygdala), individuals with this condition engage in extreme orality and will indiscriminately pick up various objects and place them in their mouth regardless of their appropriateness. There is a repetitive quality to this behavior, for once they put it down they seem to have *forgotten* that they had just *explored* it, and will immediately pick it up and place it again in their mouth as if it were a completely unfamiliar object. Although ostensibly exploratory, there is thus a failure to learn, to remember, to discern motivational significance, to habituate with repeated contact, or to discriminate between appropriate versus inappropriate stimuli or to recognize even basic feelings such as fear or love.

For example, one young man, following bilateral removal of the amygdala, subsequently demonstrated an inability to recognize anyone, including close friends, relatives, and his mother with whom he had formerly been quite close. He ceased to respond in an emotional manner to his environment and seemed unable to recognize feelings expressed by others. He became extremely socially unresponsive such that he preferred to sit in isolation, well away from others.

Among primates who have undergone bilateral amygdaloid removal, once they are released from captivity and allowed to return to their social group, a social–emotional agnosia becomes readily apparent as they no longer respond to or seem able to appreciate or understand emotional or social nuances. Indeed, they appear to have little or no interest in social activity and persistently attempt to avoid contact with others. If approached, they withdraw, and if followed, they flee. Indeed, they behave as if they have no understanding of what is expected of them or what others intend or are attempting to convey, even when the behavior is quite friendly and concerned. Among adults with bilateral lesions, total isolation seems to be preferred as they have lost the ability to feel close or to attach to anyone.

As might be expected, maternal behavior is severely affected. According to Kling, mothers whose amygdala were subsequently destroyed behaved as if their "infant were a strange object to be mouthed, bitten and tossed around as though it were a rubber ball."

Limbic Social Deprivation

The intense need for physical emotional contact remains crucially important not only during the first several years but throughout life. If physical and tactile stimulation is lacking, inadequate, or for any reason abnormal, the limbic structures that mediate these needs become affected. So too is the ability to form close relationships or to bond with others. Moreover, nerve cell growth throughout the nervous system is affected and becomes retarded. This is why infants and young children desperately require a caretaker to interact with them for a good part of every day, and why some children will accept negative attention (by acting bad) rather than be ignored. The very functioning of their brain and body requires that they receive some form of social and physical stimulation, and even negative attention is better than none at all.

So intense is the need for physical and tactual stimulation that young animals placed in social isolation where opportunities for physical and social contact are minimized initially will form attachments to bare wire frames, to television sets, to dogs that might maul them, to creatures that might kill them, and among humans, to mothers who might abuse them.[24]

Even infant animals normally reared that are later punished or even given an electric shock every time they come close to their mother will redouble their efforts and attempt to follow her even that much more closely. Although this behavior may seem paradoxical, it is due to the fact that under normal circumstances the mother is perceived as the source from which contact comfort, safety, and security may be obtained. Hence, even when mother is punishing or abusive, this behavior cannot be reconciled emotionally by the child's limbic system. This is because abusiveness from the primary caretaker is abnormal and not something the primitive limbic system is prepared to deal with. Hence, in abusive situations the child is biologically driven to seek safety in the arms of the very individual who he should be running from.

This condition has been inadvertently well demonstrated by Dr. Ronald Melzack, of McGill University in Canada, who performed some very terrible experiments on dogs,[25] a creature with a limbic system very similar to humans. Specifically, Melzack raised young dogs in isolation and when they were approximately 7 to 10 months of age he tortured them individually by sticking burning matches into their noses. He also did this with normally raised pups which of course tried to bite him and flee. However, those who were raised in isolation were so limbically starved for physical contact and social intimacy, that although they reflexively twisted away when burnt, they immediately stuck their noses right back into the flame and hovered excitedly next to him.

So pervasive is this need for physical interaction (especially among humans) that when grossly reduced or denied, the result is often death.[26] For example, the morbidity rates were typically 70% and higher for children less than 1 year of age who were raised in foundling homes during the early 1900s, when the need for contact was not well recognized. Typically, children and infants were left unattended and alone in their cribs for most of the day, and only their basic needs were cared for.[27]

Of those who survived an infancy spent in institutions where mothering and contact comfort was minimized, signs of low intelligence, extreme passivity, apathy, as well as severe attentional deficits are often characteristic. Such individuals have difficulty forming attachments or maintaining social interactions later in life. Indeed, this can even affect their ability to feel love or affection, even when they have grown and have had children.

Similarly, nonhuman primates who are raised in a motherless environment not only develop bizarre behaviors, but are unable to provide adequate mothering to their own young. As described by Harry Harlow[28]:

Figure 38. Communal bedroom of a Paris foundling home, 1846, where up to 100 infants and small children might be warehoused. From R. Dirx. *Das Kind: Das unbekannte Wessen.* Namburg: Marion von Schroder Verlag, 1964.

After the birth of her baby, the first of these unmothered mothers ignored the infant and sat relatively motionless at one side of the cage, staring fixedly into space hour after hour. As the infant matured desperate attempts to effect maternal contact were consistently repulsed. . . . Other motherless monkeys were indifferent to their babies or brutalized them, biting off their fingers or toes, pounding them, and nearly killing them until caretakers intervened. One of the most interesting findings was that despite the consistent punishment, the babies persisted in their attempts to make maternal contact.

However, even temporary separation from the mother or primary caretaker can exert deleterious consequences which become more severe and permanent as the separation continues. For example, children who are separated from their moms between the ages of 1 to 2 years of age and placed in a hospital, children's home, or what not, eventually pass through three stages of emotional turmoil, the first of which is characterized by a protest period where they frequently cry and scream for their mommy, and this may last several weeks or even months. This is followed by a stage of despair in which the child ceases to cry, loses interest in his environment and withdraws, and this, too, can persist for months. In the final stage, he ceases to show interest in others, loses his appetite, and ceases to respond to the affection offered by others and becomes quite passive and unresponsive. Instead, he may sit or lie for long periods with a frozen expression on his face, staring for hours at nothing. If the separation continues he deteriorates further and becomes physically ill and may die.[29] In general, males are more severely affected than females.

Those who were only temporarily removed, once they were returned home, would desperately cling to their mothers, follow them everywhere, and become extremely fearful when left alone even for short time periods. Within a few weeks or months, however, most seem to recover. However, those who were deprived of maternal contact for 6 months or more instead behaved in a withdrawn, depressed manner and showed no interest in and were unable to reestablish their normal attachment to their mother. According to J. Bowlby, children who suffer long-term or repeated separations during the first 3 to 5 years of life are usually permanently affected.[30]

The reason these affects are permanent is due to the catastrophic effects early deprivation induces within the limbic system. That is, limbic neurons, become smaller, and throughout the nervous system there is cell death and atrophy. As noted, when the amygdala functions abnormally, social functions become similarly disturbed.

The Fragility of the Male Ego

It is noteworthy that in regard to temporary or long-term deprivation, males are much more severely affected than females.[31] Moreover, although adults who are separated from their spouse due to divorce, death, or abandonment may show many of the same symptoms as deprived children, it is the male who is more likely to respond with limbic, infantile rage. When this occurs they sometimes stalk and kill their wayward spouse. Males are much more fragile than females, but as adults, they are much more dangerous.

This is also in large part a function of the differential manner in which men and women view the expression of dependency needs and the unwillingness, and in fact the neurological inability of many males to accept nurturance during times when a male is feeling alone, weak, afraid, or in need of comfort.[32] Even many married men with otherwise good relationships often refuse or rebuff attempts by their wives to provide them with the contact comfort and tenderness that they need such as following the loss of a job or what they perceive as a personal failure or shortcoming.[33] Instead, many males are much more willing to allow their emotions to fester and wreck their physical and psychological health.

However, in part, this is due to differences in brain structure and the greater access women are provided in regard to information exchange between the right and left half of the brain (i.e., a larger anterior commissure—a fiber pathway that links the two amygdalas), as well as evolutionary influences and selective pressures acting differentially on the male versus female limbic system. Males have a male neural pattern, and females, a female neural pattern of limbic interconnections in this region which differently effect aggressiveness and the expression versus suppression of sympathy, compassion, and connectedness, which in women is also experienced and expressed as maternal love and nurturance.

Paradoxically, many men respond as if they find the provision of comfort and nurturance as aversive, particularly when in time of need, because they do. Many men feel they must deny their hurt rather than admit to it, especially since other men would not accept this display of weakness but may in fact deride a man for it. Moreover, there have long been selective pressures on hunting males to control their feelings and stifle their fears, for those who ran away in fear from a charging wounded beast, were more likely to end up prey or the laughing stock of his fellows.

Hence, among modern males, when hurt, or feeling afraid or abandoned, like the abandoned and deprived child, many tend to withdraw or become expressionless, opting to "tough it out." In contrast, women are much more willing to talk about their feelings, exchange hugs, and receive nurturance, love, understanding, and emotional, physical support from family and their social groups,[34] which not only meets their limbic needs for contact comfort but decreases the likelihood that women will respond with murderous limbic rage.

However, men also need women and their wives as they provide their only source of contact comfort. In consequence men are much more vulnerable to the shock of divorce and the loss of their wives and often become severely depressed when they are abandoned. In fact, men are seven times more likely than women to commit suicide following catastrophic occurrences such as divorce or the loss of a job.[35]

It is due to these same tremendous dependency needs that most men refuse to divorce their wives until they have another woman waiting in the wings. When women initiate the divorce proceedings (which occurs about 35% of the time), it is not only unwanted but comes as a shock to the husband.[36] This is particularly traumatizing to a man who feels he is being emasculated and has now lost control over his own wife. As such, rather than seek contact comfort, he may fly into a rage.

Moreover, because the man feels castrated and abandoned this activates those same limbic nuclei which raged when mother increasingly ceased to meet his infantile desires and abandoned him for his father. However, now an adult, instead of simply raging helplessly, he may threaten, stalk, beat, or even kill his former spouse or girlfriend and perhaps even his own children. Or, if it's his job he's lost, he may threaten, attack, or as occurs with increasing frequency in this society, kill his boss and coworkers.

Males, be it babies, little boys, or grown men, cannot tolerate being alone or suffering prolonged separation from the primary sources of contact comfort, affection, or self-esteem, such as their wives, girlfriends, or jobs. It is for this same reason that single or divorced men have a much higher death rate than those who are married; their limbic brains are more fragile.

Deprivation and Amygdala–Septal Functioning

A baby cries when it is put down and left alone, and it ceases to cry when it is held and rocked. It is the amygdala and **cingulate gyrus** of the limbic system that cries, and it is these same limbic nuclei which are appeased by being held and comforted.[37] Tactual sensations are so vital to the limbic system and amygdala that single cells may respond to touch regardless of where on the body the person was stimulated.

Unfortunately, when there is inadequate tactile and physical–social interaction during early development, the ability of these cells to develop and function adequately is significantly reduced. The amygdala and other limbic nuclei, as well as neocortical cells, becomes environmentally and emotionally damaged. Not receiving sufficient and appropriate stimulation these cells begin to die and atrophy from disuse; just like a muscle if unused. "Use it or lose it." Once these neurons die or if certain interconnections between different regions are not maintained, they are no longer able to respond appropriately to physical, emotional, and social interaction.[38]

In fact, an individual raised in these conditions may fail to derive emotional satisfaction or even pleasure from lower-order limbic functions,

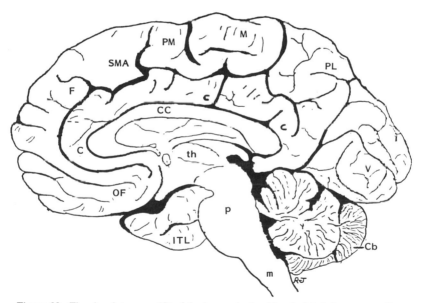

Figure 39. The cingulate gyrus (C) of the human brain. Also depicted: the corpus callosum (CC), visual cortex (V), cingulate gyrus (C), thalamus (th), inferior temporal lobe (ITL), pons (p) and medulla (m) of the brain stem, parietal lobe (PL), frontal lobe (F), frontal motor (M), premotor (PM), supplementary motor (SMA), and the medial and orbital frontal lobes (OF). From R. Joseph. *Neuropsychology, Neuropsychiatry, and Behavioral Neurology.* New York: Plenum Press, 1990.

such as eating. On the other hand, many lonely or emotionally traumatized individuals try to fill up that empty space so deep inside their soul by eating, eating, eating. Indeed, eating, just like the sense of touch and smell as well as the need for social stimulation, is also mediated by the amygdala, hypothalamus, and other limbic structures, and it is for this reason that some people substitute one limbic need for a wholly different one.[39] This is because these seemingly different needs are limbically linked and when fulfilled can cause the limbic system to respond with feelings of pleasure even when the original desire continues to go wanting.

When the limbic system experiences pleasure, regardless of its source, other needs and concerns sometimes fall temporarily by the wayside, even when the original need is never satisfied. Hence, sometimes eating can temporarily assuage one's loneliness because it induces feelings of pleasure in the same part of the brain which subserves the need for social contact, that is, the limbic system.

Food Sharing and Kissing

It is due to limbic mediation that eating is also a social function that can promote bonding. The motoric components associated with eating and food sharing have in fact become adapted for the purposes of achieving, establishing, and maintaining social–physical intimacy for adults, children, and especially the infants of many species. Hence, friends and family eat together, and many of us bite, nibble, lick, or suck on parts of the body of a loved one. Most people, regardless of culture (but not climate) make substantial mouth-to-mouth and even tongue-to-tongue contact with individuals they respect, love, or who sexually excite them, and this is called kissing.[40]

It has been argued by scholars such as Eibl-Eibesfeldt that kissing is a form of ritualized feeding behavior, which among primates has evolved into a nongustatory means of maintaining intimacy and social bonding.[41] In part, this is why humans and animals kiss, bite, nibble, suck, and even lick one another. Many mammals in fact lick their partners as a form of greeting and as a form of courtship or mating foreplay. When sex is sought, much licking occurs, particularly in the genital region.

Baboons use lip smacking as a form of courtship and will pucker up almost as if they are about to suck.[42] Similarly, so-called pygmy chimpanzees, the bonobos, not only kiss but will stick their tongues deep inside the mouth of their partner or even their human caretakers if they are receptive.[43] Dogs and wolves also utilize a considerable amount of tongue action as a form of greeting.[44]

Humans will kiss each other in greeting, will kiss the ring of a pope or bishop, and will kiss their hands so as to blow greetings and love to departing friends and relatives. Among some cultures, where on the body

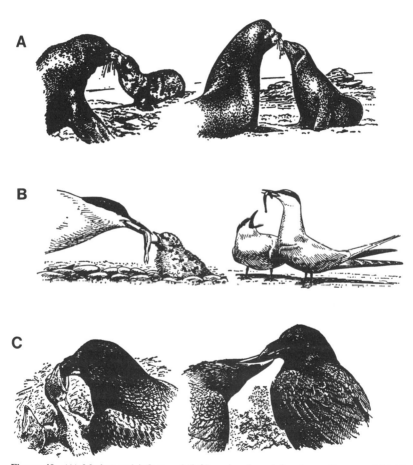

Figure 40. (A) Mother and infant seal (left), and male and female sea lions greet (right). (B) Infant bird begging mother for worm (left). A male feeds a female tern a worm as a prelude to copulation (right). (C) A raven feeds its young (left). A male is greeted by his mate by billing (right). From W. Wickler. *The Sexual Code*. Garden City: Anchor Press, 1973.

two parties kiss is a highly ritualized form of social greeting as well as an indication of rank. For example, among Iranians, kissing on the mouth is for those of equal rank. If of unequal rank they kiss on the cheek. However, if the person is far below the other in rank they will prostrate themselves.

Hugs, Kisses, and the Social Stomach

All insect states have social castes which are very rigidly structured.[45] There is only one queen, who is responsible for reproduction, and hundreds of thousands or even millions of soldiers, workers, brood tenders, scouts, and so on.

Via pheromonal analysis, in all insect states strangers are easily recognized and are driven from the nest or killed even when of the very same species. Similarly, citizens of the same insect city are able to easily recognize one another and are respectful of the duties each has to perform as they are all highly dependent upon one another.

It is perhaps this mutual dependence which maintains the insect state; otherwise its members may simply disband. On the other hand, like dogs, chimpanzees, and humans, perhaps this mutual dependence is also a dependence on social stimulation. That is, bees and ants, wasps, and so on may derive some type of insect enjoyment from being together. Nevertheless, like social mammals, insect societies are often maintained by a form of *kissing* and the maintenance of a social stomach.[46]

It has long been assumed that social insects are able to recognize strangers, even when they are of an identical species, because they have a different nest scent. In contrast, fellow citizens go unmolested about their duties because they all have the same characteristic odor. Another major factor that has been assumed to play a role in the maintenance and cohesion of the group is mutual social feeding.[47]

Social insects, such as bees, wasps, termites, and ants, often beg their fellow citizens for food which their compatriots then regurgitate. When a bee returns after having indulged in a delicious dinner, other bees will gather around and will then beg with animated gyrations of their antennas until a drop of food is disgorged. Those who immediately feed are then subject to begging from other members of the hive and will then regurgitate for them as well. Sometimes they will beg each other simultaneously and then feed each other.

This whole process begins at the larva stage, for larvae are fed mouth to mouth. For example, wasp larvae hang head down from their little cells and

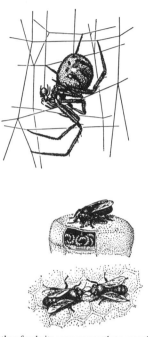

Figure 41. (top) A spider mother feeds its young mouth to mouth. (center) A bumblebee feeds larvae. (bottom) A worker bee feeds another by mouth. From W. Wickler. *The Sexual Code.* Garden City: Anchor Press, 1973.

are fed by adult wasps who premasticate grubs and other insects and then feed this to them. What is the reward for the adult? Why are they so altruistic? The larvae, when touched, will produce a drop of liquid which adults find quite tasty. Hence, they readily and eagerly feed their little larvae.

Human beings provided each other with similar rewards. Be it the ancient Cro-Magnon mother who in addition to providing the milk of her own body, and premasticated much of what she fed to her infant, or the modern woman who feeds her child mass-marketed baby food, they are rewarded by their charges who use their little mouth not only for eating and sucking but for smiling.

An additional reward that some breast-feeding women as well as some authorities find disconcerting is the pleasurable sensation she experiences

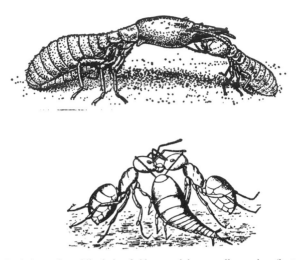

Figure 42. (top) A termite soldier being fed by mouth by a smaller worker. (bottom) Two ants and a silverfish feed each other by mouth. From W. Wickler. *The Sexual Code*. Garden City: Anchor Press, 1973.

as her vagina contracts and spasms in response to the reflexive limbic system, hormonally induced secretion of oxytocin (triggered by the sucking of her nipples), which aids in the production of milk. In one recent case of neocortical injustice and ignorance in Florida, a young mother was arrested, charged with pedophilia and sex abuse, and her baby taken away after she called the department of social services and mentioned these feelings of genital pleasure to the wrong person. It took her 2 years before she was able to get her daughter back.

Sucking, licking, and eating are all manifestations of limbic activity and are related to touch. Similarly, the social kiss may be an offshoot and modification of the original means of food sharing and physical intimacy shared by mothers and their infants. That is, tactual mouth-to-mouth contact appears to be not only the precursor to eating and kissing but serves as the means of establishing the first long-lasting emotional bond. For example, the orangutan, gorilla, chimpanzee, and even the wolf all feed their young mouth to mouth, often with regurgitated food.[48] Wolves will in fact regurgitate food for their mates and other pack members if, when approached, they are first licked around the mouth.[49] And of course their infants use their mouths to lick and suck.

Figure 43. Adult jackals (top) and dogs (bottom) nuzzle the mouths of their partners in the same manner that their young beg for food from adults. From W. Wickler. *The Sexual Code*. Garden City: Anchor Press, 1973.

Similarly, chimpanzees often greet each other likewise, and mothers often kiss their infants on their hands, face, arms, chest, and other body parts, whereas adult chimps sometimes take the hand of a fellow chimp and place the entire appendage in their mouth.[50] Indeed, this kissing behavior occurs even among a variety of birds. That is, when the parent returns to the nest, the young will all make begging actions with their mouth, and the parent will then drop them a morsel of a tasty worm or some such delicacy.

Such contact is not limited to food sharing, however. Dogs and wolves, for example, typically attempt to greet one another, and their human friends, via mouth-to-mouth contact, often in the form of licking. Dogs and wolves also love to kiss, hug, wrestle, and nibble on one another and will often engage in elaborate greetings with fellow packmates once they awake in the morning, or after a fellow packmember has returned from some adventure. Moreover, they will treat their human friends likewise if given the chance. Unfortunately, when dogs and wolves attempt to kiss a human they will jump up and sometimes smack the person on the mouth, sometimes even accidentally piercing the skin.[51] My good friend Dr. Gallagher discovered this inadvertently one day when Nietzscha (my Belgian shepherd) jumped up and gave her such a smack on the lips that she burst into tears from the sudden shock of his teeth hitting her in the mouth.

Moreover, these creatures nuzzle not merely as a means of affection, but so as to determine if one's friends or mate has had anything to eat and what it might be. Jackals sometimes even take on an infantile begging behavior, similar to what young pups do when begging for food. That is, they will greet their mate by licking the lips and then by nuzzling the jaws as if asking to be fed. However, dogs, wolves, and jackals will also lick an opponent's mouth as a form of appeasement, that is, they present a puppyish behavior to show they are submissive.[52]

Chimpanzees who hunt are subject to similar forms of begging from their compatriots, and among humans in hunter–gatherer societies this same manner of food dispersal occurs.[53] Sharing of food for acceptance, contact comfort, appeasement, and even sex is common among many species including humans and birds.[54] A domestic cock will try to attract a female to him by finding a crumb, or even a pebble, pecking it on the ground, and emitting sounds which will make her look at him. He will then drop the crumb and wait. The female, if interested, will run up and bend over in front of him so as to search the ground for his little love offering. At this point he takes his pleasure. The rascal!

Male cormorants also present their mates with food gifts, such as a starfish, or a bunch of seaweed, for if he fails to do so he will not be allowed by his female consort to return to the nest. The female may fly at him angrily, pecking and driving him away. However, once he returns with the proper respectful offering, she will allow him entry, and he will then take her beak in his own and then nibble her neck.

Among humans, some females may respond similarly by kissing her mate and allowing him to fondle her to varying degrees when she is a recipient of edible gifts or vegetable matter (e.g., flowers). A male chimpanzee, when attempting to entice an estrous female to leave with him, will in fact pick not just flowers but an entire bush, which he then offers and beckons to his lady love by shaking it as he walks away while simultaneously displaying an erect penis.[55]

Sucking, Loving, Licking, Kissing

Among almost all mammalian and avian species, the first physical contact occurs in relation to the mother, and it is the mother through whose presence that feelings of discomfort, such as hunger, are alleviated and satisfied and feelings of security are promoted. Similarly, when young are frightened, it is to the mother that most higher vertebrates immediately seek refuge. Among monkeys, apes, and humans, it is only after a number of years that young begin to seek protection from higher-ranking males or their fathers. Among older beasts and human adults, when frightened they may seek and will run toward, huddle, and embrace one another. If isolated, be it human, ape, or monkey, it may simply embrace itself.

Initially, however, flight is always in the direction of the mother. Among humans, when in a strange and novel environment, the infant will periodically explore and then quickly run or crawl back to mother for reassurance before again venturing forth. Mother represents a mobile but secure home base. However, human females return more often, more quickly, and

Figure 44. The need for contact comfort when frightened. Rhesus mother with infant and older infant, and two Sonjo children clasping one another when frightened. From I. Eibl-Eibesfeldt. *Ethology.* New York: Holt, 1975.

spend more time near the mother than males, who are more adventurous and less easily frightened. (On the other hand, boys are more severely affected by deprivation.) Baby chimpanzees respond similarly and like frightened humans will often bury their heads in their mother's chest. Likewise, when upset or afraid and seeking comfort, adult primates and humans tend to bury their head in the chest of another, who may then wrap his or her arms protectively around them.[56]

When frightened, baby humans, apes, and monkeys not only bury their head in the mother's chest but often attempt to take hold or bite onto the breast. If successful they will then suck as they fearfully keep a close eye on whatever upset them.[57] Hence, sucking serves several masters; that is, feeding, comforting, and emotional and sexual bonding. It is a product of limbic mediation and originates as a form of parental care reminiscent of being held in the mother's arms while being fed at her breast or by the bottle she held in her hand.

Indeed, touching, kissing, smiling, licking, and sucking are all derivatives and are closely related to food consumption and the need for physical interaction. Even a male shark will gently bite and nibble on his lady love during sexual intercourse. These intimate emotional activities are mediated by the limbic system and are associated with the establishment of contact comfort, sexual coupling, and as common forms of social greeting that are employed by humans and animals alike.

4

Dances with Bees
The Languages of the Body in Motion

She floats through the air on gossamer wings and with finesse, grace, and delicacy, alights with almost magical precision upon her tiny feet before the expectant and eager audience. Excitedly they surge forward together so as to behold her every movement. Eagerly she accommodates their feverish desires and performs a dance with the sweet promise of nectar and fragrant flowers, of sunlight and the lilacs of the valley. It is a dance replete with meaning that has been passed on secretly from generation to generation, from sister to sister, from mother to daughter, and for which no male was meant to behold. Its origins shrouded in mystery and veiled by time difficult to comprehend, the dance of the honeybee is easily 150 million years old.

Men and women dance for joy, in triumph and victory, when overcome by passion and eros, to appease the gods, and for the sake of art. Dance can be for ourselves, or for others, and may serve as an emotional release, or as a means of imparting joy, love, and pathos. Dance is an art and a form of communication, for it can symbolize not only emotion and intent, but a fire burning, swirling smoke, a babbling brook, a bird in flight, or aesthetic motion for art's sake.[1]

Among ancient and primitive human beings, the first dances may not have served such lofty designs. Instead the dance of early man and woman may have been a means of pure uninhibited emotional expression and sexuality, sometimes spontaneous, sometimes as part of a ritual release.[2] In fact, the recreational dancing of modern-day "rock and rollers," or the frenzied religious writhings of believers in the throes of mystical passion and ardor, are probably in many ways no different from those of our dancing ancestors who roamed the planet a million years ago. The first true "rockers" and "whirling dervishes" no doubt first danced their passions a long time ago.

Figure 45. Cro-Magnon female engaged in ritual dance and surrounded by other dancers. Painted approximately 15,000 to 20,000 years ago. Courtesy of Bildarchiv Preubischer Kulturbesitz.

Figure 46. Cro-Magnon female dancers (Australia). Painted approximately 10,000 to 15,000 years ago.

However, long before humans first writhed in rhythmic, melodious joy, or first set foot on forest floor or clung from redwood bough, the busy bee danced. Although bees may dance for whatever constitutes beelike pleasure, the primary purpose is communication.[3] The dance of the honeybee is part of a complex system of information exchange that has all the earmarks of language, even though bees have no ears and cannot speak.

Of primary concern to a honeybee is the location of a food source, and dancing serves as a complex language by which they are able to inform their sisters of the presence and location of such delicacies. A secondary concern, which arises only seasonally, is that of a good nesting site. Periodically, the colony must migrate and move to a new and accommodating abode.

Food and food sharing, however, are the social glue which binds most insect states, and it is a concern even when foraging for new nesting sites.[4] Thus, dancing scouts are continually coming and going so as to keep the hive informed as to the status of potential harvesting sites, and this they accomplish through dancing.

Those who watch her dancing body are soon informed how far away they must fly in order to reach these promised gastronomical delights, and in which direction they will be found. When the honeybee dances round and round, they know it is near; when she waggles her abdomen, the rhythm of her writhing tells them its precise distance away. By her vertical gyrations the direction is taught to them in relation to the sun. For example, if they must fly toward the light, the bee dances toward the zenith, and if it is away from the sun, they orient their dance to the nadir.[5]

Be it food or shelter, when located by a scout, not only is direction and distance indicated with a astonishing degree of precision, but the amount of effort that must be expended to reach it, such as if the bees must fly into a strong headwind. Moreover, through these complicated maneuvers honeybees are able to transpose the orientation of the sun in regard to gravity.

Those who closely observe the dance will then fly away and partake of the riches to be found, and when they return, they, too, dance. However, when all the food has been extracted and carried away, the dance stops, and the bees then wait for the next act which, like the last, will again inform them of where the promised land lies.

When Karl von Frisch first reported his remarkable discoveries regarding the dance of the honeybee over 50 years ago, most scientists dismissed them out of hand. Of course, their dismissals had nothing to do with the veracity of his startling evidence, for just as bees dance and birds fly, it has always been the knee-jerk reaction of mainstream science to deny. Nevertheless, 50 years of research has only verified and extended von Frisch's original discoveries, which are no longer considered controversial.

In general, a honeybee is able to communicate complex forms of information by engaging in one of three types of dances and through individual variations in the number of abdominal waggles and the rhythmic speed at which she prances and shakes. These three dances are called the "round dance," the "sickle dance," and the "waggle dance."[6]

The waggle dance is the most complicated and imparts a greater degree of information and is the most evolutionarily advanced. The round dance, by contrast, is the simplest and is thought to be the most ancient. Similarly, among humans the round dance, later known as the "magic circle," is probably the oldest form of group dancing.[7]

Formal dance among modern-day, Westernized humans also seems to fall within three broad categories. This includes classical ballet which first appeared around the midseventeenth century, modern dance as typified by Martha Graham and Isadora Duncan, and new modern dance as exemplified by Paul Taylor and others.[8]

Although modern dance has been for some odd reason labeled as such, it is in fact one of the oldest of all forms of dance. Isadora Duncan passionately argued that modern dance was an attempt "to recover the natural cadence of human movements which have been lost for centuries" so that "the harmonious expression of our spiritual being" is made possible.[8]

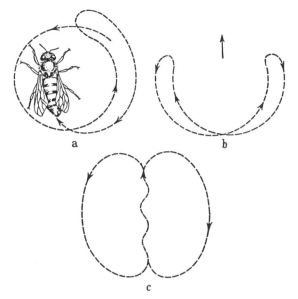

Figure 47. (a) The round dance, (b) the sickle dance, and (c) the waggle dance of the honeybee. From K. von Frisch and M. Lindauer. The language and orientation of the honeybee. *Annual Review of Entomology*, 1 (1956), 45–58.

However, to accomplish this requires that movements become free, unrestricted, and natural, similar to "the movements of the savage who lived in freedom in constant touch with Nature."[9]

Be it modern, ballet, rock and roll, or the "round" dance performed by our ancient ancestors, dance is an expression and a sometimes purposeful attempt to convey to others feelings, thoughts, ideas, hopes, desires, and even the most seemingly intangible of emotional experience.[10] Dance among insects and humans is a subtle and complex language and can be employed to artfully convey even that which cannot, or should not, be put into spoken words. Two dancing strangers, a young man and woman, may contort wildly together with licentious abandon, conveying volumes of indecent thoughts and desires that they would never dare put into words until weeks into a relationship if at all.

Movement as Language

Like the dance of the honeybee, human dancing serves as a means of imparting movement-based information. In this regard, human dance movements, like those of the honeybee, could again be considered to fall within three broad categories: narrative movement which involves the unfolding and telling of a story; depictive and expressive movement which involves the portrayal of mood, nature, and feeling; and nonthematic movements (e.g., pirouettes, glissades) which may be used for spacing and in which no particular theme is meant to be communicated.[11] Of course, be it classical, modern, new modern, or even rock and roll, all three forms of movement may be found, even within a single dance.

The point is, like the dance of the honeybee, human dancing, too, can be a form of language which is expressed via movement. Through movement, feelings of tenderness, astonishment, and compassion can be indicated, and narrative symbolism can be employed. Through depictive and expressive movements, particular aspects of the world can be portrayed, such as a wisp of smoke, a soaring eagle, the undulations of water, the wind blowing through the trees.[12]

However, the honeybee, too, dances of trees, of wind, rock, and the undulations of water; important landmarks which guide their flight. Moreover, by their dance they display what appears to be a unique capacity to make reasoned choices.

When in need of a new nesting site, several scouts are sent searching. When they return, each scout dances to announce not only her discovery, but her appraisal of its desirability. Those who especially like a certain location dance with more fervor and verve and for a longer time period than those who are not all that impressed with what they have discovered. By these actions, the most excited dancing bee is able to draw a larger crowd which will then observe and then fly off to investigate to make their own appraisal, which they advertise upon their return. If it is an especially good site, even those who had at first danced their discovery of a different location will change their mind and dance in favor of the new piece of real estate.[13]

As one might assume, although bees do not see with anything resembling the precision of the human eye, they have a good knowledge of their surroundings, including the location of landmarks such as trees, streams, and large rocks. Just as men or women can picture in their mind's eye the appearance and location of the furniture within their living room, bees, too, maintain a mental map of their environment and its terrain. It is in this way that by watching the dance that bees are able to deduce with an amazing degree of precision exactly where they are to travel and around which landmarks.

"Dirty Dancing"

Bees, however, are not the only species to dance. Birds, monkeys, and apes engage in dancing displays.[14] As based on paintings left in ancient and forgotten caves some 15 to 25,000 years ago, it is also apparent that our human ancestors, like our apelike cousins, were dancers as well. Among ancient humans it is likely that dances sometimes may have served purposes related to the procurement of food. However, with the exception of the busy bee, among all other animals who dance, including humans, the procurement of a mate is a major driving stimulus to dance. For example, male egrets dance before a potential partner to indicate sexual interest, and if she is agreeable, they dance together. Even the common male sparrow puts considerable effort into dancing for his lady love, and sometimes two or more males may dance simultaneously, competing to win her affection.

Humans dance for a variety of reasons, such as the success of a hunt or the vanquishing of an enemy. Still the sexuality of many forms of human dancing is undeniable. Perhaps, this is why the Catholic church and other Christian sects expended so much effort trying to stamp out dancing in the fourth century and again during the Middle Ages; an effort that was doubly resumed in the thirteenth century due to the outbreak of dance epidemics and the spread of dancing diseases which sometimes caused the afflicted to flock dancingly to various religious shrines so as to obtain relief.[15]

Interestingly, these dance epidemics are thought to be the result of widespread ergot infections that damaged portions of the basal ganglia in those who ingested contaminated bread. The writhing movements associated with disorders of this ancient motor center are referred to as St. Vitus' dance.

Nevertheless, although food and shelter were of primary importance to the honeybee, as well as that of ancient humans who danced so as to gain some control over nature, the writhing, rhythmic movements of the dancing bee appear to be devoid of sexual enticement. Bees, however, were probably the first species to dance together so as to increase generalized feelings of excitement and arousal, as well as the first to perform a jitterbug. It is precisely this latter type of close, wild dancing which for thousands of years has so outraged the guardians of our moral character.

When engaged in what some scientists have referred to as the "vibration dance," jitterbugging bees will approach a partner, grasp them by the legs and literally start "shakin' all over." They not only vibrate and grind against each other, but sometimes search out and dance in this fashion with the queen.[16]

This latter form of dancing, however, does not seem to serve as a source of complex information exchange. Rather, when bees perform their dancing jitterbug, it appears to act more like a generalized stimulant of sorts, causing an increase in general excitement and arousal. Because mutual vibration dances occur predominantly in the morning, some scientists have in fact referred to this as an "awakening dance."[17]

Possibly bees perform their morning dances in preparation for a day of foraging. Humans, too, have long used dance as a means of preparation for a successful hunt, the accomplishment of some goal, and also as a preparation for war. In this regard we note that dancing not only acts as a stimulant, it can act to intoxicate and produce even trancelike states.

Humming in Harmony

One could ask of course what is a dance without music? However, some ballets are performed in the absence of music as are some modern and new modern dances. Like humans, bees sometimes dance without musical accompaniment.

For the most part, however, bees in fact dance to melody, the hum of their wings. When dancing bees hum, the sound they produce also serves to inform by imparting information regarding distance.

Hence, just as music can inform modern-day humans as to mood, emotion, intent, desire, as well as instructing as to the speed at which one might move and sway while dancing the night away, the musical hum of the honeybee is similarly instructive and informative, that is, to a bee.

Apollo and Dionysius

For the ancient Greeks, as voiced by Plato, dance, too, was thought to be of extreme importance as a formative instructive element in education and which contributed to one's becoming a good citizen.[18] Athenaeus claimed during the third century A.D. that certain forms of dance instruction made men into better soldiers. This warlike dancing today would be recognized as marching.

Although, the ancient Greeks classified dance as one of four types, they essentially viewed the dance as belonging to one of two categories, only one of which was acceptable.[19] According to Plato, only "noble dancing," or what Nietzsche called Apollonian dancing,[20] was to be allowed, and anything suggestive of a "Bacchic nature" or of a lascivious character was unfit for good citizens to observe. This would include fertility dances, war dances, rain dances, sun dances, and all forms of orgiastic dancing. Anything of a Dionysian spirit, including ancient dances performed since before the time of Babylon, was deemed base. Plato referred to this latter type of dancing as primitive and sexual and not worthy of being considered art.

In contrast, the dancing of Apollo represents the mind, ideas, and thoughts, and reveals our presumably "higher" and more "noble" nature as expressed through movement. It was this form of movement that Plato valued most highly and as expressive of that which is most lofty.

And yet, as a communicative device, dancing can hide and deceive as well as reveal, and movement can be restrictive and act to limit and conceal. Nietzsche, Isadora Duncan, and many modern-dance enthusiasts have decried the type of dancing which Plato cherished so dearly and instead have argued that it is only the most "Dionysian," natural, uncontrolled, and unfettered forms of movement which are the most revealing and the most meaningful.

According to Nietzsche, when engaged in this type of dance one is in a frenzy of melody and is able to "tear asunder the veil of Maya, to sink back into the original oneness of nature; the desire to express the very essence of nature symbolically."[21]

Similarly, Isadora Duncan repeatedly decried such restrictions as stifling not only dance, but consciousness and creative thought. For her, dance should only be of free and natural movements which are closer to our original nature and best reflective of the upwellings in our soul.

> The movement of the free animals and birds remains always in correspondence to their nature, the necessities and wants of that nature, and its correspondence to the earth nature. It is only when you put free animals under false restrictions that they lose the power of moving in harmony with nature, and adopt a movement expressive of the restrictions placed about them. So it has been with civilized man.[22]

Drinking and Dancing

Although the cultural elite of ancient Greece or the guardians of today's public morals might disagree, some might ask, What is dancing without refreshments such as alcohol so as to liven things up even more? Among humans, dancing is often accompanied by drink, and the imbibing of alcohol was a central feature of the Dionysian movement. Indeed, Dionysius was also the god of the vine.

Humans are not the only ones to partake of the pleasure and misery of alcohol. Some dancing bees, such as northern bees who have migrated or been transplanted below the equator, not only imbibe, they get drunk, a consequence of their feeding on fermenting nectar which they quickly metabolize.

Unfortunately, like humans, when bees get drunk this not only disrupts their foraging and dancing but can result in accidents and even death. They crash into trees, get lost, and when they return to the hive they are usually turned away due to their altered social behavior. Those that are allowed back into the hive while drunk pass on drunken messages. Moreover, those that survive generally appear to suffer a very debilitating hangover.

Dances of Power

Dancing may well have appeared on the human scene long before any other type of expressive art and, along with singing, eating, and sex, may have been one of the first and earliest pleasures. Dancing no doubt evolved from spontaneous expressions of emotion so as to eventually convey mystical and magical ideas, which in turn became ritualistic and religious in content and symbolism.[23] All along, however, dance has served as a means of expression, information exchange, and a form of potential entertainment for others to observe. Among our ancient ancestors, however, dance was probably also recognized as a form of potential power.

The world of the ancient human was filled with power; the power of nature, of animals, of fear, death, hunger, thirst, and sexual desire. The powers of the unseen, however, were the most awesome of them all. The unseen, the unknown, of course, had no form and thus could not be objectively represented, not even by name even after complex spoken language had been invented. However, the forces contained within and one's feelings regarding such powers could be displayed and communicated in the abstract, and this was accomplished through movement and dance.[24]

If one can create an image of the nameless and of that which has no form, then one's potential power over these forces is increased. They are now subject to one's potential control. When that image can be created in the mind of another, this is power that can be utilized to control other human beings as well.

When humans were first observed to dance, those who watched as well as those who participated were probably infected with similar feelings, fears, hopes, and desires. In this way, even long after words and speech had been invented, dancing assumed a magical and mystical quality as it became a powerful means of communication.

An ancient Cro-Magnon might think it magic if one were to write down something secret he had said, and then passing it to someone who could read, had it read aloud. Dancing would be viewed with just as much awe and excitement, and its power enhanced even further by the frenzied participation of other tribe members. Not only thoughts and desires could be transmitted by dancing, but the feelings, hopes, and fears of others could be altered and possibly controlled. People felt compelled to dance and participate, the effects of which were always potentially hypnotic and even narcotic.

It was possibly in this manner that dance evolved into a medium involved in religion and magic ceremonies as its power was seen as being able to influence the thoughts, feeling, and actions of those who observed and who felt desirous or compelled to participate. However, if it could affect the dancers and the audience, could it also affect the outcome of events or affect the lives and futures of individuals or animals the dancers depicted? Certainly such thoughts must have occurred to our ancient ancestors. In this way, dance was probably seen as exerting an influence on the world itself.

Humans mimic their environment through movement and sound, and one of their early pleasures may well have been the mimicry of animals through song and dance. Humans would dance like a tiger or a bear, wear animals' heads and skins, and did so, so as to not only identify with nature, but to exert some controlling influences over their environment. One could dance out the hunt, the slaying of prey, or the ensuing victory over a foe. By dancing ancient man was able to exert control over the world and the outcome of all important events.

The Dance of the Gods

According to Curt Sachs, in his *World History of Dance*,[25] the oldest dance is that of the circle dance. Human beings would form a circle and dance round and round and soon become enveloped in a frenzy of emotional, mystical, and spiritual excitement (feelings mediated by the amygdala and temporal lobe). This has been referred to as the "magic circle."

The circle, thought divine by many ancient societies, was seen as magical as it is the embodiment of perfection and symmetry and is thus in harmony with the universe. The sun is round, the moon is round, and the earliest forms of magic are the circle. Hence, the circle dances were probably also danced to alter events, increase or give thanks for success of the hunt, and so on.

Thus, dance over eons of time evolved beyond limbic sex and wild emotional abandon to encompass the mystical, magical, and spiritual worlds, including those of the gods and loved ones long dead. Soon, shamans, witch doctors, magicians, and medicine women and men evolved and performed their dances so as to affect the audience and the Unseen.[26]

Indeed, dancing continues to play a significant role in the rites of many modern religions. Japanese Shinto religious dances are performed for the enjoyment of ancestors and as a means of making contact with their god. The whirling Dervishes, a religious Muslim sect, use dance ecstasy as a major element in their religious worship. So do some Hasidic Jewish sects in which they dance in order to increase their contact with the mystical powers of God. Dancing also plays a major role in Hindu religious practices as even the gods are known to dance.

Dance of Art

Magical dances and those involved in fertility rites are not usually meant as a form of individual emotional expression but often serve a particular purpose benefiting the tribe. Early in the course of human social–cultural evolution, dances became ritualized and were performed in the same manner with certain proscribed and prohibited movements. Many modern forms of dance are in fact ritualized, including ballet and even rock and roll. In contrast, those forms of dance which attempt to defy ritualized constraints instead rely on natural movements and gestures which in turn are

Figure 48. The Hindu god, Shiva, dancing. From the 12th century.

sometimes the most pregnant with individual meaning. These movements are under the auspices of the basal ganglia, which, like the amygdala, provide emotional coloring and meaning to movements. Indeed, it is via the amygdala that even spiritual feelings of the mystical come to be generated.

As dancing evolved from purely emotional to ritualized, it gained in its potential power to influence and communicate beyond the natural and innate. In the evolutionary process it also became art. However, it is quite possible that dance preceded not only the more formalized means of communicating such as through spoken and written language but the development of the plastic and representational arts as well. Long before the development of visual art, dancing may have already become an art as well as a later subject of ancient as well as modern artists. Hence, we find that in the earliest depictions of art in the recesses of ancient caves and beneath ancient cliffs that the Cro-Magnon people painted their fellows and the women in the act of dancing.

Figure 49. Sketches of Isadora Duncan dancing. From I. Duncan. *The Art of the Dance*. Drawing by Maurice Denis. New York: Theatre Arts Books.

Figure 50. Cro-Magnon dancers. Engraved in walls of Cave of Addaura, Palermo, Spain.

Be it the expression of emotion, magic, art, spirit, or soul, dancing thus represents an important form of intellectual activity. Indeed, be it modern human or dancing bee, dance is not only a language, for it speaks of the past and the future and time and space and that which is ineffable.

The Language of Movement

Dance is language through motion and is a language of movement.[27] Through dance, a panoply of meanings and events can be conveyed and displayed. Dance accomplishes this, in part, by utilizing natural movements and gestures to convey that which needs not words in order to be understood. These movements, insofar as they are unconscious yet meaningful, are under the auspices of the limbic system and the limbic portion of the old cortical motor center—the basal ganglia.

Unfortunately, among modern-day humans, many of these signals are seldom attended to, and many of these movements are suppressed, hindered by clothing, and in the process of "growing up" are forgotten or greatly modified by the neocortex. Still, even when not dancing, the movement of the body speaks volumes.

Communication through movement preceded the development of language and thought as is evident from studying the language of bees. From an evolutionary perspective, regardless of species, it has always been through observation or the detection of movement that one is able to make a multitude of judgments regarding the motivation and intent of those who are moving, such as a stalking predator about to strike.

Movement, in fact, is a main source of communication for modern human beings and is one of the earliest and most primitive as well as one of the most advanced form of communication. Even spoken language is based on movement and gesture. That is, in order to speak, words must be formed, articulated, and expressed via a complex synergy of movement involving not only the lips, larynx, lungs, and tongue, but via complex programming which takes place in the cerebellum and motor neocortex of the brain.

Human dance, language and movement, is of course much more complex than that of the bee and involves a considerable degree of flexibility and maneuverability a bee is not capable of. Complex movement and complex communication are linked.

One need only compare the dance maps of von Frisch and his honeybees, and Laban's "icosahedron," a spatial map of all potential human movements, to see this most dramatically. The "icosahedron" is basically a geometric, spatial map which indicates all the points of movement orientation in space including the dynamic tensions arising between different movements oriented in different directions.[28] Via this scale, which also takes into account the force of gravity, any and all of the potential movements that can be made by woman or man can be identified.

Based on this geometric analysis of movement in space, Laban was able to develop what he called a "vocabulary of movement" and established a system of movement notation called Labanotation (or kinetography). Labanotation allows for the precise recording of all aspects of three-dimensional movement including its timing, continuity, and extension in space.

Given the multiplicity of potential human movement, perhaps it is no surprise that, in comparison to bees, humans are able to symbolize, abstract, and represent that which has no representation. This is because with increases in the ability to make complex movements, the ability to communicate complex messages increases as well. However, to accomplish this required that the hands and limbs become adapted for purposes other than walking, wing flapping, or holding on for dear life.

In fact, as the evolution of complex movement preceded the develop-

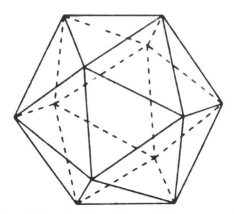

Figure 51. Laban's icosahedron, depicting all the possible movement which can be made in space.

ment of spoken language, certain movements are able to effectively communicate and express some forms of information that cannot be handled well by spoken language. For instance it is much more effective to describe a "spiral" by gesture than by spoken words.

Gestures can be executed more rapidly than speech and can convey concepts that are quite cumbersome to describe verbally. Indeed, it has been estimated that the human hand is 20,000 times more versatile than the mouth in producing comprehensible gestures.

Nevertheless, human beings have grown accustomed to listening to spoken language and paying less conscious attention to these movement-based cues. Although potentially linguistically complex, due to the imposition of civilization and formal learning, movement as a means of communication has become diminished, more repetitive in nature, and quite restricted. Clothing restricts movement as do social expectations. For Isadora Duncan, movement is most communicative and spiritual when it is most natural and unhindered by clothing. However, when movement is restricted, so too is human consciousness. She lamented:

> Very little is known in our day of the magic which resides in movement, and the potency of certain gestures. The number of physical movements that most people make through life is extremely limited. Having stifled and disciplined their movements in the first stages of childhood, they resort to a set of habits seldom varied. So too, their mental activities respond to set formulas, often repeated. With this repetition of physical and mental movements, they limit their expression until they become like actors who each night play the same role.[28]

Dancing on the Right Side of Your Brain

As noted, dance is emotional, spiritual and religious, and intellectually attuned. In many ways, dance as a language parallels certain aspects of emotional functioning associated with the limbic system and the right half of the brain. As noted, even with severe left-cerebral injuries, the right half of the brain is still able to swear and utter religious oaths, and it is the limbic, amygdala–basal ganglia association which not only produces emotional movements but can infuse them with religious and mystical awe as well as generate narcotic mystical states via the secretion of opiate neurotransmitters, the enkephalins.

Are there other links as well between the limbic system and the right half of the brain and dancing? Yes. But not all forms of dancing are mediated by these regions of the brain. Like music, the left brain contributes as well, particularly where the rhythmic, temporal–sequential aspects of movement are exaggerated or pronounced, the most extreme of which is marching.

Dancing and Geometric Space

Neuroscientists and neurologists know that the ability to control and direct fine motor skills is under the auspices of the motor cortex and the left half of the brain in the majority of the population. This is also demonstrated via hand preference as in the right hand being controlled by the left half of the brain.

In contrast, the right hemisphere exerts controlling guidance over the trunk and lower extremities so that successful movement of the whole body in space is made possible. For example, the capacity to guide movements in space and to take into consideration factors such as gravity, distance, depth,

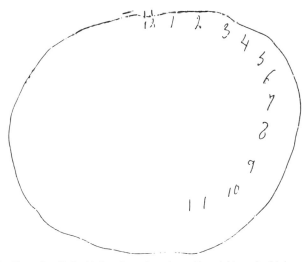

Figure 52. Example of left-sided neglect. A patient with a right cerebral injury was asked to draw a clock with all the numbers and to make it say, 10 after 11. From R. Joseph. The right cerebral hemisphere. *Journal of Clinical Psychology*, 44 (1988), 630–673.

and so on, so that one does not trip and fall, is dominated by the right half of the cerebrum. This includes the perception of direction, shape, orientation, position, perspective, and figure–ground, the detection of complex and hidden figures, route finding, maze learning, as well as locating targets in space.[29] The determination of the directional orientation of the body as well as the positional relationships of different body parts is also under the dominant domain of the right half of the brain. Of course, the basal ganglia and cerebellum also make major contributions in this regard.

It is the right hemisphere which enables us to find our way in space

Figure 53. Examples of left-sided neglect and distortion. Patient with a right cerebral injury was instructed to copy the above figures. Note preservation of right-sided details. From R. Joseph. The right cerebral hemisphere. *Journal of Clinical Psychology*, 44 (1988), 630–673.

without getting lost, to walk and run without tripping and falling, to throw and catch a football with accuracy, to drive a car without bumping into things, to draw conclusions based on partial information, and to see the forest when looking at the trees. If not for a right hemisphere, a dancer could not dance as it is this half of the cerebrum which makes it possible for us to run, skip, jump, leap, throw, hit, and to catch a football tossed long and wide via the analysis of body–visual–spatial relations.[30]

If the right half of the brain were injured, distance and depth perception would be altered, half of the body image may be erased, and our ability to maneuver our body about in space would be compromised. There would be no more dancing.

Among humans, it may well be the right half of the brain, coupled with limbic influences and that of the basal ganglia which provides and extracts emotional meaning from movement. However, over the course of human evolution, neuronal organization has become more complex, and new brain structures have evolved that have made possible the translation of movement into meaningful gestures and then, finally, into complex human speech. In fact, it appears that the development of complex gesturing was a precursor to the development of spoken language, and that, conversely, spoken language required that movement be modified and imposed on sounds. However, before the first word, there was the sign, and the sign still accompanies and gives meaning to the spoken word, even when nothing has been said.

5

Knowing Chimpsky
The Phylogeny and Ontogeny of Gesture
and the Language of Signs

Hans was born in Berlin, Germany, early in this century and long amazed the local citizens with his ability to rapidly perform mathematical calculations. It was not the difficulty of the problems which was so astounding, but the fact that Hans was a horse. For example, after a math problem was written on a blackboard, Hans would count out the answer by tapping his left forefoot to indicate 10, 20, 30, and so on, and with his right forefoot to indicate 1–9. In this way, he was able to indicate if the answer were 17, 24, or whatever.

Trickery was ruled out because clever Hans could solve the problems even when his owner was out of sight. However, when the owner and the audience were not shown the problem or the answer, Hans suddenly lost his amazing prowess. This is because the audience tended as a body to move forward as they watched Hans's feet and would signal him via their movements as to how to respond. For example, they would tense as he approached the correct answer and would then move their heads when he reached it, signaling him to stop. When he did as their subtle body language suggested, he was rewarded by a nice treat. Hence, it was well worth his while to act on their unconscious bidding.

If a horse can detect unconscious, seemingly unintentional body signals made by humans and then determine what they mean so as to act on them, then why not humans? The fact is, humans are also keenly attuned to

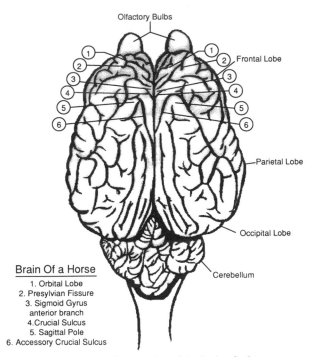

Brain Of a Horse

1. Orbital Lobe
2. Presylvian Fissure
3. Sigmoid Gyrus
   anterior branch
4. Crucial Sulcus
5. Sagittal Pole
6. Accessory Crucial Sulcus

Figure 54. Superior, top view of the brain of a horse.

receiving as well as transmitting similar body signals.[1] However, at a conscious level this nonverbal form of language when subtle does not make a significant impression. Nevertheless, at an unconscious level a sometimes wholly unintentional form of communication takes place with behavior being correspondingly affected.

Although body language may contradict or express views and attitudes that are not consciously intended, often these cues provide multiple layers of meaning and significance to what is being produced verbally as well. Sometimes, in fact, nothing need be said as the language of the body says it all.

Sara and Lisa were talking animatedly as they crossed the campus quad when Jesse waved to them and began strolling their way. Immediately Lisa hunched up her shoulders and clasped her books in front of her chest. Sara, however, threw her shoulders back and lifted up her chin, a pleased smile on her face.

Jesse strolled up confidently and standing with legs slightly spread apart grabbed hold of his belt buckle and knowing that Sara was watching him allowed his eyes to wander boldly up and down her body. Sara laughed, rolled her eyes, and with one hand absently played with her necklace. Lisa, meanwhile, remained hunched over, arms crossed close to her chest, her eyes on the ground. Jesse bent down and picked up a twig and then stuck it between his lips. Giving Sara a wicked smile, he leaned close, and taking it in hand pretended to inspect her necklace.

Lisa took a step back, whereas Sara held her ground, smiled back, and licked her lips very slightly with a darting movement of her tongue. Letting go of the necklace, he again took hold of his belt and then suddenly with his other hand tapped her shoulder with the twig. Sara laughed, hit him lightly on the shoulder, then brushed her hair away from her face with a sweep of her hand and arm which caused her breasts to jut out promisingly. However, although Jesse intended to wave his arms about and to keep invading her space by touching her, she held her ground, sometimes leaning close, sometimes leaning back, and only occasionally touching him in return.

Nevertheless, for the next 10 minutes, whenever Jesse said something funny, boisterous, or crude, she would roll her eyes, bat her eyelashes, and pucker up her lips and then smile or laugh. Occasionally, she reached out and hit his arm. Lisa, however, her body still in the same cramped position, had taken hold of the edge of her sweater and absently played with it.

Sex Differences in Body Language

Some young girls learn to carry their bodies in a provocative manner so that their developing breasts are proudly displayed. Young males often tend to stand with a wider base and to make gestures toward their pockets (or rather genitals?).[2] Nevertheless, be it male or female, their movements and posture speak volumes as to their attitude, self-confidence, and even hidden desires. For example, some, like Lisa, will hunch up so as to hide themselves or absently cover select regions of their body as if they were a source of embarrassment.

It is clear that the Sara feels good about herself, whereas by her body language we know Lisa has a troubled self-concept and is insecure about her self and her body. Moreover, although we do not know what was said, it certainly seems that Sara and Jesse are very attracted to one another. Jesse almost seems aggressively interested due to his willingness to repeatedly invade Sara's personal space.

Males in general, however, are much more likely to invade the space of others and to gesticulate more away from the body. They thus tend to take up more space when sitting or standing as their arms and legs are more likely to be extended out and as they tend to move around more when they sit or stand. Women tend to keep their arms and legs closer together and to move about less. Females also tend to gesture toward their body and to engage in more self-touching.[3] In fact, when it comes to self-touching, both sexes are more likely to use the left hand, whereas the right is more frequently used for gesturing during speech.[4] However, when women touch, it is likely to be much more gentle and much more caressing than a male who tends to use much more force.

In general, men touch both females and males more than women touch males. Women and girls are much more likely to hug, touch, hold hands, and make repeated and long-lasting physical contact with one another but are quite reticent to behave in the same manner with a man unless they are lovers or family. Indeed, even in regard to simple touching, a woman is almost four times more likely to be touched by a man than vice versa. However, when men touch men, this contact is usually very brief and consists of friendly slaps, slugs, punches, and pushes.[5] Unfortunately, due to female reticence and the male proclivity for repeatedly making brief physical contact, when a female is a recipient of such attention there is some likelihood that it may be misinterpreted as threatening or sexual even when it is not.

Females also begin smiling soon after birth and at an earlier age than males. Females also smile on average about 30% more than males and are far more likely to maintain sustained and direct eye contact with a woman than a man is with a man.[6] Males often perceive direct eye contact as a threat (which is true for most primates and many other male animals) and are thus less willing to provide it.[7]

When females make eye contact, this is also much more likely to be coupled with appeasement facial expressions such as smiling and head bowing or nods.[8] The smile serves so as to reduce social tension and threat. Again, these are common appeasement gestures for our cousins, the apes, as well as for wolves, dogs, and similar creatures.

Insofar as a smile may be mistaken for appeasement, men are less likely to employ a smile even when making direct eye contact. This reluctance regarding eye contact is even evident during infancy and is even suggested in the types of toys boys versus girls prefer to play with.[9] That is, females are more drawn to toys that have an easily identifiable and humanlike face, such as teddy bears and dolls, whereas boys like toy guns, trucks, blocks, hammers and so on, including action figures where their essential humanness is deemphasized.

Male and Female Brains

These sex differences are in part a function of differences in hormones and the structure of the male versus female limbic system. Specifically, the pattern and density of growth and interconnections within the hypothalamus and the amygdala of male and female brains are sexually dissimilar.[10] This is a consequence of genetic programming and the presence or absence of the male hormone, testosterone, within the first few fetal months of life. Those whose baby brains are bathed in testosterone subsequently develop the male pattern of neural growth. This is important, for the female limbic system is called upon to perform many functions, such as involving the monthly menstrual cycle, pregnancy, and lactation, that male brains simply are not wired to accomplish. Hence, their patterns of limbic neural growth and interconnections reflect these different concerns.

As noted, the ability to experience pleasure or aversion is a function of the limbic system, the hypothalamus in particular. Hence, one consequence of this difference in limbic structure and function is that males versus females experience and express pleasure in response to different stimuli, thoughts, and acts. For example, the hypothalamus and amygdala generate feelings of pleasure when males act aggressively and when they kill. In contrast, the female limbic system is wired so that they experience positive sensations and feelings when presented with creatures that are helpless and dependent, such as their own smiling babies. Thus females are more nurturing and less aggressive, which is also a function of males but not females being in possession of one of the most dangerous chemical substances found in nature—testosterone. It is testosterone which induces the male pattern of neural development and which later in life acts to promote aggressive and even murderous behavior, a trait that would not be to the advantage of the individual or the species if possessed equally by females.

Only females who feel compassion and tenderness, who love (or can at least tolerate) the sensation of dependency, and who can lovingly nurture her baby even as it is biting and gnawing on her sore body, or waking her repeatedly night after night, make good mothers. This nurturing tolerance comes more naturally to them as their brains are wired so as to generate feelings of pleasure in response to the helpless and loving dependency of her baby. Of course, there is a wide range of variability in this regard and cultural factors also play a major role. The female must be able to suppress the desire to strike out or to kill, for otherwise her kind would soon be weeded out.

In contrast, over the course of evolutionary history, males who were unable to suppress feelings of compassion and nurturance were also weeded out. They would have made very poor hunters, for when confronted by those dear, sweet Bambi eyes on some small or helpless creatures, they would have felt compassion and, more often than not, might have let it go. As such, except for the meat brought home by their mate, their diet would probably have been largely vegetarian.

Males who derived little pleasure from killing and instead chose other means of procuring food would probably also have had trouble finding breeding partners due to the higher premium placed on men who brought home meat from the kill. Meat, which they would use to share and barter, added to their stature in the bargain.

In the ancient hunter–gatherer societies, successful hunters probably had many wives and even more extramarital affairs, and thus many, many children. Hence, those who experienced pleasure when killing and those with limbic systems wired differently from the female so as to generate positive feelings in this regard may well have had some advantage over those less inclined. Hence, similar to those females who respond with love and compassion to those who are helpless and dependent, these ancient hunter killers were more likely to have thrived, survived, and passed on their murderous limbic traits to their forebears.

Contradictory Body Language and Sex

Most people reveal a great deal of personal information inadvertently via body and facial cues, posture, eye contact, hand gestures, inflectional vocal nuances, and their clothing. However, because this nonverbal form of communication was developed long before the advent of consciousness or even the evolution of the ancient hominids that gave rise to man and woman, most of it continues to occur and make impressions at an unconscious level. As such, it continues to influence human behavior sometimes in a manner that neither the receiver nor the sender intends, particularly if a contradictory message is being conveyed through a different modality—speech. For example, the woman who says "no," but with her body and eyes says "yes."

In fact, females of many species provide contradictory signals to males regarding sexual availability, one form of which is flirting.[11] Moreover, it is a pronounced tendency of sexually receptive females (e.g., lizards, birds, dogs) to make their availability obvious and to then run away only to be actively pursued by one or several sex-crazed males. Among these species, it is only the dominant male who does not accept these seemingly contradictory messages who is able to successfully breed. On the other hand, among some animals, such as rabbits, young males who fail to heed the otherwise receptive female's adamant disinclination for sex end up as castrates, for she is likely to bite off their testicles. Modern human males need to be similarly wary and may be best advised to interpret "no" as meaning exactly that.

Body Language and Inner Natures

Gesture is not just an adjunct to speech, for humans are able to converse quite fluently through the motions and movement of their body even in the absence of the spoken word.[12] The language of the body is a veritable treasure trove of information as to a person's intentions, desires, fears, emotional state, and self-confidence. The manner in which it is employed can greatly aid or diminish one's ability to successfully navigate the stream of socializing that makes up our everyday world, be it business, love, or political or sexual conquest.

Through gesture and facial expression, emphasis, context, elaboration, and depth are added to the content of all that is uttered and can open our eyes to possibilities not conveyed by speech. This is because gesture is a completely different and separate language by which human and animal speak through their movements.

Body Language and Fashion

Our human ancestors were no doubt at one time covered with a dense, thick coat of hair and ran about naked for much of the season. As such, many social signals could be conveyed via changes in one's coat, by one's hair standing on end or puffing out such as in displays of fear or anger. Signals could also be transmitted via facial and hand gesture and manipulation of exposure of the genitals, such as an erect penis or a swollen red vagina.

With the loss of much of man and woman's hair and the consequent covering up of the genitals, dress and other adornments took on considerable importance in making a social statement and thus body language. Hence, body language can be conveyed through clothing and in this regard, clothing can reveal as much as it hides.

A woman wearing a tight skirt, cut short, with high heels and a blouse that reveals a good part of her cleavage is definitely sending a signal which reeks of sexuality regardless of what her claims to the otherwise might be. A man who shows up at an informal gathering wearing a suit and who refuses to relinquish his coat may be perceived as quite stuffy, conservative, and reserved. Certainly not the life of the party. Conversely, the man who shows up at a formal party wearing sneakers, T-shirt, and bluejeans may be perceived as inappropriately casual, scruffy, and rebellious if not rude.

Attitudes and feelings conveyed via adornments, decorations, and dress in fact have biological underpinnings. Many mammals, birds, fish, and reptiles use body decorations as a means of signaling status and intention, particularly males as their coloration and coat also serve to attract willing female sex partners.[13] In this regard, males in general tend to be much bigger, more muscular, colorful, and sport the more luxurious coat of hair, whereas females are generally quite drab in appearance. Females are nevertheless inordinately affected and attracted by these signals and are more likely to mate with those who are the most colorful, biggest, and so on.

Indeed, due to this tremendous power coloration and body adornments exert over females in general, human females eagerly apply numerous cosmetics and dyes to their body and hair, and dress in a variety of colors and fashions, and these interests consume a considerable amount of their time and attention. The human female remains biologically predisposed to respond to such cues, but unlike females of other species, she is no longer dependent on the male to provide her with this stimulation. She can apply it herself and instead responds to the dress and fashion of other women.

This is why many women indicate that they do not dress for men, but for other women, and why females are inordinately concerned with the dress and cosmetics worn by other females.[14] Indeed, in contrast to many Westernized women who may sport and own dozens if not hundreds of outfits and shoes, many males are content to dress the same from day to day. For a long period of human history, the clothes one wore and the decorations that adorned one's body also conveyed volumes as to one's character and capability. The ancients would clothe themselves in the skins of animals and would wear their teeth, claws, and feathers, all of which indicated one's strength, status, and accomplishments. Hence, if you wore the teeth of a lion, bear, or tiger, you must be pretty ferocious. If your hair was decorated with the feathers of an eagle, you were not only an agile hunter but also wise and cunning. Moreover, not just the material but also the manner in which it had been fashioned spoke volumes as to the talents of the seamstress. One was clothed in his or her conquests and accomplishments, or those of their mate, and was able to parade his or her abilities and say to all who he or she is and what he or she is capable of.

Sadly, almost ridiculously so, among modern humans the nature and meaning of these displays has completely broken down. Now all one needs to do is buy certain mass-marketed clothes and assume an identity via identification. In this manner, one may look "cool" or "bad" or like a rock star, cowboy, lumberjack, fashion model, movie star, or gang member, or worse, one can parade the name of some fashion designer, like a brand on his or her butt or chest. Many modern Westernized humans have become walking billboards advertising their own psychic impoverishment, or in the case of children, that of their parents.

Before Babylon: The Universal Language

Originally (and beginning in prehistoric, preverbal times), infants and adults communicated many of their likes, dislikes, desires, and needs with little reliance on words but instead used gestures and facial expressions which were innate and natural in origin and understood across cultures.[15] Indeed, children are said to employ over 150 natural gestures and signs, many of which they unlearn or modify as they grow older.[16]

Due to their innate origins, many forms of body language serve as a silent, nonverbal means of complex signaling that are universally understood. Regardless of culture, people smile the same, and a smile accompanied by raising of the eyebrows and a nod of the head is a universal greeting gesture.[17] Of course, a smile does not necessarily indicate friendliness but may mask and even express hostility or appeasement. A similar gesture passed between men and women is often seen as flirtatious. Conversely, the balled fist, flexed arm, furrowed brow, and open-mouthed or tight-lipped expression are all indications of anger be it monkey, ape, or human. Of course, not all signs are universal.

Because many natural signs are a manifestation of the hardware of the brain and have little or nothing to do with learning, we see that even children who are born blind and deaf still smile and emit appropriate sounds when happy. Similarly, when sad or upset they will frown, stamp their feet, and clench their fists. Severely retarded children with no language capabilities, who are extremely limited in their ability to learn or even mimic what they see, can still laugh, smile, weep, and show signs of rage by yelling and screaming and stamping their feet in anger. Chimpanzees and gorillas respond likewise when angry, including baring the teeth in an open-mouth posture of rage.

Many complex depictive gestures which are employed predominantly by adult human beings have common biological roots as they are related to the functioning of the body. It is these gestures which are more or less expressed similarly or have the same meaning across cultures. For example, touching the ear with a single finger to indicate that one does not understand or hear what is being said, holding one's nose and wrinkling one's face and mouth may indicate the presence of a disagreeable odor, rubbing the tummy may indicate hunger or a stomachache, touching one's temple or forehead with the index finger may indicate thinking, turning the same finger in a corkscrew fashion may indicate craziness (a "mixed-up" mind), and slicing a finger across the throat may indicate an upcoming death by decapitation, or the need for the opposing party to shut up.

Natural (biologically rooted) signs are first and foremost rooted in emotional displays and thus have common limbic origins.[18] However, many gestures, although originally natural and seemingly shaped by innate physical and biological predispositions, also come to be shaped by the cultural and social environment in which one is raised. Natural signs and gestures, like spoken language, are subject to modification and abbreviation, or they become stylized to various degrees as they become part of a common currency of exchange.[19] For example, in a military culture the open (empty, nonthreatening) hand above the eyes (which decreases the threatening nature of direct visual contact) has become a military salute.

Just as dialects and languages have evolved from common roots, many

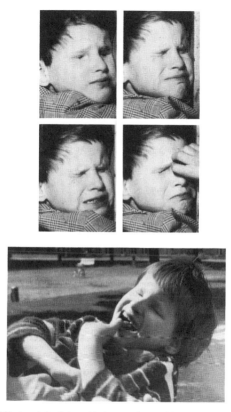

Figure 55. A girl, blind and deaf since birth, weeping and laughing. From I. Eibl-Eibesfeldt. *Ethology.* New York: Holt, 1975.

gestures have also differentially evolved so as to form distinct gestural languages as well. Nevertheless, many gestures and facial expressions such as feelings of puzzlement, doubt, contempt, anxiety, insecurity, or haughtiness are easily understood between most cultures.[20] Many gestures in fact circumvent language and illuminate or make unnecessary an exchange of dialogue; for example, blowing a kiss, shaking hands, kissing on the cheek. In fact, certain gestures convey meanings that are very hard to put into words, such as demonstrating versus explaining how a corkscrew works and what it is shaped like. In many ways, spoken language, being a more recent acquisition, still lags behind gesture and specific body movements as a form of intellectual and social currency. In many situations, it is clearly not as efficient as gesture for communicating.

Natural Signs and Gestures

Signs and gestures, albeit natural and innate, are not limited to emotional exclamations but can impart precise information about one's experiences, intentions, hopes, and desires. This form of communication is possible even among individuals raised thousands of miles away who speak completely different dialects, or who cannot speak at all having been born deaf.

For example, MacDonald Critchley described a number of "deaf and dumb" children from France, Austria, Czechoslovakia, and Italy with no formal training in sign language or lip reading who were temporarily gathered together at a special school for a weekend.[21] Critchley noted that within an hour or two they were all communicating freely with each other and understanding each other almost completely via the use of natural signs and facial expressions alone.

Similarly, among the American Indians, although over 65 linguistic families were still in existence by the mid-1800s, even those who lived thousands of miles away and spoke a completely different dialect had little difficulty communicating via the use of natural and imitative signs. For example, to indicate snow or rain they might hold their palms up and then slowly wiggle their fingers as they dropped their hands. To indicate cold, they might clench both fists then fold them against the chest followed by a trembling motion. Via gesture and the use of natural and imitative signs, complete and detailed conversations could be held between individuals who knew nothing of the other's native tongue.

Nevertheless, unlike spoken language, this natural gestural language of signs does not possess the "deep-structure" grammatical organization that supposedly typifies spoken language. This is true if it is employed by the deaf or by the various tribes of American Indians, all of whom are able to communicate quite effectively via the use of these agrammatical natural signs. In contrast, formal gestural systems such as American Sign Language are grammatically complex.[22] Being more complex in turn makes possible greater complexity and precision in communicating and inquiring and thus greatly expands upon the horizons of what is more natural. Natural and formal gestures, in fact, complement one another.

The Origin and Evolution of Gestures

Gestures are the result of tendencies to shape naturally occurring behavior patterns so as to convey and represent one's feelings and internal emotional and motivational states.[23] It is in this manner that body movement becomes infused with meaning. Similarly, ontogenetically and phylogenetically, the earliest movements, be they of a modern infant or a primitive mammal scurrying about in the darkness of the jungle, are representative of diffuse and rather global feelings and emotions such as pleasure, rage, and fear, or they are produced as a product of self-stimulation or reflex. Although expressive of a feeling or motive, the early movements and gestures of an infant do not consist of depictive signs but occur within certain contexts which give them meaning. That is, a smile is just a smile and in the absence of context is of limited meaning. The same could be said of an adult smile, which may convey contempt, anger, disbelief, as well as happiness.

Infantile movements themselves are initially very global, sprawling, and unfocused, involving the entire body musculature or, at a minimum, body parts irrelevant to the action being explored. It is only as the child ages that movements become more distinct and specific such that bodily gestures eventually become increasingly localized to the arms, hands, and face.[24] As the infant and its nervous system mature, the child increasingly adds to its repertoire of movement patterns. This in turn can be increasingly modified for the purposes of interacting with the environment and grasping some desired object. Many movements and gestures, in fact, come to be associated with the use of certain objects, tools, toys, and utensils.

With the rapid differentiation of the limbic system and the right-cerebral mental system, not just movements but emotions come to be more refined, specific, and less global as the child ages.[25] Now specific emotional states are experienced, and these in turn come to be expressed via specific movements as the infant screams, babbles, cries, or coos. Soon the lips and vocal tract also come to be specifically shaped so that certain words are formed.

Similarly, with the maturation of the neocortical motor areas in the frontal lobes, movements become more precise. From this point on, these and related motor programs begin to receive a temporal–sequential stamp as they are acquired and repeated endlessly. Finally, these movements become stylized and habitual as they are repeatedly relied upon to perform certain tasks.

When pretending to perform a certain task or using some imaginary tool or object, the behavior is then recognized as mimicry and pantomime, a form of gesture that lies intermediate between natural gestures and depictive gestures including what is recognized as sign language.[26]

Over time there is increasing differentiation such that innate and imitative gestures become more and more spatially and temporally distinct from that which they are representing as well as less and less naturalistic. That is, gestures and movements become increasingly distinct from their biological roots.

The Ontogeny of Imitative Gestures

Human beings sometimes use imitative gestures when painting and drawing, when pretending, playing, mocking, or storytelling, as well as when attempting to describe. This is a capacity that is not only informative and amusing but which is also fairly unique to apes, monkeys, and humans.

Children begin to make imitative moments almost from the time of infancy.[27] However, because of their poor motor control they may mimic with the wrong muscle groups. For example, they may open and close their mouth when mommy is blinking her eyes. To merely imitate, however, communicates very little and may in fact communicate nothing at all and serve only to reflect.

True depictive movements and gestures do not begin to appear until between the ages of 2 to 3. At this age, children can depict by gesture, size, form, and then later, objects, animals, and activities. Although this, too, is made possible via imitation, it is imitation meant to depict and to inform. For instance, they may depict a flickering light bulb by rapidly blinking their eyes, or they may make stirring motions with their finger to indicate a spoon, or they may outline the shape of the item or object via hand movements in the air.

As children increasingly refine their ability to imitate particular patterns of movements, they are also increasingly able to separate the signs and gestures they use from the objects they are trying to depict. In the early stages of imitation, they pretend to drink out of a glass that is full. Then, as they age, they pretend to drink out of an empty glass, then without a glass while making smacking and swallowing motions.

Similarly, consider the use of a pair of scissors. One can depict the use of scissors by bumping her thumb and forefinger together as if she were holding and cutting with it. Or she can imitate the scissors by taking the index and middle finger and snapping them together as if cutting. However, as pointed out by Werner and Kaplan, in the latter instance, the movements are functional, and more reality oriented so as to serve the purpose of imitating events and activities of everyday life.[28]

Initially, however, imitative acts involve movements which mimic the actual use of objects, with the exception that part of the body, the hand or arm, is substituted for the missing object, such as in pretending that the hand is a hammer instead of pretending to hold a hammer. In all other respects, these imitative acts involve the repetition of the exact motor sequence which makes up the natural act. Similarly, symbolic movement and gestures evolve from imitative acts.

One might suppose that the ontogeny of symbol acquisition has followed a similar path from ape to human during its long evolutionary journey. In fact, when we consider the most recently acquired and advanced form of making depictive gestures, that is, writing, it is clear that its development followed a similar evolutionary course being preceded by drawing. This appears to be true for modern-day children as well as for the evolution of pictorial and written verbal thought. Writing evolves from picture making.[29]

Gesture and Children's Speech

The use of gesture expands in accordance with refinements in motor control and seems somewhat independent of speech. Indeed gesture is organized much differently from speech and retains its natural, nongrammatical organization and is not always employed in coordination with what is being said. In part this is a function of speech being mediated by the temporal and frontal lobes, and gesture being comprehended by the visual areas within the parietal lobe.[30] Different brain regions mediate different language systems. However, these differences in expressions are also a function of the different rates of maturation for different regions of the child's brain, such as those controlling speech versus gestures.

For example, a 4-year-old will pantomime the use of a hammer before describing it. A 10-year-old will pantomime while he is explaining, where-as a 14-year-old will gesture selectively and employ gestures only in relationship to certain words so that the two systems of communication are coordinated.[31] Hence, with increasing age, gesture and speech become more closely associated and organized together, and gesture becomes more grammatical as well.

This developmental sequence is neurologically determined and further indicates that speech and gesture are in fact different communication systems that are interlinked but which also rely on different brain tissue for their expression and comprehension.[32] The developmental progression of gestures, as they are modified from natural and innate to imitative, to depictive and temporal–sequential and symbolic, are dependent on the neocortical maturation of the parietal lobe, the inferior regions in particular which sit at the junction of the visual, auditory, and tactual regions of the brain. The inferior parietal lobe is one of the last neocortical regions to functionally mature, taking over 10 years.[33]

Gesture, Sign Language, and the Inferior Parietal Lobe

The limbic system acts to analyze sensory and tactual information in regard to its potential emotional and motivational significance. It is via the neocortex of the parietal lobe that one comes to analyze the external sensory–physical properties of various objects and stimuli and to associate them together in the form of ideas. It is also the six- to seven-layered neocortex which enables us to achieve knowledge of the physical world as well as discern its potential motivational and intellectual attributes. The neocortex, however, does not replace the old cortical nuclei. Rather it acts semi-independently to analyze the emotional as well as those nonemotional characteristics of the world which the limbic system is not concerned with. It is in this manner that we come to know and know that we know.

Physical and tactual sensations are transmitted to the amygdala as well as to the primary receiving area for somesthesis, which is located within the neocortex of the parietal lobe. The parietal lobe is sensitive and responsive to tactual stimuli regardless of where on the body it is applied. In fact, via the reception of these signals from the sensory surface of the body, the entire body surface comes to be spatially represented in the sensory neocortex.

When part of the body is touched or stimulated, cells in the primary tactual receiving area fire. Conversely, if select regions of the primary

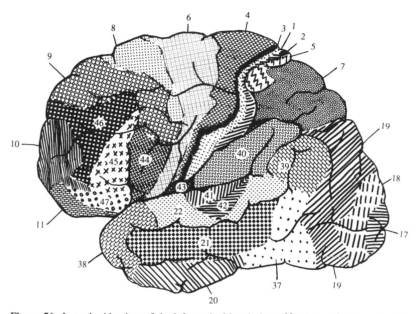

Figure 56. Lateral, side view of the left cerebral hemisphere. Numbers refer to anatomical analysis performed by Brodal and cells that are similar in composition receive the same numerical designation. Primary frontal motor (4), premotor (6), and Broca's expressive speech area (44) are depicted. The primary tactual receiving areas (3, 1, 2) in the parietal lobe, the supramarginal (40) and angular (39) gyrus of the inferior parietal lobe, Wernicke's receptive and auditory association areas (42, 22), and the primary auditory (41) and visual (17) areas, and visual association areas (18, 19, 7, 37). From R. Joseph, *Neuropsychology, Neuropsychiatry, and Behavioral Neurology.* New York: Plenum Press, 1990.

receiving area were electrically or abnormally stimulated, the person would experience a tactual sensation occurring on a circumscribed portion of the body; that is, that body part which projects via a series of cellular relays to the neocortical cells being stimulated. Electrical stimulation of these neocortical cells can elicit well-localized sensations as if they were occurring on the opposite half of the body such as numbness, pressure, tingling, itching, tickling, and warmth, when in fact it is only these neocortical cells which have been activated.

The parietal lobes are responsive to a variety of divergent stimuli. This includes tactual, kinesthetic, and proprioceptive information, sensations

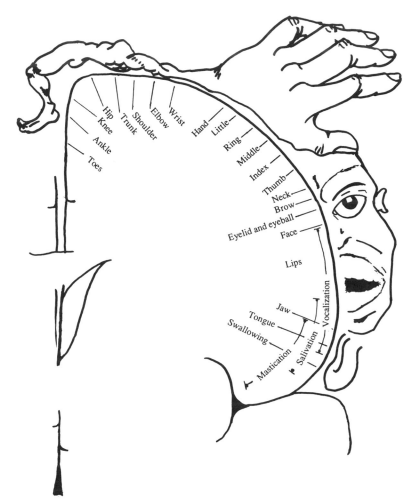

Figure 57. The neocortical map of the body image. The body parts are distorted as some regions receive more neocortical representation due to their sensory importance. From R. Joseph. *Neuropsychology, Neuropsychiatry, and Behavioral Neurology*. New York: Plenum Press, 1990.

regarding movement, hand position, objects within grasping distance, audition, eye movement, as well as complex and motivationally significant visual stimuli.[34] Moreover, the parietal lobe contains neurons which are visually sensitive to events which occur in the periphery and the lower visual field, regions where the hands and feet are most likely to be viewed. Some neurons in this area also become highly active when an individual reaches for some item, whereas other fire when the item is grasped. These latter neurons are referred to as hand-manipulation cells.[35]

Analysis and guidance of the body's position and movement in visual space, the hand in particular, is a primary concern of the parietal lobe. That is, it guides the movement of the arms and hands as they move through space regardless of purpose or object of desire. In this regard it mediates eye–hand coordination. The parietal lobe, however, does not keep its "eye" on the ball (or the net, hoop, hole, or whatever) but on the hands and arms. It does not receive visual information from the fovea of the retina of the eye. Rather it has only black-and-white peripheral vision. Again, these are the visual areas where the hands and legs are most likely to be viewed and in this regard the parietal lobes are responsible for comprehending the significance and meaning of hand movement and gesture as well as coordinating the feet as a person runs, dances, or walks about in space.

It is via the perceptual activity of the parietal lobule that we come to know that a wave of the hand means "come" or "goodbye," or that someone has balled up his fist and may punch us in the nose. In their most refined form, these movements and gestures are expressed in the form of sign languages including the writing in pictures and in script, these actions are highly dependent on feedback from touch and are made possible via select regions within the parietal lobe, specifically, the inferior parietal lobule, a structure which is largely unique to human beings.

By receiving visual as well as information regarding the limbs of the body, the parietal lobe, especially that of the right half of the brain, enables us to run, jump, do somersaults and perform gymnastics, as well as guide and program the hands so that the skills of carpentry, bricklaying, drawing, painting, and other fine arts are made possible. It is also via the parietal lobe that complex, skilled temporal–sequential tasks can be performed, such as preparing a pot of coffee or brushing one's teeth.[36]

Through its control over body and limb movements, the parietal lobe, over the course of evolution, has become increasingly involved in gestural communication. Indeed, this brain area in fact mediates the ability to

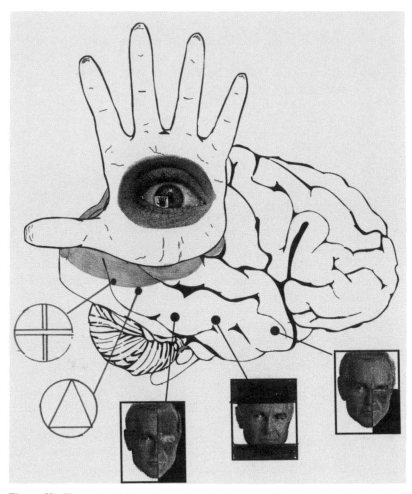

Figure 58. The parietal lobe is frequently referred to as a "lobe of the hand" as the hand receives extensive neocortical sensory representation here. The parietal lobe also receives lower and peripheral visual information, the region where the hands are most likely to be viewed, as this area mediates and comprehends hand movements and gestures. The neocortex of the inferior temporal region contains feature-detecting cells which respond selectively to faces, body parts, and geometric forms.

perform not only simple and "natural" signs, but complex, grammatically based, gestural sign systems such as **American Sign Language** (ASL). In fact, if this area of the brain were injured, the ability to comprehend ASL as well as natural signs would be compromised, although a person could still speak and understand what was said to him or her.[37] Speaking and verbal comprehension are a product of the left frontal and temporal lobes, respectively.

Knowing Chimpsky and Dances with Wolves: Communicating via Signs

There once was a girl who could neither hear, nor speak, or write, nor communicate by signs. It was only through a long terrible struggle that occurred completely within her world of touch that the mind of Helen Keller was suddenly illuminated by the world of language.

> Earlier in the day we had a tussle over the words "m-u-g" and "w-a-t-e-r." Miss Sullivan had tried to impress it upon me that "m-u-g" is mug and that "w-a-t-e r" is water, but I persisted in confounding the two. I became impatient of her repeated attempts, and seizing my new doll, I dashed it upon the floor. I was keenly delighted when I felt the fragments at my feet. I had not loved the doll. In the still, dark world in which I lived there was no strong sentiment or tenderness. . . .
>
> We walked down the path to the well-house. Someone was drawing water and my teacher placed my hand under the spout. As the cool stream gushed over one hand she spelled into the other the word water, first slowly, then rapidly. I stood still, my whole attention fixed upon the motions of her fingers. Suddenly I felt a misty consciousness as of something forgotten—a thrill of returning thought; and somehow the mystery of language was revealed to me. I knew then that "w-a-t-e-r" meant the wonderful cool something that was flowing over my hand. That living word awakened my soul, gave it light, hope, joy, and set it free! . . .
>
> I left the well-house eager to learn. Everything had a name, and each name gave birth to a new thought. As we returned every object which I touched seemed to quiver with life. . . . I saw everything with the strange new sight that had come to me. On entering the door I remembered the doll . . . and picked up the pieces. Then my eyes filled with tears; for I realized what I had done, and for the first time I felt repentance and sorrow.[38]

Humans have employed gestures and movements to mime and dance, to pantomime and to convey the symbolic, the sublime, the abstract, the mythical, magical, and mystical (which is why many religions utilize symbolic gestures in their rituals and liturgies) long before the development of written or complex grammatical spoken language. Gesture may have not only been a forerunner to spoken language, but it provided the context in which it could develop.

Because gesture is a precursor to the development of speech, it is commonly employed by those without speech but who possess highly developed social brains. Apes, monkeys, dogs, wolves, and other animals are able to gesture through eye and body movement, and facial expression, and are even able to mime and dance.[39] Even unrelated species make several of the same gestures which essentially have the same meaning; a puppy will sometimes naturally raise its paw as if to shake and show friendliness when confronted appropriately, and apes and monkeys commonly make hand-to-hand contact as a signal of appeasement and friendliness.

Given that so many postures, facial expressions, and arm and hand movements are understandable not only between cultures but between animals and humans certainly raises the possibility that the neurological foundations of gesture are similar between species. These similarities could also be due to the possibility that they were subject to similar environmental pressures in order to survive.

This is why, perhaps, that dogs and humans both feel guilt, shame, depression, anger, jealousy, and love. Both share a very similar limbic system, have similar brains, and have lived quite comfortably together for at least 10,000 years.

However, dogs don't rhyme, curse, and use foul language or put much effort into describing their world, though they are certainly not lacking in curiosity. In contrast, as has been shown with the chimpanzees Washoe and Lucy, Koko the gorilla, and other apes (see below), these capacities are not limited to human beings as these primates possess similar aptitudes.

How do we know this? First via the extensive observations of Jane Goodall on wild chimpanzees living in the Gombe field reserve in Africa compiled over a 30-year time period, and through the efforts of the Gardners of the University of Nevada, Francine Patterson of Stanford University, and many other scientists.[40] Indeed, the Gardners and Dr. Patterson were able to teach these creatures American Sign Language and were thus able to "talk" with them.

American Sign Language and the Acquisition of Temporal–Spatial Grammar

American Sign Language (ASL) is a complex gestural language utilized by the deaf as their primary means of communicating with others. It is composed of natural as well as artificial signs which have been forged into a grammatical gestural language that is thus visual and verbal, but not auditory.

In general, there are two structural levels to signing.[41] The first level includes the rules which govern the relations between signs within sentences. The second level is concerned with the internal structure of the lexical units.

ASL makes use of space patterns and the contours of movement, that is, hand shape, movement, and spatial location, which are actively manipulated in regard to indicating syntax, nominals, and verbs. For instance, nominals are assigned locations in the horizontal plane, and verb signs move among these spatial loci so as to indicate the grammatical relations between the subject and object. Moreover, the same hand shape presented in different motions or at different spatial locations conveys different grammatical and semantic information.

Although highly grammatical and dependent on temporal contrasts, ASL is very sensitive to spatial contrasts. However, these spatial contrasts are heavily dependent on verbally described relationships such as "right" versus "left," "up" versus "down." Space and movement become subordinate to verbal labels and coordinates. Unlike gestures, be they cultural or natural in origin, ASL employs linguistic, temporal–sequential, and visual referents, and thus represents a multidimensional as well as grammatical means of complex communication via movement.

The Inferior Parietal Lobule and Temporal Sequencing

To be capable of learning and producing this formalized system of gestural interrelationships, considerable evolutionary adaptation and development in the parietal lobes was required. Some of this development increasingly occurred in the superior parietal lobe, which became more concerned with the movement of the hands and arms in visual space and less so with the feet. This is because, as the hands became adapted for grasping versus walking, they began to take up more neocortical space. With the evolution of parietal tissue at the juncture of the occipital (vision), temporal (auditory), and inferior frontal (motor) areas, the development of a complex grammatical gestural language including the temporal sequencing of sound was made possible.[42] The appearance of the angular gyrus of the inferior parietal lobule helped make possible the creation of human speech and all associated nuances of language, for example, reading, writing, and arithmetic.

Grammatical, denotative spoken language is in part a secondary acquisition which follows the evolution of the hands and the development of the superior and inferior parietal regions, all of which made the ability to gesture in temporal sequences possible. Nevertheless, animals that cannot speak but whose hands have evolved and who possess some of the same brain structures as humans are also able to learn and employ complex systems of gesturing and signing so as to communicate their own interests and desires.

Apes and humans are all blessed with hands and well-developed parietal lobes and limbic systems. Apes, however, possess only the first hint of what in humans is referred to as the angular gyrus which along with the marginal gyrus makes up the inferior parietal lobule. These neocortical structures are extremely important in the acquisition of language. Sitting at the junction where visual, auditory, and tactual sensations are processed, they are able to integrate these different signals so that multiple categories can be assigned to a single sensation or idea.[43]

Primates are able to make considerable use of their face, hands, and arms for the purposes of gesturing, and this has been made possible via the development of the hands and the tremendous evolutionary development that has occurred in the parietal lobes. Thus the ability to gesture with the hands and arms as a means of communication is most developed in primates. Similarities in gestures and the brain, however, are also a function of both species being subject to similar evolutionary pressures early in their history, especially in regard to the development of the upper arm and hand, and the ability to grasp and manipulate objects. It is also for these reasons that humans and apes are in fact capable of acquiring and communicating via complex systems of gesturing, such as via ASL.

Washoe

The Gardners began training in ASL in 1966. They used the method of molding the hands while he was also looking at the appropriate action or

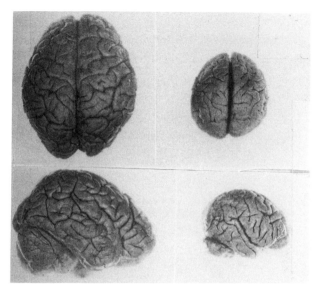

Figure 59. (top) The superior and (bottom) the lateral, side view of the human and chimpanzee brain which are proportionally depicted. From F. Tilney. *The Brain from Ape to Man.* New York: Hoeber, 1928.

object, followed by giving him a rewarding treat.[44] Given the complexity of ASL, he made surprising progress.

After 2 years of training, Washoe had learned over 30 signs, which he was able to use in appropriate two-word combinations. In fact, he was capable of using two-word gestures before his first year of training was up, such as "give food." Washoe picked up signs through imitation, was able to correctly invent certain signs such as a gesture indicating a "bib," and made up gestural combinations to accurately describe items he had no name for. For example, for some time, he mystified his caretakers by repeatedly asked for a "rock berry," which later they discovered meant a Brazil nut. He also called Alka-Seltzer a "listen drink," and a cucumber a "green banana." After 3 years of training he was able to use correctly over 85 signs, and after 4 years he had progressed to over 130 signs.

Washoe, in fact, not only learned signs and could communicate proficiently, but actively engaged in tutoring a young chimpanzee, Louis, in how to sign as well. On one occasion, when a human presented a bar of

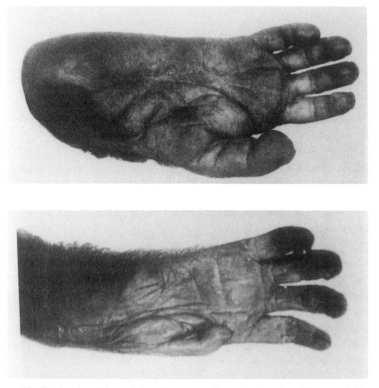

Figure 60. The hand and foot of the chimpanzee. From the American Museum of Natural History.

chocolate, Washoe became very excited and started signing "food." However, noticing the eager and curious look of Louis, Washoe walked over and took his hand, pointed at his mouth, and then molded the child's hand in the form of the appropriate sign, which was then readily learned.

Nevertheless, when the Gardners first presented their evidence on Washoe, they were attacked by many "scientists" who ridiculed their findings. Indeed, great efforts were extended to deny their evidence and even ruin their reputation. Unfortunately, throughout history, it has not at all been uncommon for a few scientists, like temple priests, to strive at all costs to maintain the faith of the status quo. Nevertheless, the efforts of the Gardners have been replicated by many others.

Lucy, another chimpanzee who was taught ASL and raised like a child in the home of the Temerlins, not only learned to communicate, but in some ways became more humanlike than chimpanzeelike.[45] She learned how to turn on the TV to watch her favorite shows (which says something of the intellectual level television is aimed at), and learned to go to the cupboard and then the refrigerator so as to mix herself a gin and tonic. Lucy also liked to flip through magazines and comment through ASL on what she saw.

Such humanlike behavior in a chimpanzee is probably not all that surprising given their proficiency for aping humans in so many other contexts. In this regard, they are just like children. Indeed, Washoe's proficiency at signing has often been favorably compared to the language capability of a young child. On the other hand, chimpanzee behavior is also very complex.

As Dr. Goodall and her colleagues have demonstrated, chimpanzees in the wild live very multifaceted lives, and many of their social interactions involve complex gesturing.[46] They use elaborate facial, hand, arm, and body gestures to convey their needs and goals, develop long-lasting best-friend relationships with other chimps, learn strategies for achieving dominance including the formation of coalitions, and engage in deception so as to do things in secret without the interference of more dominant chimps. They also engage in prolonged child care with mother–son and especially mother–daughter bonds lasting a lifetime.

Chimpanzees greet each other with hugs and kisses, hold hands, pat each other affectionately, engage in long periods of mutual grooming, seek reassurance by hugging and embracing each other, and will even risk their own lives to help friends or family members who are in distress or danger. They will punish estrous females for faithless sexual indiscretions, reconcile after an argument, will care for sick family members, and will cooperatively hunt monkeys or small animals which they kill, eat, and share.[47]

Moreover, when chimpanzees are frightened or alarmed they will clasp hold of one another. Humans often behave similarly, and wolves and dogs, too, will seek the close proximity of each other or their master, sometimes even climbing into laps.

When a lower-ranking chimpanzee wants to pass by a more dominant animal, he may reach out his hand and wait for the dominant male to reach out and touch it so as to provide reassurance. The dominant chimp will open up his hand, stretch it out, and sometimes take the hand of the less dominant animal and give it a squeeze.

Male chimps also often form coalitions, which enables them to assume

Figure 61. (top) A chimpanzee reaching out to make hand-to-hand contact with a dominant male. (bottom) Two other monkeys kiss and greet one another. From I. Eibl-Eibesfeldt. *Ethology*. New York: Holt, 1975.

dominant status. However, if they fail in these endeavors or when harmony is broken the troublemaker may eventually make conciliatory gestures by grinning nervously, stretching out his hand and begging to make up, sometimes by bowing and grunting submissively while the dominant male may stare or approach with lips pressed together firmly and with the body inflated. Frequently, the dominant will take the offered hand and either touch or kiss it, sometimes taking the whole hand in his mouth. Then the dominant male and the troublemaker might approach and kiss each other and then embrace.[48] Sometimes, however, the dominant will mount the subordinate as if he is about to engage in sexual intercourse. Similar sexual mountings to indicate dominance or submission are employed by gorillas, monkeys, dogs, and humans, and this is one of the reasons that human males respond aversively or quite aggressively in response to the sexual overtures of homosexuals. It is not a learned prejudice but a biological predisposition, as homosexual overtures are likely to be perceived by heterosexual males not as sexual, but as attempts to dominate or indicate submission.

Among apes, when dominance or appeasement is not forthcoming and a fight ensues, it has an extremely excitatory effect on the others, who hurry over to watch the scene, and give high-pitched barks of encouragement. Spectators may also be drawn into the fight by one of the adversaries who begs for reinforcement. Those who are fighting draw attention by screaming at the top of their lungs; they may put an arm around a friend's shoulders to get him or her to join in; with open hand they may beg for help from bystanders; and if they start to lose the fight, they may flee to a protector and, in safe proximity, shout and gesticulate at their opponent. In consequence, large-scale confrontations between different sections of society may result.

Moreover, in attempting to build a coalition so as to increase their status and power, males will seek female support, which is accomplished by grooming the females and playing with their infants, much like a politician feels a need to kiss and hug babies. This is valuable, for if his position is challenged, those females whose support he has won may assist him. However, females are likely to provide support only for their closest of friends or for kin. Male chimps, however, will form coalitions even with former enemies with whom they've fought, if perchance it will help them gain power. In contrast, females are less interested in power and status and instead will engage in cooperative behavior not for the purposes of obtaining power or some advantage but to maintain a cooperative relationship with kin and those they like. The parallels with human male versus female behavior are surprisingly similar. According to Dr. F. Waal:

> Adult male chimpanzees seem to live in a hierarchical world with replaceable coalition partners and a single permanent goal: power. Adult females, in contrast, live in a horizontal world of social connections. Their coalitions are committed to particular individuals whose security is their goal . . . for them it is of paramount importance to keep good relationships with a small circle of family and friends. [However,] a female chimpanzee may, for example, instigate an attack by a male friend on another female. She will sit next to the male, arm

around his shoulder, directing a few high pitched barks at her rival until the male obliges by charging at the other female.[49]

The Killer Apes

Humans are not the only species that derive enjoyment from killing one another. Chimpanzees show a surprisingly humanlike propensity for murder, mayhem, and gang attacks on both male and female chimps who are not or no longer part of their group. This latter behavior occurs while the males of a group, sometimes accompanied by a female, patrol the outskirts of their territory.[50]

Chimpanzees will viciously assault other chimps who stray into their territory, including former friends, strange females, or infants, even when the stranger attempts to be friendly by holding out a hand or trying to touch them. If a strange male or female actually makes contact, this may be completely rejected by the patrolling chimpanzee gang member, who not only moves away but will take leaves and scrub the spot where he has been touched. The stranger is then usually severely beaten or killed. Chimpanzees also sometimes engage in sneak attacks on their leaders, sometimes castrating them and biting off fingers and toes.[51]

These types of behavior, however, should be completely distinguished from those of a natural predator. Although a predator such as a wolf or a lion may find even the prospect of the hunt exciting, hunting for humans and apes seems to be a derivative of the desire to kill. In my opinion, humans do not engage in the killing of other humans because they are hunters and predators. Rather, they became hunters because they enjoyed killing. The same, I believe, can be said of chimpanzees who kill not for the sake of obtaining meat, but for the thrill of killing, which has a terribly excitatory effect on the whole troop, who may then beg and beg for just a tiny morsel of the still-warm and bloody flesh of the victim.

Like humans, chimpanzees also engage in tool use and construction and will often utilize tree branches, rocks, or other handy items in their displays. For example, they may sneak up on their compatriots, and then suddenly charge while dragging or waving huge branches which they rattle or strike against the ground. They may also pick up rocks and throw them (sometimes with punishing accuracy), stamp on the ground or drum loudly on a tree trunk, as well as pick up cans or other noisy objects that one can bang on and even use to swat others who happen to be in their way.

Moreover, they show a surprising capacity to think about and plan their attacks, often spending much time looking for the right-sized rock and then waiting again for the right moment to strike.

These similarities between humans and chimpanzees (and the other apes, the orangutan, and the gorilla) should probably be not all that surprising particularly in light of chimpanzees being the closest living relative to human beings. In fact, chimpanzees are genetically more similar to humans than they are to gorillas. Indeed, human DNA differs from chimpanzee DNA by only 1%. Moreover, biochemically, chimps and humans possess an identical order of amino acid molecules.

On the other hand, some gestural and behavioral similarities also exist between humans and dogs (which is why they form close relationships), as well as between humans and gorillas. In fact, perhaps about 100 million years ago, a predatory canide line and a herbivorous line may have branched off from the end of the therapsid, reptomammalian line, with some taking control of the ground and the others the trees. There they lurked and lived, hiding from the dinosaurs, until a cataclysmic event 65 million years ago turned the world to both their advantage.

Dances with Wolves

Dogs and wolves are extremely social animals who maintain a home base, or den, which they defend, and where they rear and care for their young. Wolves and wild dogs often develop monogamous pair bonds and bond for life, though divorces sometimes occur.[52] They are possessive, act with jealousy, and attempt to prevent other members of their pack from establishing similar bonds with their mates.

Among a pack of wolves, although there may be several members of both sexes, usually the dominant alpha male will bond exclusively with his alpha female. She in turn refuses to be mounted by other males (who seldom make such advances), and when other lower-ranking females go into heat they will be harassed by male and female alike and sometimes driven away or to the periphery of the group until they are no longer sexually receptive. The other males, although no doubt desirous of sex, nevertheless respect the authority of their leaders and strictly adhere to pack discipline, avoiding sexual contact with the temporary outcast whom they will harass as well.

Dogs and wolves place a high premium on olfactory communication via pheromones. But, status, anger, aggression, playfulness, curiosity, defense,

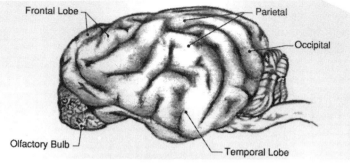

Figure 62. The lateral, left side of the dog's brain.

and sexual interest and arousal are also indicated through body posture, facial gestures, muscle tension, and, of course, tail wagging. They also engage in considerable body contact. For example, courting pairs will flirt with their heads by tossing them coyly to the side, often with their tongues out, and often they will throw themselves together and wrap their forelegs around each other's necks while they playfully bite at the neck and engage in a sort of wrestling. They will also nibble at one another's mouth and will pretend to bite one another on various regions of the body.

Dogs and wolves both have very expressive faces. They smile when happy, and like apes and humans, they also have a smile that indicates fear or anxiety, or the desire to appease aggressive actions of a more dominant male or female. By their facial expression and body language they also indicate shame, guilt, disappointment, expectation, excitement, depression, and irritation.

Similarly, although most people think that when a dog barks, that a bark is a bark, this, too, can communicate different meanings which is reflected by variations in the manner in which they produce their sounds. The most common bark, such as when one dogs sees another, could be roughly translated as "Hey! Hey! Hey!" However, by subtle shifts in intonation to a higher pitch, this same bark can indicate "Come on, come on," such as when they are excited about the possibility of going for a walk or a ride in the car. If they are extremely happy and excited, their bark may instead sound like a scream of joy. In contrast, a lower-pitched bark, often accompanied by a growl, can indicate irritation such as when someone dares trespass on their territory.

Similarly, not all growls are the same. A deep rumbling growl often emphasizes aggression and feelings of anger and belligerence. However, a softer, higher-pitched growl may indicate frustration or a desire to obtain something, such as if they want your attention so that they can come in or go outside.

Wolves will sing and howl when they are happy or sad, and they usually howl before the onset of a hunt. When traveling alone they often howl to communicate and keep in touch, or to communicate with neighboring packs, or sometimes to advertise the fact that they are alone and looking for a mate, or just to advertise their presence and to *talk* with other wolves who might also be lonesome.[53]

Dogs and wolves are also quite mischievous and enjoy teasing and playing tricks on one another or on their human owner. For example, Sara, my 2-year-old Belgian shepherd, often likes to sneak up or run past me or Jesse, my 6-year-old Belgian shepherd, and give a little nip on the butt or steal certain objects from my hand (like a sock I am about to put on). When I give them both a cookie, she often likes to wait until he has finished his, and then she will take hers and prance back and forth to tease him with it. However, Jesse has started to simply take the cookie away, so she does not do this much anymore. Wolves, too, like to tease and play pranks, such as leaping and trying to steal the cap someone is wearing or a handkerchief hanging from a coat pocket. If given the chance, dogs also are quite capable of at least trying to mimic human beings.

Dogs and wolves do more than simply smile, growl, and kiss, so as to indicate their intentions or feelings but rely on a considerable degree of body language. An alpha male, for example, when approached by a subordinate or strange male will stand very erect in a very regal fashion, with his ears up and head erect, mouth held firmly together or in a tight grin. If he needs to prove to an intruder that he is dominant, he may give him a shove, mount him, and if that fails to be convincing, will wrestle and throw him to the ground. Usually, a subordinate will respond when so treated by grinning in submission or by throwing himself on his back, exposing his stomach and neck but with his tail firmly pressed against his tummy so as to protect his genitals.[54]

If two strange male dogs meet for the first time, they tend to approach one another slowly in a very stiff-legged manner with their tail and head held very erect and the mouth closed in a very tense, tight-lipped fashion. The more dominant of the two may then take his forepaws and, like a boss to his subordinate, place them across the shoulders or the back of his opponent, or give him a shove with his shoulders and then either mount him or escort him off the property. Or the subordinate member will voluntarily tuck his tail between his legs and slink away with an embarrassed grin of appeasement on his face. Dominant males will respond in this manner even when not on their own territory.

Such behaviors are easily recognized and understood by humans, and if a dog were to approach a man in this manner, it would be obvious that the dog is not being friendly. No words need be spoken, nor would a growl or a bark need be uttered.

Dogs and wolves, like chimps and humans, often form coalitions. Sometimes a dominant male initiates attack against others so as to enforce discipline or to maintain his status. Sometimes when others are being harassed by other subordinates, he simply watches, and in some cases he may come to the aid of a subordinate and even allow him to get a few good whacks in before breaking up the fight. Moreover, sometimes subordinates will approach the alpha male so as to solicit his support if they feel they have been wronged in some manner. They are also capable of shifting their coalitions such that even subordinate males and females may temporarily join together so as to prevent or dissuade a dominant from engaging in a certain action.

Dogs and wolves, like humans and chimps, often share the spoils of the hunt. Indeed, dominant male wolves not only share food with their mate, the sick, injured, or the young, but have been observed to distribute food among fellow pack members, sometimes by even throwing it through the air to a compatriot.[55] In general, however, alpha males and females tend to eat first and are then followed by the other pack members according to their status.

Fathers also take an active part in the rearing and training of the young, whereas male chimpanzees, like many modern-day humans, are content to have sex and then simply abandon the mother and her infants once they are born. Moreover, although male chimps can become excited and harm any infant or female who happens to get in their way, male wolves and dogs are very solicitous of pups and especially their own offspring and would almost never harm them. Other pack members, usually females, will also baby-sit pups while the others are out hunting. Such behavior is not a product of "instinct" per se, but is maintained via cooperative communication in the form of body language and is dependent on the manner in which the mother was raised.

Monkeys and chimpanzees do not seem to mourn to deeply when a friend dies, though they are obviously highly disturbed if they witness the death of a friend. A chimpanzee mother will eventually drop her dead infant on the ground and leave it there without a look back. In contrast, wolves and dogs mourn for their dead and will sometimes bury their young. And although these creatures love to dig, once buried the site will not be molested.

Moreover, like human beings who for the last half-million years up until the present have spent at least a third of their life engaged in cooperative hunting, an activity which places a premium on silence and gestural communication, dogs and wolves are essentially cooperative group hunters who, while engaged in the hunt, communicate in a silent fashion.

It is because of these similarities that dogs and humans were able to form and establish bonds that have persevered for eons. That is, since they hunt in a similar fashion, humans and dogs are able to hunt together. Since they both maintain a home base which they defend, they have been able to live together. Since they both establish and are respectful of dominance relationships, dogs were able to cohabitate with humans and accept a subordinate position. Since both are capable of forming loyal monogamous relationships, close social and emotional bonds could be formed between a single human, or his family, and a single dog. Given all these similarities and the ease at which they communicate and live together, dogs became useful not only to hunters, but also to herdsmen, farmers, and modern humans as both protector and loyal friend.

Dogs, however, are extremely limited in their ability to use their limbs for gesturing, a consequence of their line taking to the ground, whereas ours took to the trees and developed a precision grasp. In fact, our ancestors were probably hunted and chased up those trees by the ancient predatory ground dwellers, who would one day become one of our best friends.

Hence due to these limitations, although dogs and wolves are capable of performing complex actions, and are extremely sensitive to social signals, much of their gesturing occurs via vocal tone, posture, and facial expression. On the other hand, the same could be said in regard to our second closest living relative, the gorilla. However, like the chimpanzees, gorillas also display an amazing propensity for learning human sign language.

Gorillas and Gesture

Gorillas are generally very peaceful creatures who maintain their massive bulk predominantly through a vegetarian diet, although they also like to eat meat on occasion, including bird eggs and termites. Like dogs and humans they are very social and roam through Central Africa in bands of up to 30 individuals which are led by a single dominant male. Their predominant mode of auditory communication is through barking, roaring, grunting, whining, whimpering, grumbling, humming, and purring. Many of their sounds, however, can convey different meanings depending on the context in which they are elicited, as well as through variations in pitch and quality. It has been estimated that their calls range from 20 to more than 30. Often, however, their vocalizing seems to have no effect on other gorillas, who may not even bother to look up.[56]

Another primary means of communication is through complex gestures. This includes facial expressions and a variety of postures, including chest beating, and tearing limbs from trees or waving branches in a threatening manner.

Chest beating is one of the more obvious and most easily observed

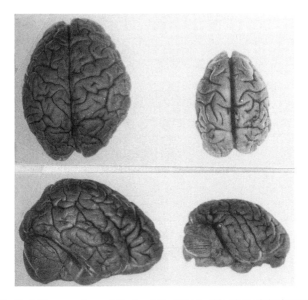

Figure 63. (top) The superior and (bottom) the lateral, side view of the human and gorilla brains, which are proportionally depicted. From F. Tilney. *The Brain from Ape to Man.* New York: Hoeber, 1928.

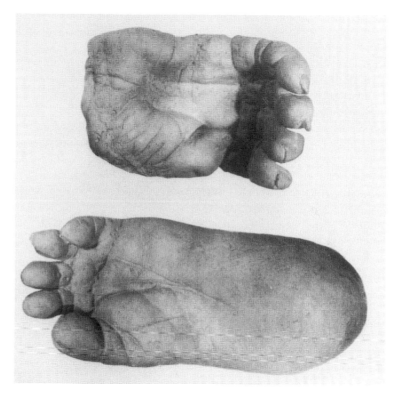

Figure 64. The hand and foot of an adult male gorilla. From the American Museum of Natural History.

gestures they employ, and it usually is part of a temporal sequence of movements, calls, and actions. That is, before they beat their chest, they begin to "hoot" and will then suddenly grab some leafy vegetation and stuff it in their mouth while just as suddenly they will rise up and throw the remainder into the air. Then the chest beating occurs, the sound of which can carry up to a mile away. While so engaged they will kick one leg into the air, and then begin running sideways while they snatch a tree limb and wave it menacingly. This whole sequence is then brought to a conclusion by the gorilla holding both hands clasped together over his shoulder like a boxer who has just won his prize fight. He then slaps and thumps his hand forcefully against the ground. One could only imagine that if he held a football, he would substitute spiking for thumping.

It seems any number of events can trigger the chest-thumping display, such as the presence of a human, or that of another gorilla, in response to the movement of some animal or object the identity of which they cannot completely discern, or in response to displays by other gorillas. They also engage in this display in play and in fun, or just for the heck of it. In this regard this behavioral sequence can express irritation, anger, lonesomeness, or even boredom.

Another means of communicating to the group is walking in a very rigid and stiff-legged fashion, moving very abruptly or lunging with the body, staring, head shaking, cowering, and even mock biting. However, what these gestures mean depends on context and the status of the gorilla who employs them.

In comparison to chimps, the tool-making capabilities of the gorilla are sorely lacking. In part, this may be due to their larger, more clumsily shaped hands, which reduces their ability to perform fine motor tasks. Indeed, the thumb is very small, and they still use their upper extremities for knuckle walking.

Nevertheless, they frequently handle and carry objects, will construct elaborate nests for resting during the day and sleeping at night, and will tear up vegetation or yank apart tree branches as part of their display. On the other hand, those in the wild seem to lack any inclination to explore.

Koko

One might assume from these observations that gorillas may not seem to be the brightest of our fellow apes. However, this may be due in part to problems inherent in the observational method. As demonstrated by Dr. Francine Patterson, gorillas possess tremendous mental powers as well as the ability to converse in depth via gesture or to even play practical jokes, such as with a plastic alligator.[57]

For example, Koko the gorilla apparently likes to sneak up on some (supposedly) unsuspecting human with a little toy alligator hidden behind her back. When she is near her victim, Koko will abruptly spring up, brandishing the alligator wildly. The human is expected to assume a terrified look, scream, and run. Koko knows the whole thing is a charade, but she thinks it is vastly funny. Nevertheless, as pointed out by Dr. Patterson, it is rather amazing to think that "a gorilla might even for a moment think that it needs a prop like a toy alligator in order to scare a human."[59]

Koko the gorilla began her formal training in 1972 at the age of 1 and for the next several years underwent a grueling full-time educational regimen in ASL acquisition. This was accomplished by molding her hands to form the appropriate signs, and then by giving her a reward if she did so correctly. Before giving her a drink from her bottle, Dr. Patterson would show her the sign for drink, which is shaping the hand as if hitchhiking and placing the thumb in the mouth.

Surprisingly, Koko began to learn signs almost immediately, although admittedly they were sloppily made and often consisted of demands for food. Indeed, by her second month of training she was able to use about 16 different two-word combinations. "Pour that hurry drink hurry . . . me me eat . . . you me cookie me me . . . gimme drink thirsty," and so on. Like Washoe, she invented and utilized natural signs, such as "give me" which looks like a beckoning gesture. Again this is a gesture that not only primates and humans understand, but which is comprehensible to a dog as meaning "come here."

In any case, after 3 years of training Koko had acquired and was regularly employing over 125 words and had utilized an additional 150. This included words such as "pillow, love, necklace, baby, hot, up, down, in, out, finished, yes, no, don't, can't."

Koko could also understand many spoken words as well and sometimes would eavesdrop on conversations, or answer questions that were asked out loud. For example, one day a visitor asked Dr. Patterson what might be the appropriate sign for the word "good." Before Dr. Patterson could respond, Koko made the sign for her.

Sometimes she would also demand some item, like candy, that the humans had been talking about. To keep Koko from knowing what they were discussing, these experimenters had to often resort to spelling and even then Koko figured out, for example, that c–a–n–d–y spelled her favorite treat.

Many of Koko's comments and questions were spontaneous, and she soon learned to use words in an insulting fashion. For example, Koko did not like birds and hated the bluejays who screeched outside her window. Hence, when she dealt with someone or something she didn't like, she would call it "bird." Later she progressed to calling certain individuals "stupid devil," or "Devil head," and began using filthy language to insult Dr. Patterson or her friends, by signing at them, "dirty toilet," "stupid toilet," or "Penny dirty toilet devil."

Figure 65. An irritable Koko the gorilla, using American Sign Language to call Dr. Francine Patterson "You, dirty, bad, toilet." From F. Patterson and E. Linden. *The Education of Koko*. New York: Holt, 1981.

Koko, like Washoe, also learned to combine signs to describe animals or objects, the signs for which she had not yet been taught. This includes, "barefoot head" to describe a bald man, "Giraffe bird," to describe an ostrich, "finger bracelet" to describe a ring, "eye hat," to describe a mask, and "fine animal gorilla" to describe herself.

Talking Apes and Talking Dogs

Given some of the social and neurological similarities between apes, dogs, and humans, as well as their advanced methods of expressing and comprehending complex gestures, it may seem curious that no other species save humans has learned to converse as do man and woman. If these creatures can gesture, why can't they learn to talk or understand complex spoken commands?

There have been attempts to teach dogs and apes to speak, but they have failed. For example, Keith and Catherine Hayes spent many years trying to teach spoken language to a chimpanzee by the name of Viki. Although seemingly as intelligent as many young children, Viki could only speak six words, most of which were not all that easily understood.

Similarly, Alexander Graham Bell attempted to teach his dog to speak and was able to make it say, "How are you, Grandma?" This was accomplished by training it to growl at a steady rate while moving its jaws and throat in a certain fashion. That is, he did it by shaping. However, what the dog said actually sounded more like "ow ah ooh gwahh mahh."

Apes and dogs are not well equipped for producing the sounds of speech because of the shape of their mouths and throats. In addition to lacking the mechanisms necessary to make speech sounds, they also lack the temporal–sequential perceptual capabilities so as to hear the individual units and sounds of speech and are unable to string words together so that they are grammatical and make syntactical sense. Although apes are able to sign, their ability to employ grammar even in their gestures is all but nonexistent.

Washoe was never able to understand the rules of grammar and his ability to use grammatical relationships did not even approximate that of a 3-year-old human child. For example, if Washoe wanted to ask to be tickled, he would sign, "tickle you, me tickle, or you tickle" as if all meant the same thing. In this regard, Washoe's understanding of grammar may be similar to that of the right hemisphere of an adult human. Grammar is a left-brain function.

However, David Premack showed with his chimpanzee named Sara that she was able to learn some aspects of grammar and to read and write simple words using a system of language based on variously colored and sizes and shapes of plastic chips which she would arrange. For example, she could manipulate these chips to say: "Put the apple in the pail and the banana in the dish." However, she would make errors on one of four trials indicating that her understanding of grammar remained rudimentary and incomplete.

Dogs have also been trained to manipulate plastic chips and various objects in order to communicate their needs. In fact, a German investigator, H. Subbok, early in this century taught his dog to read, or at least to recognize certain written words. This particular dog would be shown four cards with different words written on them and when that particular word was uttered by his master, he would reach out and point with his paw to the correct word regardless of the order in which it was arranged. Nevertheless, this dog was unable to rearrange these written words to make meaningful sentences.

Grammar and the Inferior Parietal Lobe

The reason apes do not understand grammar or use spoken language is because they are almost completely lacking a recent neurological evolutionary acquisition. Indeed, evolution did not stop with the apes, and humans possess a significantly increased expanse of highly evolutionarily advanced parietal tissue near the juncture of the occipital and temporal lobes. This expanse of tissue is referred to as the inferior parietal lobule and contains the famous landmark, the angular gyrus, a structure almost wholly unique to humans. It is the inferior parietal lobule which makes not only fully formed ASL possible, but the evolution of grammatically correct language, be it spoken or gestural in the form of drawing, painting, or writing.

Situated at the junction of the temporal (auditory), occipital (visual), and superior parietal (hand movement) lobes, as well as inferior frontal motor areas, the inferior parietal lobule became able to sample and integrate multiple messages to create categories and concepts. It was also able to impose temporal order on what was heard and seen as well; the same temporal sequencing that had long been refined in the course of developing skilled hand movements and gestures.

The appearance of the inferior parietal lobule was an important evolutionary advance not only in language. This made complicated tool construction possible as well as the ability to engage in complex sequential actions, such as in making a tool, constructing or sewing clothing, or even putting them on and wearing them.

Take the simple steps necessary to make a pot of coffee. From obtaining the coffee container, to heating the water, to pouring it, and so on are just a few of the many steps that must be performed in a highly interrelated sequence. Take just one of these steps and perform it out of order, that is, pouring cold water on the coffee, then heating up the empty coffee pot, and one destroys the overall integrity of what one was attempting to accomplish. This condition can occur as a function of an injury to the parietal lobe. The affected person is said to suffer from **apraxia**.

Apraxia is a disorder of skilled temporal–sequential movement that in most cases is due to strokes or tumors that have invaded the inferior parietal lobule of the left half of the brain.[59] Depending on which half of the brain has been injured, the individual may be unable to make a pot of coffee or put on his or her clothes, much less sew them together.

Creatures such as dogs and apes, as well as our own ancient ancestors who roamed as recently as a million years ago, are without such capabilities and are unable to make complex tools or fashion or wear clothes. This in turn is a function of the evolutionarily advanced angular gyrus of the inferior parietal lobule, which they, and our ancient ancestors, lack. Indeed, like the bulk of the frontal lobe, the inferior parietal lobule may not have appeared until following the evolution of the Cro-Magnon people, for they were the first not only to draw, paint, and construct truly complex tools, but to sew clothes and to make the first sewing needle.

Apraxia

Usually apraxia patients show the correct intent but perform the movement in a clumsy or disorganized fashion. This can also result in gross inaccuracies as well as clumsiness when making reaching movements or when attempting to pick up small objects.

Apraxia is usually mildest when the sufferer uses the actual object, and performance deteriorates the most when he or she is required to imitate or pantomime the correct action. The patient may be asked to show the examiner how he would use a key to open a door, or hammer a nail into a piece of wood. In many cases, the patient may erroneously use the body, such as a finger, as an object (e.g., a key). That is, rather than pretending with the finger and thumb to hold an imaginary key, they instead might stick out their index finger as if it were the key. Or he may pretend his hand is the hammer rather than he is holding and swinging a hammer with his hand. In other words, with destruction of the parietal lobe of the left half of the brain, the individual loses the capacity to make depictive movements involving temporal sequences and instead reverts back to a more primitive form of imitative object description such that the sign, symbol, and object become fused, the body part and the object become one. That is, their ability to gesture is similar to that of young children, who, it turns out, have a very immature inferior parietal lobule.

Moreover, patients with apraxia may demonstrate difficulty properly sequencing their actions. If you were to pretend to place a cigarette and matches in front of the patient and ask them to demonstrate by pantomime how they would light it and take a drag, they may pretend to hold up the match, blow it out, strike it, and then pretend to suck on the cigarette and then light it. That is, they would incorrectly produce the sequence, though the individual acts may be performed accurately. They'd just be in the wrong order.

It is noteworthy that apes lack this most recently acquired neocortical tissue, the angular gyrus of the inferior parietal lobule, and that complex tool-making capabilities, the capacity to draw, as well as their ability to make even the simplest of clothing, is completely lacking. Only humans make or wear clothing which in turn is also a product of this recent evolutionary neocortical acquisition. However, it is the right and not the left inferior parietal lobe which makes possible the ability to wear clothes, for when damaged or injured, these abilities suffer. Some apraxic disorders, such as **constructional** and **dressing apraxia**, are often due to right parietal injuries.[60]

As the name implies, when patients suffer from dressing apraxia they have difficulty putting on their clothes. For example, a patient may attempt to put a shirt on upside down, then inside out, and then backwards. Those with constructional apraxia lose the ability to draw accurately, and they may in fact fail to attend to or notice the entire left half of visual space.

It is thus important to note that although we possess two parietal lobes, they do not perform the exact same functions. The left parietal is more concerned with temporal sequences, grammar, and language including writing, spelling, and the production of signs such as in ASL. The right parietal lobe is more concerned with guiding the body as it moves through space as well as the manipulation and depiction of spatial relations such as through carpentry, masonry, as well as painting and art or even sewing together or putting on one's clothes. As such, the two parietal lobes are concerned with different aspects of language and the art of gestural communication. Indeed, it was the evolution of these two differently functioning regions of the neocortex and their subsequent harmonious interaction that made possible complex, grammatically correct written and spoken language.

II

THE ORIGIN OF SPEECH, LANGUAGE, AND THOUGHT

6

The Knowing Hand
The Evolution of Reading, Writing, and Arithmetic

Language and the Left Half of the Brain

It is now well known that among over 90% of the right-handed population, and 75% of those who are left-handed, the left cerebral hemisphere provides the neural foundations for the comprehension and expression of grammatically complex spoken language. The left half of the brain dominates in the perception and processing of real words, word lists, rhymes, numbers, Morse code, consonants, consonant vowel syllables, nonsense syllables, the transitional elements of speech, and single phonemes. It is also dominant for recognizing phonetic, conceptual, and verbal (but not physical) similarities, for example, determining if two letters (*g* and *p* vs. *g* and *q*) have the same vowel ending.[1]

The neocortical mantle of the left half of the cerebrum mediates most aspects of expressive language functioning such as reading, writing, speaking, spelling, and naming. This includes the comprehension of the grammatical, syntactical, and descriptive components of language, as well as time sense, rhythm, verbal concept formation, analytical reasoning, and verbal memory.[2]

Perceiving, organizing, and categorizing information into discrete temporal units or within a linear and sequential time frame are also left-hemisphere dominated activities. Indeed, the left half of the brain is sensitive to rapidly changing acoustics be they verbal or nonverbal. It is also specialized for sorting, separating, and extracting in a segmented fashion, the phonetic and temporal–sequential or articulatory features of incoming auditory information so as to identify speech units.[3]

Language and Temporal–Sequential Motor Control

The left cerebral hemisphere is specialized in regard to the temporal–sequential control over hand movement as well as perceiving and expressing units of information in regard to sequences. In fact, language and the production of verbal thoughts are related to and in part are an outgrowth of left-hemisphere-mediated motor activity.[4] This is why individuals often gesture with the right hand when they speak and why injuries to the left half of the brain disrupt the performance of actions requiring a set series of sequences—apraxia.

Temporal sequencing is of course a fundamental property of language as demonstrated by the use of syntax and grammar. That is, syntax is a system of rules which govern the positioning of various lexical items and their interrelations to one another. This allows us to do more than merely name but to describe and to analyze how various parts and segments of speech interrelate. We can determine what comes first or last (e.g., "point to the door after you point to the window"), and what is the subject and object. When the left hemisphere is damaged, expressive and receptive aspects of syntactical information processing suffer.

In contrast, the right half of the brain has great difficulty utilizing syntactical or temporal–sequential rules.[5] In studies of individuals who have undergone the surgical separation of the two cerebral hemispheres (i.e., split-brain operation), when the isolated right brain is given a command, such as "pick up the yellow triangle after you pick up the blue star," it will simply (via the left hand) pick up the yellow triangle. The right hemisphere responds to the first item in the sentence regardless of grammatical relationship.[6] In fact, the ability of the right half of the brain to understand vocabulary and the denotative aspects of language is also limited to emotional and just a few concrete words.

The reason that only the left half of the brain is able to understand and perform a simple task such as the above, whereas the right half is unable to do so, is because the ability to extract denotative meaning from complex spoken language is dependent on the ability to organize and coordinate speech into temporal and interrelated units. This is a function at which the left hemisphere excels, and an ability which is at least in part strongly related to the predominant usage of the right hand for repetitive sequential activities for much of human history.[7] Among the majority of the population, motor control in turn is dominated by the right hand. Hence, the left parietal lobe would be the more neurologically wired to perform these actions as compared to the right parietal region which (in conjunction with the right frontal motor area) mediates left-hand functioning.

Handedness and Language

There is considerable evidence that the grammatical and syntactical components of spoken language are directly related to handedness and the evolution of the left inferior parietal lobule. Among the majority of the population, it is the right hand which is dominant for grasping, manipulating, exploring, tool making, sewing, writing, creating, destroying, and communicating. Although the left hand assists, it is usually the right hand which is more frequently employed for orienting, pointing, gesturing, expressing, and gathering information concerning the environment. We predominantly use the right hand for activities such as throwing, hitting, writing, drawing, and so on, and are more likely spontaneously to activate the right rather than the left half of the body. While engaged in nonemotional, explanatory, or descriptive speech, most individuals are more likely to gesture with the right arm and hand, which appears to accompany and even emphasize certain aspects of speech.[8] Hence, the right hand appears to serve as a kind of motor extension of language and thought insofar as it often accompanies or acts at the behest of linguistic impulses.

Given the preponderance of right-hand activity while talking or gesturing, it is apparent that gesture, like speech, is linked to the functional integrity of the left half of the brain and that both share or are tightly interlinked with some of the same neurons that support speech activity.[8] In fact, immediately above the neocortical frontal lobe motor areas where the neurons which represent muscles of the face and mouth are located, can be found neurons which represent the hand, and Broca's expressive speech area is adjacent to both.This same close proximity is maintained in the sensory areas of the parietal lobe which in turn are tightly interlinked to the frontal neocortical motor area via extensive axonal–dendritic interconnections.

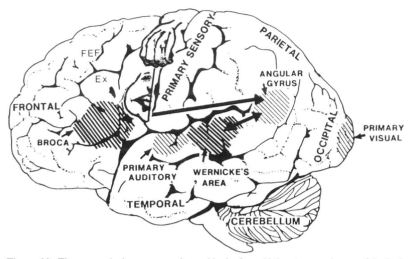

Figure 66. The neocortical motor areas located in the frontal lobe. A motor image of the body is represented in regard to motoric importance and fine motor skills. Hence, the fingers of the hand have a greater representation than does the thigh. Broca's area activates the oral musculature so that the sounds of speech may be articulated.

When one speaks, due to the spreading of excitation to neighboring groups of cells, hand neurons become excited as well including those involved in controlling the muscles of the mouth. Hence, speaking triggers hand movement. Conversely, as is evident from working with the deaf, signing also often triggers sound production. The deaf often make sounds while signing because presumably activation of those neurons involved in hand control and gesturing stimulate neurons involved in vocalization. As in normally speaking adults, the neurons subserving both functions are adjacently located.

Because the hand and speech area partially occupy closely adjacent neuronal space, when both are simultaneous activated there results competitive interference. While speaking, the ability to simultaneously track, manually sequence, or maintain stabilization of the arms and hands is concurrently disrupted.[9] In addition, as the phonetic difficulty of the verbalizations increase motor control decreases—a function presumably of simultaneous activation of (and thus competition for) the same neurons.

Motor functioning, however, is dependent on sensory feedback which in turn is provided by the parietal lobe.[10] That is, if not for the sensory feedback provided by the muscles and joints, information which is transmitted to the parietal lobes, movement would become clumsy and uncoordinated since persons would not know where their limbs were in space and in relation to one another. Since the parietal lobes and the motor areas in the frontal lobes are richly interconnected, they serve in many ways as a single neurocortical unit—sensorimotor cortex.

It is in part due to this interrelationship among sensory feedback, motor control, and gesture that language production is greatly dependent on the parietal and frontal lobes where the motor centers and Broca's speech area are located. Although language serves to question and describe, initially it is through touch that one first comes to know the world. Knowing is first made possible via the hand which in turn initially, during the early stages of development, tends to pick up everything within reach and place it in the mouth where it is then orally explored (a function of the immature amygdala). Hence, the hand and the mouth are linked almost from the outset in regard to the acquisition of knowledge. Be it a human infant or an adult ape, it is the hand which is first employed for the purposes of information gathering and classification.[11]

The Knowing Hand

As noted, the parietal lobe is sensitive and responsive to tactual stimuli regardless of where on the body it is applied. In fact, it is via the reception of these signals from the sensory surface of the body that the entire body comes to be spatially represented in the neocortex. However, body parts are represented in terms of their sensory importance, that is, how richly the skin is innervated.[12] For example, more cortical space is devoted to the representation of the fingers and the hand than to the forearm as there are more sensory receptors located in these regions of the skin. Because of this, the neocortical body map is very distorted, and the hand receives extensive representation.

When tactile sensations are transmitted from the body surface to the amygdala and thalamus, and finally to the primary receiving areas for somasthesis (e.g., body sensations), located within the neocortex of the parietal lobe, one is able to determine not only the qualities of the object being touched (e.g., hard, round, wet, sharp, heavy, pitted) but what part of the body is being stimulated (e.g., hand, elbow, foot). It is within the parietal area of the brain where conscious, perceptual, and tactual associations are formed.

It has been said that the parietal lobule is an organ of the hand.[13] As noted, the hand appears to be more extensively represented than any other body part, and parietal neurons mediate temporal–sequential hand control and become highly activated in response to objects which are within grasping distance.[14]

It is also by the hand that the parietal lobe gathers information regarding various objects (stereognosis), and about the self and the world, so that things and body parts come to be known, named, and identified. Ontogenetically, the hand is in fact primary in this regard. The infant first uses the hand to grasp various objects so they may be placed in the mouth and orally explored. As the child develops, rather than mouthing, more reliance is placed solely on the hand (as well as the visual system) so that information may be gathered through touch, manipulation, and visual inspection. It is at this stage where still-immature parietal-lobe neurons begin to join the amygdala in information acquisition.

As the child and the neocortex of its parietal lobe mature, instead of

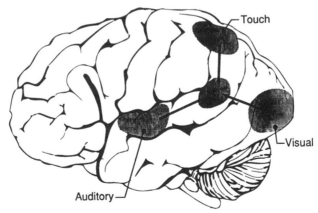

Figure 67. The inferior parietal lobule is situated at the boundary where complex auditory, visual, and tactual associations are formed. Neocortical cells located in this vicinity act to receive and assimilate these associations so that multiple names, descriptions, classifications, and concepts may be formed.

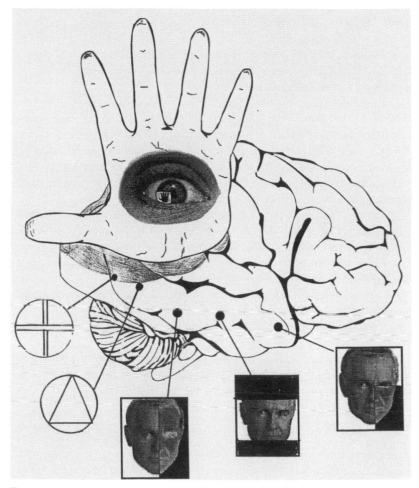

Figure 68. The parietal lobe is often referred to as a "lobe of the hand." Parietal neurons guide hand and reaching movements via sensory feedback to the motor areas in the frontal lobe. Many parietal neurons also receive visual information from the periphery and lower visual fields, the regions where the hands are most likely to be viewed. Parietal neurons act not just to guide but to comprehend hand movements and gestures.

predominantly touching, grasping, and holding, the fingers of the hand are used for pointing and then naming. It is these same fingers which are later used for counting and developing temporal–sequential reasoning; that is, the child learns to count on his or her fingers, then to count (or name) by pointing at objects in space. It is at this point that the latest and slowest region of the brain to mature, the inferior parietal lobule, begins to play an increasing role in cognitive development.

Counting, naming, object identification, finger utilization, and hand control are ontogenetically linked and all seem to rely on the same neural substrates for their expression; that is, the left inferior parietal lobule. Hence, when this portion of the left hemisphere is damaged, naming (*anomia*), object and finger identification (*agnosia*), arithmetical abilities (*acalculia*), and temporal–sequential control over the arms and hands (*apraxia*) are frequently compromised.[15]

Hence, a variety of symptoms are associated with left-inferior-parietal-lobe damage. However, when this particular constellation of disturbances, that is, finger agnosia, acalculia, *agraphia* (loss of the ability to write), and left–right disorientation occur together, they have been referred to as *Gerstmann's syndrome*.[16] Gerstmann's symptom complex is most often associated with lesions in the area of the inferior and superior parietal lobule. However, because this symptom complex does not always occur together, some authors have argued that Gerstmann's syndrome, per se, does not exist.

We will not take issue, pro or con, on this minor controversy but instead will focus on those aspects of Gerstmann's syndrome which have not yet been discussed (finger agnosia, acalculia, left–right disorientation).

Finger Agnosia

Finger agnosia is not a form of *finger blindness*, as the name suggests. Rather, the difficulty involves naming and differentiating among the fingers of either hand as well as the hands of others.[17] This includes pointing to fingers named by the examiner or moving or indicating a particular finger on one hand when the same finger is stimulated on the opposite hand, or if you touch their finger while their eyes are closed and ask them to touch the same finger, they may have difficulty.

This may not seem like a terribly disabling disorder. However, when one considers the importance of the hand and especially the fingers in exploring an object in order to comprehend its character and potential, finger agnosia, particularly if it occurs in a child, can drastically reduce and interfere with the very foundations of knowledge acquisition.

Acalculia

By the time a child has reached the age of 3 she or he is capable of making simple calculations, and this is a function of the early stages of maturation of the inferior parietal lobe. Counting is important in knowledge acquisition because it aids in the ability to determine what is, versus what isn't, and thus to form categories consisting of abstract notions. The ability to count requires and relies on the ability to make fine distinctions and to differentiate not only what is "4" versus "8," but to classify so as to determine what is 4 *apples* versus 8 *bananas*, and how that adds up to a *dozen* pieces of *fruit*.

When the parietal lobe has been injured, or is slow to mature, an individual may in consequence have problems not only in classifying, but in adding and subtracting.[18] How this problem manifests itself depends on whether the right versus left parietal lobe has been injured.

A individual with a right-parietal disturbance, when performing arithmetical operations may misalign numbers when adding or subtracting (referred to as *spatial acalculia*). With left (or right) parietal injuries he or she may erroneously substitute one operation for another, that is, misreading the sign " + " as " × ," such that he or she multiplies rather than adds. Or he or she may reverse numbers, that is, "16" as "61," substitute counting for calculation, that is, 21 + 6 = 22, or inappropriately group: 32 + 5 = 325.

On the other hand, with injuries localized to the vicinity of the left inferior parietal lobe, patients may have severe difficulty performing even simple calculations, for example, carrying numbers, stepwise computation, borrowing, and they may in fact be unable to recognize numbers. For example, they may be unable to write out or point to the number "4" versus the number "8" versus the letter "*B*." Hence, in some cases it is the spatial nature and in others it is the temporal sequential deficit and the inability to recognize or write out abstract signs and symbols which causes them difficulty.

Right–Left Disorientation

When reading and writing, or when performing gestures involved in the production of ASL, the ability to correctly orient oneself and one's movements and perceptions in regard to right versus left and up versus down, is very important. Gestures are made in space, and the movement of the hands in space is made possible via the parietal lobes. Hence, when this structure is injured, the ability to orient to the left or the right may be impaired.[20] Nevertheless, although the right half of the brain is clearly dominant in regard to analyzing and expressing spatial relationships, it is the left half which is concerned with the temporal–sequential aspect of movement in space, including orientation to the right or the left.[19]

In part, it seems somewhat odd that right–left spatial disorientation is more associated with left rather than right cerebral injuries, given the tremendous involvement the right half of the brain has in spatial synthesis and geometrical analysis. However, orientation to the *left* and *right* transcends geometric space as it relies on language. That is, "left" and "right" are designated by words and are defined linguistically. In this regard, left and right become subordinated to language usage and organization. Hence, left–right confusion is strongly related to problems integrating spatial coordinates within a linguistic framework.

It is for this reason that when the left parietal lobe has been injured, an individual who is proficient in ASL may become aphasic for signs. They may lose not only the ability to comprehend them but to make them appropriately. However, if they were able to speak and converse before their injury, they retain these abilities although they can no longer sign.

Right–left orientation is vital not only in the production of signs and gestures, but in the evolution of reading and writing.

Reading, Writing, Naming, Spelling, and the Inferior Parietal Lobule

The development and acquisition of language seems to be related to manual gesturing and fine motor exploration, complex sequential processing, and the ability to form concepts and to classify a single stimulus in a multiple manner. For example, classifying and labeling a chair according to the type of fabric, cost, color, comfort, style, and to describe it as such using words. Or, conversely, hearing the word *chair* and being able to consider it from multiple perspectives. This is made possible via the inferior parietal lobe.[20]

The development and evolution of the inferior parietal lobule enabled

complex gestures to be learned and performed. Through its rich interconnections with the immediately adjacent neocortical areas in the temporal
lobe which are responsible for the perception of sound, the inferior parietal
lobe is also able to impose and stamp temporal sequences on auditory input
and thus break it up into units. This includes sounds that are transmitted to it
from the auditory (Wernicke's) association areas located in the immediately
adjacent superior and middle temporal lobe of the left hemisphere. This
relationship in part also made possible the development of grammar.

The acquisition and expression of complex grammatically correct
spoken language is not dependent only on grammar but the ability to develop
and assign labels and multiple categories so that what was heard could be
quickly analyzed for meaning. Again, in order to comprehend that the word
chair indicates an object that you can sit in, the person must be capable of
associating completely different sensory experiences into an interlinking
category so that multiple possible meanings can be assigned and extracted.
It is in this manner that we come to know what words may signify as well as
their grammatical relationship. Otherwise a sound would be merely a sound
having little more than emotional or motivational significance.

This ability to form multiple categories and to impose grammatical
structure on all that is said is made possible by the same structure which
provides temporal sequencing to movement and which guides and controls
the ability to gesture and perform complex actions—the inferior parietal
lobule and the angular gyrus.[21]

The Inferior Parietal Lobule: The Multimodal Assimilation Area

Developmentally, of all cortical regions, the inferior parietal lobule is
one of the last to functionally and anatomically mature.[22] In this regard,
ontogeny replicates phylogeny in that humans are the most advanced species
to appear on this planet and the only one to possess an angular gyrus of the
inferior parietal lobe. It is due to its slow maturation that many capacities
mediated by this area (e.g., reading, calculation, the performance of
reversible operations in space) are late to develop appearing between the
ages of 5 to 8. Because humans are the only species to possess this
neocortical region, no other animal is capable of these cognitive feats.

The inferior parietal lobule contains many hundreds of millions of
neurons most of which are multimodally responsive. A single neuron may
simultaneously receive highly processed somesthetic, visual, auditory, and

movement-related input from the neocortical association areas located in the parietal, occipital, temporal, and frontal motor areas. Hence, many of the neurons in this area are multispecialized for simultaneously analyzing auditory, somesthetic, and spatial–visual associations, have visual receptive properties which encompass almost the entire visual field, and respond to visual stimuli of almost any size, shape, or form.[23]

Due to their extensive interconnections with the adjacent neocortical areas, inferior parietal neurons are able to assimilate and create multiple associations. The inferior region thus acts to increase the capacity for the

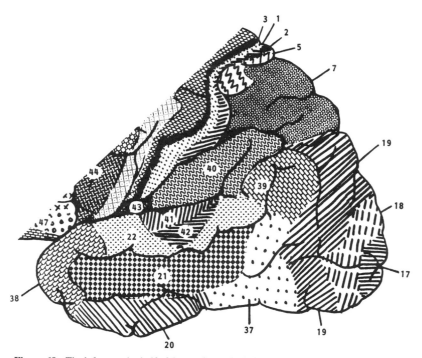

Figure 69. The left posterior half of the cerebrum depicting the primary tactual receiving areas (3, 1, 2) in the parietal lobe, the supramarginal (40) and angular (39) gyrus of the inferior parietal lobe, Wernicke's receptive and auditory association areas (42, 22), and the primary auditory (41) and visual (17) areas, and visual association areas (18, 19, 7, 37). From R. Joseph, *Neuropsychology, Neuropsychiatry, and Behavioral Neurology.* New York: Plenum Press, 1990.

organization, labeling, and multiple categorization of sensorimotor and conceptual events. One can thus create visual images, somesthetic, or auditory equivalents of objects, actions, feelings, and ideas simultaneously, and can conceptualize a "chair" as a word, visual object, or in regard to sensation, usage, and even price. Similarly, when we see an object such as a chair, we are then able to name it.

Spoken Language and the Inferior Parietal Lobe

Because of its involvement in functions such as those described above, one effect of damage to the left angular gyrus is a loss of the ability to correctly name things, or to find words, a condition called *anomia*. Patients so affected cannot recall the names of things even when looking right at them such that the words they search for seem to be perpetually on the "tip of the tongue." However, this condition is much worse for if told the word they immediately lose it again. These individuals have difficulty naming objects, describing pictures, and so forth.

On the other hand, if the area of tissue destruction is quite circumscribed so that the inferior parietal lobe is disconnected from just one adjacent neocortical area, for example, the visual cortex, the loss of naming ability will be only for things the person looks at, though it could be named if described verbally or felt in the hand. If instead the damage disconnected the inferior region from the rest of the upper parietal lobe, then they would be able to name the object if they looked at or heard it but not if they just felt it in their hand.[24] These are called *disconnection syndromes.*

Moreover, lesions involving the angular gyrus, or when damage occurs between the fiber pathways linking the left inferior parietal lobule with the visual cortex, there can also result *pure word blindness*, and the person loses the ability to read words or sentences. This is due to an inability to receive visual input from the left and right visual cortex (located in the occipital lobe). Although these people can see without difficulty, they are unable to recognize verbal symbols, such as words and even letters because this information is no longer linked up with Wernicke's area and the inferior parietal lobule. Like anomia, pure word blindness is due to an inability to transmit visual–linguistic information to Wernicke's area so that auditory equivalents may be called up. This is due to disconnection. One can no longer link or match the sound or verbal name to the visual image and can no longer read.[25]

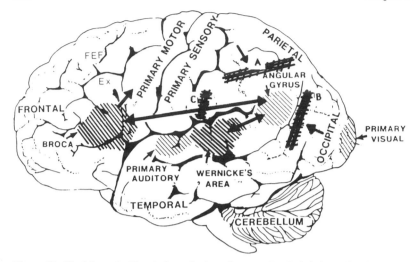

Figure 70. The left cerebral hemisphere. Lesions disconnecting the inferior parietal lobe (IPL) from the (A) remainder of the parietal lobe result in an inability to name or recognize what is held in the hand. However, they can be named if they are looked at or if it makes a sound. (B) Disconnection of the IPL from visual cortex results in inability to name objects or letters that he sees. However, he can name the same object if he touched it or if it makes a sound. (C) Disconnection of the IPL from Broca's area results in conduction aphasia.

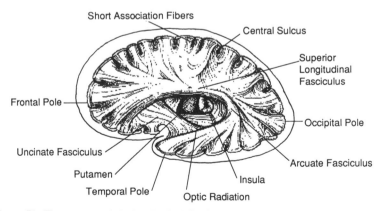

Figure 71. The arcuate and the longitudinal fasciculus contain axonal fibers which link the amygdala, Wernicke's area, inferior parietal lobule, and Broca's expressive speech area. The arcuate fasciculus is part of the axonal white matter fiber pathways which travel beneath and between neurons located in the neocortex and the old cortical centers.

Depending on the extent of the injury, some patients may experience only a mild disturbance involving word and letter recognition, and this is referred to as *dyslexia* as they have difficulty reading. Dyslexia, however, can result from a number of conditions including the slow or incomplete maturation of the inferior parietal lobe. Those who are completely unable to read and recognize words are described as suffering from *alexia*.

The inferior parietal lobe acts not only to assimilate different associations but to relay this information via a bundle of axons (called the arcuate fasciculus) to Broca's expressive speech area in the frontal lobe. It is via Broca's area that words come to be articulated.

Destructive lesions involving this axonal fiber pathway between the inferior parietal and Broca's area can result in *conduction aphasia*.[26] That is, individuals would know what they want to say and they might even be able to hear the word in their head. However, due to destruction of this fiber pathway the information is cut off from Broca's area and cannot be expressed. Nor would someone so affected be able to repeat simple statements, read out

Figure 72. Cro-Magnon art. The life-sized head and chest of a bull and the head of a horse. Aurignacian-Perigordian. Picture gallery. Lascaux. Courtesy of Bildarchiv Preubischer Kulturbesitz.

Figure 73. Cro-Magnon art. Running horses. From the picture gallery. Lascaux. Courtesy of Bildarchiv Preubischer Kulturbesitz.

loud, or write to dictation. This is because Broca's area is disconnected from the posterior language zones.

The Gestural Foundations of Reading and Writing

The evolution of the inferior parietal lobule not only made language possible but provided the neurological foundations for tool design and construction, the ability to sew clothes, and the capacity to create art and pictorial language in the form of drawing, painting, sculpting, and engraving. Finally it enabled human beings not only to create visual symbols but also to create verbal ones in the form of written language.[27] In fact, with the exception of written language, all these abilities, including, perhaps, spoken language, may have appeared essentially within the same time frame, the first evidence of which is approximately 40,000 years old and was left in the deep recesses of caves and rockshelters throughout Europe and Africa.[28]

Although we have no recordings of what people may have sounded like,

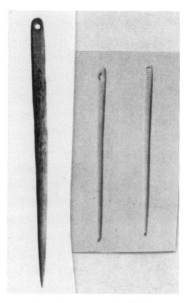

Figure 74. Cro-Magnon eye needles for sewing animal skins and making clothes. From G. Clark. *The Stone Age Hunters*. London: Thames & Hudson, 1967.

or any evidence regarding the presence of grammatical sequences in their speech, complex and detailed paintings left in the deep and forgotten recesses of ancient caves indicate that human beings, the Cro-Magnon people, were capable of telling stories and making signs which are still comprehensible and pregnant with meaning 40 millennia later. However, it is likely that they were speaking and sharing thoughts and desires since they first appeared on the scene about 130,000 years ago, or longer.

Although we can debate the merits and purposes for which these pictorial displays were created (magic, religious, instructive), they nevertheless represent the first form of written language. Body movements and gestures had now become adapted for reproduction in the form of pictures, the result of well-crafted, delicate, and precise finger, hand, and arm movements.

As can be seen based on photographs of Cro-Magnon cave art and that of ancient Egypt, Sumer, and Babylon (about 4,000 to 6,000 years ago), this

Figure 75. Cro-Magnon art. Courtesy of Bildarchiv Preubischer Kulturbesitz.

form of pictorial language remained essentially similar for almost 35,000 years. It is also apparent that although much of what was depicted during this early time frame was without any grammatical foundation, occasionally temporal elements were imposed. For example, on a dollar-sized emblem created perhaps 20,000 years ago, there appears a deer standing on one side, whereas on the other the same deer is lying on the ground, either asleep or more likely wounded or dead. Hence, art became adapted for not just depicting but also telling a changing story.

In fact, by 15,000 years ago stories regarding the exploits and conquests of humans began to be told in the form of paintings. Specifically, fearsome battles and the stalking and killing of men in combat. In fact, for the following 15,000 year similar stories of mayhem, murder, and conquest would be told to eager and receptive audiences, transmitted by satellite and movie screen directly to that ancient limbic tissue that thrives on witnessing and engaging in such acts.

Nevertheless, these first stories, which for the most part concerned

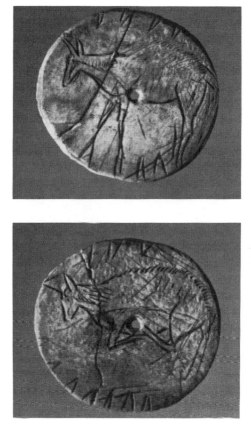

Figure 76. Cro-Magnon art. Possible temporal–sequential picture writing from 25,000 years ago. Carved on either side of a 1-inch disk of bone, an animal is standing on one side, and on the other is lying on the ground asleep or dead. From *Prehistoire de l'Art Occidental*, Editions Citadelles & Mazenod, photo by Jean Vertut.

killing and conquest, were not presented in temporal–sequential dimensions. Rather, as is evident from the "writings" of the ancient Egyptians and the Sumerians, these pictorial displays were very gestalt in formation and arranged in regard to spatial relationships. That is, they were produced in a manner typical of the right half of the brain.

Nevertheless, these ancient stories are for the most part comprehensible

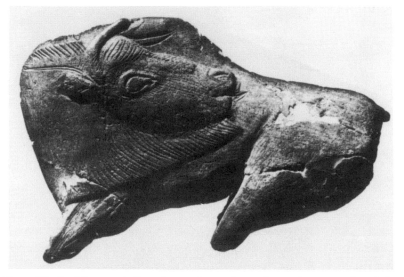

Figure 77. Cro-Magnon art. Hewn from and carved around natural contours of a reindeer antler, a bison turns and licks its flank. This 4-inch ornament is thought to be part of a spear thrower and was carved perhaps 25,000 years ago. Compare with Figure 78, of Babylonian ox carved 20,000 years later. From *Prehistoire de l'Art Occidental*, Editions Citadelles & Mazenod, photo by Jean Vertut.

Figure 78. Babylonian art. Hewn from ivory, an ox turns and licks its flank. Carved approximately 4,000 years ago.

Figure 79. Cro-Magnon bow-and-arrow hunting scenes. Drawn approximately 15,000 years ago.

Figure 80. Cro-Magnon bow-and-arrow hunting scene. Drawn approximately 15,000 years ago.

Figure 81. A bison is wounded and impaled by arrows. From *Prehistoire de l'Art Occidental*, Editions Citadelles & Mazenod, photo by Jean Vertut.

Figure 82. Detail from the "Great Lion Hunt." King Ashurbanipal hunting lions with bow and arrow. (Post-Babylonian) Neo-Assyrian. Carved approximately 3,000 years ago.

Figure 83. Wounded lion. Detail from the "Great Lion Hunt."

Figure 84. Wounded lioness. Detail from the "Great Lion Hunt."

6,000 years later by people from wholly different cultures and who speak completely different languages, including, I presume, those of you reading this book. The same, of course, was true 6,000 and even 25,000 years ago, which is what must have made this form of "picture writing" extremely effective as a communicative device, particularly in an educated society surrounded by the uneducated masses.[29]

Figure 85. Cro-Magnon warriors.

Figure 86. A Cro-Magnon warrior falls wounded. Drawn approximately 10,000 to 15,000 years ago.

Figure 87. Cro-Magnons from separate tribes in battle.

Figure 88. Cro-Magnon warriors celebrating the killing of the man in the lower right who has been impaled by half a dozen arrows.

Figure 89. Detail from a relief depicting victorious military campaign and siege led by King Ashurnasirpal II. Neo-Assyrian.

Figure 90. King Ashurbanipal's victory over Te-umman, King of Elam, at the River Ulai. Neo-Assyrian.

Take for instance, this stela erected by an Egyptian pharaoh who probably ruled Egypt around 6,000 years ago (Figures 91 and 92). Hence we see at the top the heads of two bulls which symbolize strength and below which are pictures of men, one of whom in particular is quite large and who we might assume is the leader, or a god, or the king. It is apparent that he is marching in procession with many subjects and standard bearers and that many men were decapitated in consequence.

Below that we see two giant, long-necked leonine creatures who have been captured, and below that we see the bull is besieging a city fortress. Looking to the other side, we see again a very large man who is wearing a different hat but who otherwise is dressed the same as the first man on the other side. Hence, this again must be the king, and he is smiting an enemy that has been captured. What this then tells us is that the King of Egypt went to war against and conquered a city and took many captives including this "mythical" lion-headed beast which was the symbol of his human foe who happened to live in ancient Sumer.

If perchance we had lived in Egypt 6,000 years ago and had some cultural knowledge and a context in which to interpret this stela, the story would not change much except that we would know that the attacked city was probably ruled by ancient Sumer, for the captured beasts were a motif popular among the Sumerian royalty.

We would also know that the two crowns worn by this pharaoh stand for Upper and Lower Egypt, that behind him are his sandal bearer and foot washer, and that in front are his priest and standard bearers. Since those who are decapitated appear to be Egyptians, it is possible that this king took action against some rebels or perhaps Sumerian soldiers holding a city of Lower Egypt. His success is further indicated by his initially wearing the crown of Upper Egypt only later to replace it with a second crown of Lower Egypt after defeating the Sumerians the corpses of whom lay below his feet. In this regard the stela may in fact tell two stories, one of which concerns the uniting of Upper and Lower Egypt and the other an attack on the Sumerians. The hawk and rebus above the head of the captive about to be struck basically reads, "Pharaoh the incarnation of the hawk-god Horus, with his strong right arm leads captives."

Who is the pharaoh that is depicted? Some scholars have assigned him the name of King Narmer, others believe him to be King Menes. In actuality it may well be neither and may represent a king who ruled much further back

Figure 91. Victory tablet of ancient Egyptian pharaoh, possibly King Nar-mer or Menes (reverse of Figure 92).

Figure 92. Victory tablet of ancient Egyptian pharaoh, possibly King Nar-mer or Menes (reverse of Figure 91). Compare the long-necked lion-head beasts with Figure 93.

Figure 93. Cylinder seals and impressions from the earliest Sumerian periods, 6,000 or more years ago. Note resemblance between the long-necked and tailed lions and those harnessed by victorious Egyptian soldiers in Figure 92. These ring cylinders were used by Sumerian royalty as a form of signature stamp.

in antiquity than many scholars are willing to consider, that is, about 6,500 years ago.

The Evolution of Visual Symbols and Written Signs

When, where, how many times, and how many places the art of writing was invented, no one knows. The earliest preserved evidence detailing the evolution of written symbols comes from ancient Sumer, around 6,000 or more years ago. In Sumer as elsewhere, the first forms of "writing" were pictorial, a tradition that was already in use at least 20,000 years before their time.

Indeed, in examining the pictorial representations left by the Cro-Magnon people, they, just like modern Westernized humans, utilized single pictures, such as a red hand to indicate a clear, readily understood message, such as "stop," or "do not enter." In most American cities, this same "hand" symbol is used at intersections to indicate if and when someone may cross a street. Moreover, the Cro-Magnon people also utilized abstract symbols, the meanings of which are not at all clear, though some scholars have associated them with sex, or fertility, or male versus female signs, and so on.

However, the Cro-Magnon people also liked to produce art for the sake

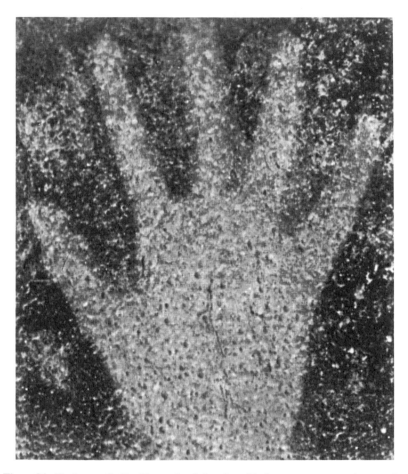

Figure 94. The image of a Cro-Magnon hand placed outside the entrance to an ancient cave in France. Hands were commonly left in caves, and many of the hands were missing fingers. Left hands tended to be depicted as frequently as right hands though in some caves left hands predominated. This hand was formed by placing the hand against the wall and then paint was blown through a tube so as to form a negative image.

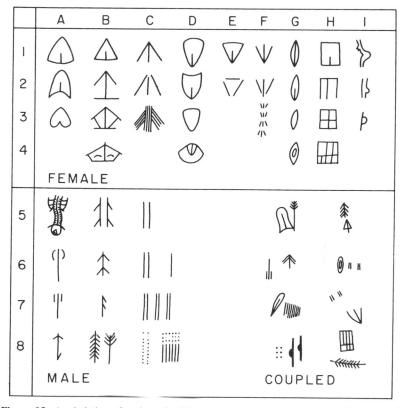

Figure 95. A tabulation of various Cro-Magnon signs and symbols found in the caves of France. From G. Clark. *The Stone Age Hunters.* London: Thames & Hudson, 1967.

of art and were the first people to paint and etch what today might be considered "cheesecake," that is, slim, shapely, naked, and nubile young maidens in various positions of repose.[30]

The Sumerians, too, initially utilized single pictorial symbols to convey specific messages, sometimes such as a "lion" to indicate "watch out, lions about," or that of a grazing gazelle so as to inform others that good hunting could be found in the vicinity.[31] Although the Cro-Magnons sometimes used single two-step pictorial displays to indicate sequence of action, as based on available evidence, it was not until the time of the Sumerians that people began skillfully to employ a series of pictures to represent not just actions, but abstract as well as concrete ideas.

For example, prior to the Sumerians, a depiction of a lion or a man might indicate the creature itself, or the Nature God that was incarnate in its form. The Sumerians took this to its next evolutionary step. For example, by drawing a "foot," or a "mouth," or an "eye," they could indicate the idea to "walk," or to "eat," and to "look" or "watch out." By combining these pictures, they were then able to indicate complex messages, such as "so and so" had to "to walk" to a certain place where "gazelles" might be found in order "to eat," but they would have to keep an "eye" out for "lions."[32]

This was a tremendous leap, for earlier in Sumerian civilization, just as in Egypt and elsewhere, to indicate a complex message such as the above would have required elaborate pictorial detail of entire bodies engaged in particular actions. Although beautiful to behold, and easily understood by all, this was a very cumbersome and time-consuming process.

The next step in the evolution of writing was the depiction not just of actions and ideas, but sounds and names. As argued by Edward Chiera,[33] if they wanted to write a common Sumerian name such as "kuraka," they would draw objects which contained the sounds they wanted to depict, such as a "mountain" which was pronounced "kur," and "water" which was read as "a," and then a "mouth" which was pronounced "ka," and by combining them together one was able to deduce the sound or name that the writer desired.

This was a tremendous leap in abstract thinking and in the creation of writing, for now visual symbols became associated with sound symbols, and one could now not only look at pictures and know what was meant, but what the symbols sounded like as well. In this manner, writing, although still in pictorial form, became much more precise. Now ideas, actions, and words, and thus complex verbal concepts, could be conveyed.

Nevertheless, although this was a highly efficient means of communication, it still remained quite cumbersome which required that further steps in symbolic thought be invented. The Sumerians met this challenge by inventing wedge-form **cuneiform** characters which gradually began to replace the older pictographs and ideographs.

Initially these characters resembled the pictures they were destined to replace. However, as the use of cuneiform continued to evolve, pictorial details were gradually minimized, until finally these characters lost all pictorial relationship to that which they were meant to describe. Among the

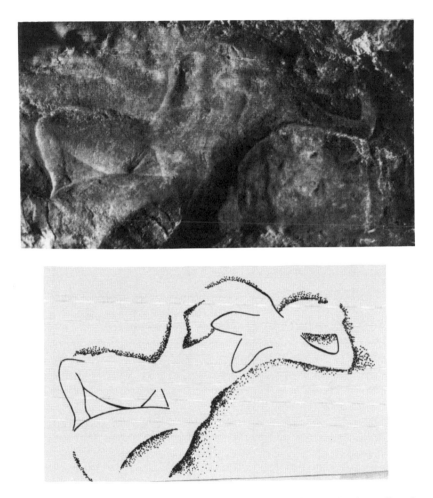

Figure 96. Remnants of Cro-Magnon cave rock carving of naked female lying down. Carved approximately 20,000 years ago. From *Prehistoire de l'Art Occidental*, Editions Citadelles & Mazenod, photo by Jean Vertut.

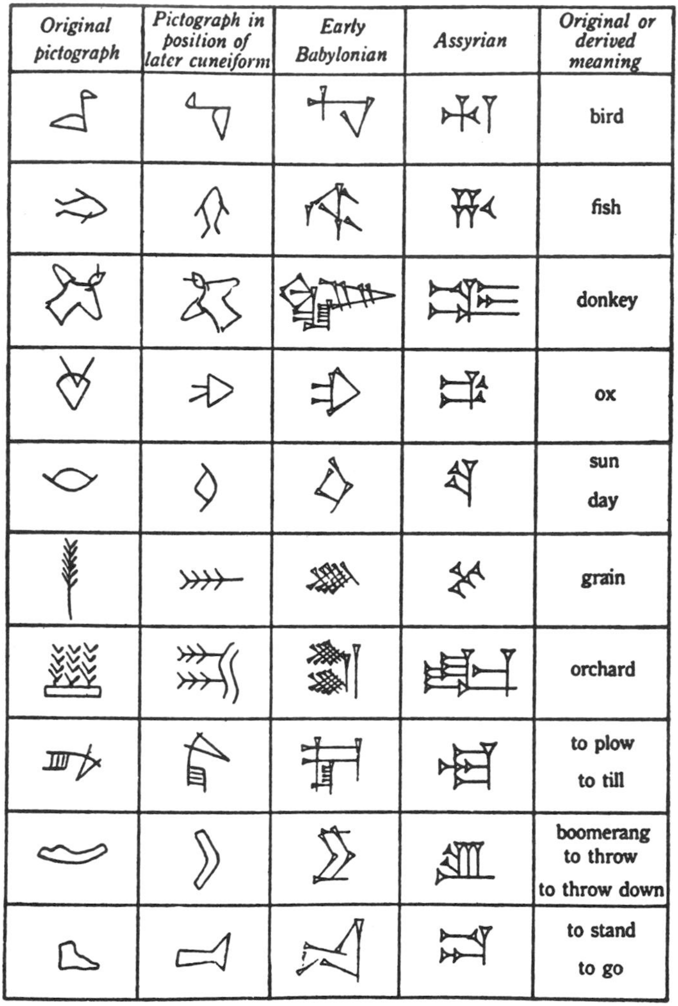

Original pictograph	Pictograph in position of later cuneiform	Early Babylonian	Assyrian	Original or derived meaning
				bird
				fish
				donkey
				ox
				sun day
				grain
				orchard
				to plow to till
				boomerang to throw to throw down
				to stand to go

Figure 97. The origin and evolution of writing from pictures employed by the people of Sumer and then Babylon to develop cuneiform characters. The people of Sumer realized that pictures could represent not only images and ideas, but sounds, which gave their writing greater precision in meaning. Reprinted with permission from E. Chiera, *They Wrote on Clay*. Chicago: University of Chicago Press, 1966.

ancient Egyptians, a similar process occurred with the exception that with the invention of hieroglyphics, pictures remained an essential feature of their writing until the very end of their civilization about 2,000 years ago.

The Sumerians, Babylonians, and the later-appearing Assyrians, however, although able to depict vowels, consonants, and complex ideas and sounds via cuneiform, did not develop an alphabet. In fact, even when they were introduced to foreign-devised alphabets, they resisted this innovation. Even so, pictures came to be placed in temporal sequences, and then the pictures themselves became represented by sequences as well, a series of wedge-shaped lines which were read from left to right.

Nevertheless, be it the writing of the Egyptians, the Sumerians, or modern-day Americans, the process of reading and writing remains essentially similar. Both require the interaction of brain areas involved in visual and auditory analyses, as well as the evolution of the inferior parietal lobule which enabled visual signals to be matched with sounds so that auditory equivalents could be conjured up. In this manner people are able to not only look at a word but to know what it sounds like.

Reading and Writing: Alexia and Agraphia

Maxwell glanced up from his book and the notes he was taking and smiled at the ugly little man who had nervously walked right up to his counter.

"Studying?" the skinny man asked as he reached into his trenchcoat pocket.

"Yeah, got a big test."

"Well, I don't think you're going to pass," the man replied as he pulled out a little .25-caliber semiautomatic and shot Maxwell point blank, as he tried to turn away, striking him in the left parietal region of his skull.

When Maxwell woke up in the hospital he immediately felt overcome by a fearsome headache and reaching up he was terribly surprised to find that part of his head was bandaged and most of his hair gone. He was "lucky" though, or so said his doctors. Although the bullet had cracked his skull, it did not penetrate but instead went round and round his head underneath his skin. Moreover, although he had suffered a fracture, the doctors felt that the amount of blood that had developed on the inferior parietal lobule of his brain, directly beneath the site of impact, was negligible. To prove this to their satisfaction they asked him a few questions, tested his memory, and then released him the next day as he seemed fine.

But Maxwell wasn't fine. On the way home, he noticed that some of the billboards along the roadway didn't make sense, and when his headache subsided enough for him to pick up his schoolbooks, he discovered that he couldn't make out any of the words or sentences, nor did any of the letters make sense.

"What's wrong with me?" he asked in panic. Had he forgotten how to read? Why did the words look so weird? Picking up a pen, he decided to write out a few words just to make sure he hadn't lost his mind. But then he wavered, his hand poised just above the paper. He couldn't quite remember how to write.

Reading Abnormalities

The process of reading involves the reception of visual impulses in the primary receiving areas located within the occipital lobe. It is within the visual neocortex where the initial forms of perceptual analyses are initiated. This visual information is next transferred to the immediately adjacent visual association cortex (areas 18 and 19) where visual associations are formed. These visual associations are next transmitted to a variety of brain areas including the inferior parietal and temporal lobes, and Wernicke's area. It is in these latter cortical regions where multimodal and linguistic assimilation take place so that the auditory equivalent of the visual stimulus may be matched and retrieved. That is, via these interactions, visual grapheme clusters become translated into phonological sound images. In this manner we know what a written word looks and sounds like. It is also possible, however, to bypass this phonological transcoding phase so that word meanings can be directly accessed (i.e., lexical reading).

Although it is likely that most individuals utilize both lexical and phonological strategies when reading, in either case the angular gyrus of the inferior parietal lobule is involved. As noted, with lesions involving the angular gyrus, or when damage occurs between the fiber pathways linking the left inferior parietal lobule with the visual cortex (i.e., disconnection), a condition referred to as Pure Word Blindness sometimes occurs. Patients can see without difficulty but are unable to recognize written language. Written words evoke no meaning because their auditory equivalents cannot be retrieved due to destruction of the interlinking fiber pathway.

There are, however, several subtypes of reading disturbances which may occur with left cerebral damage. These include **literal, verbal,** and **global alexia,** and **alexia for sentences.**[34] In addition, alexia can sometimes result from right-hemisphere lesions, a condition referred to as **spatial alexia.** All these disorders, however, are acquired and should be distinguished from developmental dyslexia, which is present since childhood.

Spatial alexia is associated predominantly with right-hemisphere lesions. In part this disorder is due to visual–spatial abnormalities including neglect and inattention. That is, with right cerebral lesions the patient may fail to read the left half of words or sentences and may in fact fail to perceive or respond to the entire left half of a written page.

Right parietal–occipital injuries may also give rise to spatial disorientation such that patients are unable to properly visually track and keep place, their eyes darting haphazardly across the page. For example, they may skip to the wrong line. Spatial alexia may also result from left cerebral injuries in which case it is the right half of letters, words, and sentences which are ignored.

Agraphia

Just as one theory of reading proposes that visual graphemes are converted into phonological units (visual images into sound), it has been proposed that in writing one transcodes speech sounds into grapheme clusters, that is, a phoneme-to-grapheme conversion. In the lexical route, there is no phonological step. Instead the entire word is merely retrieved.

Regardless of which theory one adheres to, it appears that the angular gyrus plays an essential role in writing. Indeed, it has been argued that the sensory motor engrams necessary for the production and perception of written language are stored within the parietal lobule of the left hemisphere.[35] Hence, when this part of the brain has been injured, patients sometimes have difficulty writing and forming letters due to an inability to access these engrams; that is, they suffer from **agraphia,** an inability to correctly perform the gestures necessary to form written letters or to even gain access to these symbols due to disconnection of the inferior parietal lobe from adjacent cortical tissue.[36]

Presumably the angular gyrus provides the word images (probably via interaction with Wernicke's area) which are to be converted to graphemes. However, it is possible that the initial graphemic representations are formed in the inferior parietal lobule. Nevertheless, these representations are then transmitted to the Broca's expressive speech area and adjacent Exner's writing area for grapheme conversion and motoric expression in the form of writing.

Thus, it appears that there are at least two stages involved in the act of writing, a linguistic stage and a motor–expressive gestural stage. The linguistic stage involves the encoding of information into syntactical–lexical units. This is mediated through the angular gyrus and Wernicke's area, which provides the temporal–sequential and linguistic rules which subserve writing.

The motor–gestural stage is the final step in which the expression of graphemes is subserved. This stage is mediated presumably by Exner's writing area (located in the left frontal lobe above Broca's area) in conjunction with the inferior parietal lobule. Exner's writing area sits immediately adjacent to the primary motor area that controls hand movement as well as Broca's expressive speech area.

Hence, disturbances involving the ability to write can occur due to disruptions at various levels of processing and expression and may arise secondary to lesions involving the left frontal or inferior parietal cortices.

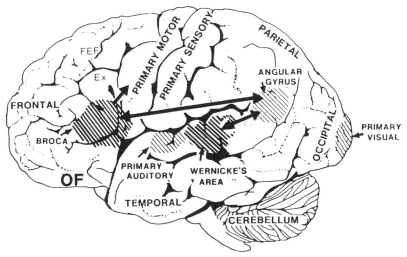

Figure 98. The language axis of the left hemisphere.

Thus, similar to alexia, there are several subtypes of agraphia which may become manifest depending upon which level and anatomical region is compromised. These include **frontal agraphia, pure agraphia, alexic agraphia** (the condition that Maxwell suffered following his injury), **apraxic agraphia**, and **spatial agraphia**.[37]

Exner's Writing Area

Within a small area along the lateral convexity of the left frontal lobe lies Exner's writing area. Exner's area appears to be the final common pathway where linguistic impulses receive a final motoric stamp for the purposes of writing. Exner's area, however, also seems to be very dependent on Broca's area with which it maintains extensive axonal interconnections. That is, Broca's area acts to organize impulses received from the posterior language zones and relays them to Exner's area for the purposes of written expression.

Lesions localized to Exner's area have been reported to result in disturbances in the elementary motoric aspects of writing, that is, **frontal agraphia**. In general, with frontal agraphia, grapheme formation becomes labored, uncoordinated, and takes on a very sloppy appearance. Cursive handwriting is usually more disturbed than printing, and there are disturbances in grapheme selection and the patient may seem to have "forgotten" how to form certain letters. Or they may abnormally sequence, that is, write letters out of order or even add unnecessary letters when writing. In general patients are unable to write because the area involved in organizing visual letter organization (i.e., the inferior parietal lobe) is cut off from the region controlling hand movements in the frontal lobe.

In addition, frontal agraphia can result from lesions involving Broca's area. When Broca's area is compromised, patients cannot speak or write. When Broca's area is spared, but Exner's area has been damaged, then they can speak but cannot write.

Spatial Agraphia. Right cerebral injuries can secondarily disrupt writing skills due to generalized spatial and constructional deficiencies. Hence, words and letters will not be properly formed and aligned even when copying. There may be difficulty keeping lines straight and letters may be slanted at abnormal angles. In some cases, the writing may be reduced to an illegible scrawl.

In addition, patients may write only on the right half of the paper such
that as they write, the left-hand margin becomes progressively
larger and the right side smaller. If allowed to continue,
patients may end up writing only along the
edge of the right-hand
margin of the
paper.
Patients with right-hemisphere lesions may tend to abnormally segment
the letters in words when writing cursively (i.e., *cu siv e ly*). This is due
to a failure to perform closure and form a single gestalt, as well as a release
over the left hemisphere (i.e., left-hemisphere release). That is, the left
acting unopposed begins to abnormally temporally-sequence and thus produce segments unnecessarily.

Math and Geometry

Three hundred thousand years ago, someone took a piece of red-ocher
pigment and sharpened it presumably so as to mark something.[38] On what
surface did it draw, and what the nature of the composition may have been,
we do not know. We can only guess that it served some symbolic purpose,
or it may have merely served only to make a mark.

Three hundred thousand years ago, someone took the rib of an ox and
carved a series of geometric double arches on it.[39] Was he or she just
doodling, or was this a common form of artistic expression even in those lost
days and forgotten nights? Again, we do not know.

Sixty thousand years ago, Neanderthals were painting their caves red
and by the time they were overrun by the Cro-Magnon people 20,000 years
later, geometric patterns, designs, and doodles soon graced many a wall.[40]
However, it was not until about 20,000 years ago that people began leaving
marks on rocks and walls that suggested that may have been keeping track
of or counting something. Perhaps the phases of the moon, or the number of
animals killed? No one knows.

Just as we have no idea when the first complex sentence was spoken, or
the first words were written, the point at which human beings first began to
count or to measure the geometric properties of the land or the universe
surrounding remains a mystery.

Geometry and the first forms of spoken and written pictorial language appear to be naturally related to the functional integrity of the right half of the brain. Conversely, however, it is likely that the first mathematical concepts were promulgated by the left cerebral hemisphere, and like writing, were related to hand use. That is, individuals first count on his or her fingers, and then they learn to count by pointing with their fingers at that which they wish to sum, and then later they grasp a pen or pencil and make marks and signs which indicate the numbers they used and their summations.

The decimal system is clearly an outgrowth on this reliance on our digits, for this system is based on the concept of tens. Even the decimal system employed by the ancients of Mesoamerica was digitalized, with the exception that they used a base of 20 as they apparently counted their toes.

It has been postulated that human beings first became concerned with geometry and numbers with the advent of agriculture (around 10,000 years ago, after the last great flood—the consequence of the ending of the last Ice Age), as apparently they wished to count their crops as well as survey their fields. However, geometry may well have first been employed to survey the heavens. It was due to its heavenly association that many of the ancients considered geometry to be the math of the gods and of divine origin. Perhaps this is why almost all ancient temples and buildings (including those of ancient Sumer) were oriented in regard to certain celestial configurations. The Sumerians were very knowledgeable about complex geometric principles.

Although it is apparent that the Sumerians were also familiar with and utilized a decimal system, and by 4,000 years ago the Babylonians had developed the fundamental laws of mathematics, both cultures nevertheless relied on a sexagesimal system for their complicated calculations because it was far superior to the decimal system.[41] For example, whereas the decimal system can be factored by 2, 5, and 10, the 60-unit sexagesimal system can be factored by 2, 3, 4, 5, 6, 9, 10, 12, 15, and so on.

However, this system, like the first forms of writing, may also have been based on right-hemisphere mental functioning as it is this half of the brain which is dominant in regard to the understanding of geometry and the constructive, gestalt, and pictorial elements of art.

The sexagesimal system is also clearly related to the geometry of space and the cosmic, or divine circle divided by the equally divine four quadrants of the universe, that is, north, east, west, and south, which in turn forms the sign of a "cross." This is the same cross that most cultures have also deemed to be divine and celestial in origin.

A circle can be divided into $360°$. An individual or object sitting opposite is 180° away, half a circle. Hence, via the complicated permutations made possible via the sexagesimal system, the Sumerians, the Babylonians, the Egyptians, the Greeks, and those living in ancient Mesoamerica were able to make very precise calculations of angular, object, and mathematical relations, and to create temples and buildings, the likes of which today could only be designed, built, and fitted together using extremely precise tools and advanced, computerized measuring devices.

It is this same sexagesimal system that is employed in the measurement of time—60 seconds and 60 minutes. Similarly, the first calendars were created in the same manner, the Sumerians dividing the circle into 12 parts in accordance with their beliefs regarding the sacred celestial nature of the number "12" and the composition of our solar system.

That is, the Sumerians and the Babylonians were well aware that the Sun, and not the Earth, was at the center of the solar system. They also realized that the Earth was one of several planets and that they all traveled around the Sun. As is evident from the engraving shown below, they postulated the presence of 10 planets plus the Moon that circles the Earth, which, when coupled with the Sun, equalled the sum of 12.

Somehow, this knowledge as to the planetary composition and structure of the solar system was lost to subsequent generations as the existence of planets beyond Jupiter and Saturn has only been rediscovered in the last

Figure 99. Star map from ancient Sumer (upper left). The sun is surrounded by 10 planets and the moon yielding the cosmic origins of the divine "12."

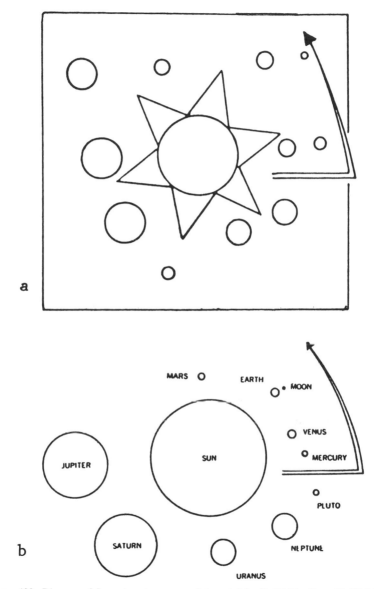

Figure 100. Diagram of Sumerian star map as interpreted by Z. Sitchin. From Z. Sitchin, *Genesis Revisited*. New York: Avon, 1990. (See Figure 99.)

century. Scientists are still searching for the elusive tenth planet that some believe exists outside the orbit of Pluto, or which may have once orbited between the Earth and Mars. It was only a few centuries ago that the Pope and the Catholic Church ordered Galileo to recant his pronouncements regarding the Sun's central position in the solar system, rather than the Earth, and to destroy his telescopes or face a public burning at the stake for heresy.

It is this same Sumerian "12" which makes the 12 hours of the day and the night (the 24-hour day), and is retained in the form of the 12 months and the 12 houses of the Zodiac. The ancient Egyptians essentially adapted this system for designing their own calendar, and in Mesoamerica an almost identical calendar system was devised.

However, the Babylonians (and probably the Sumerians before them) took the decimal and 60-unit sexagesimal system one step further and invented a way to write these numbers in a temporal sequence, the grammatical order of which revealed the value of the sum. In this manner, thanks to the Sumerian–Babylonians, when one writes 4254, it is clear that the first "4" is a thousand times greater than the last "4." It was not until the ancient Hindus and Mayas appeared on the scene that the concept of "nothing" and thus "zero" came into being.

Just as written language soon came to be organized in a nonpictorial series of temporal sequences, so, too, did the understanding of the cosmos, geometry, time, and numbers. These tremendous intellectual and creative achievements, however, like language, were dependent on the functional integrity of the inferior parietal lobe (as well as other neural structures such as those located within the frontal and temporal lobe and the thalamus), for with the destruction of this tissue, one's sense of time, space, geometry, written language, and math is abolished.

With the advent of language and the ability to analyze events in regard to temporal sequences, not only linguistic usage, but nature and reality itself has in consequence become fragmented. This is a consequence of the development of grammar which requires that all that is formulated within a linguistic framework be sequenced and arranged in a certain order. The continuity and flow of events and actions has become subject to segmentation into distinct things. A tree ceases to be part of the forest where birds and animals roam and live, but a word, distinct, separate, and isolated, much in the same manner that most human beings have become separate and psychically isolated from nature and their environment. Reality has come under the yoke of language and the ability to label and sequence, and so, too, has a good part of the human mind and brain.

7

Language and Reality
Universal Grammars and Limbic Language

Long before amphibians left the sea to take up life on land and in the air, fish had already learned to detect and analyze sounds made in the water. This was made possible by a structure located along both sides of the fish's body, called the **lateral line**. The lateral line is very sensitive to vibrations including those made by sound. The mammalian auditory system, however, did not evolve from the lateral line but from a cluster of nerve cells located within the brain stem, the vestibular nucleus.[1]

The human brain stem vestibular nucleus is very sensitive to vibrations and gravitational influences and helps to mediate balance and movement, such as in swimming and running.[2] Over the course of evolution, it is through the vestibular nucleus and the amygdala where the first neuronal rudiments of "hearing" and the analysis of sound also took place. Among all mammals this is a function the vestibular nuclei continues to serve via the signals it receives from the inner ear; the brain's outpost for detecting certain vibratory waves of molecules and their frequency of occurrence.

Indeed, it is via vibration and tactual sensations that individuals who are deaf are able to "hear" sound and music. This is also why it is possible to feel music, that is, the vibrations emanating from a high-amplitude speaker. Vibrations, of course, are a function and a result of movement, and movement and physical contact are prime sources of sound. In fact, the auditory system appears to have evolved so as to detect sounds which arise when something else moves, that is, a food source or a predator.

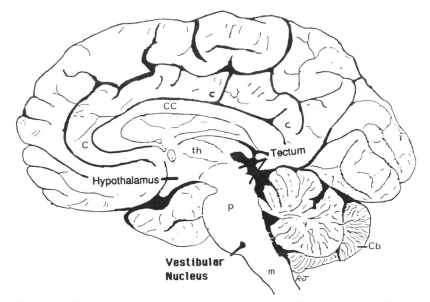

Figure 101. The human brain stem with the vestibular nucleus depicted, and the midbrain depicting the (visual and auditory) colliculi (tectum).

Be it mammal or human, the auditory system remains specialized to perceive sudden transient sounds which are typically made by the movement of other animals (such as prey or predators). The auditory system, however, remains tightly linked with the limbic system as well as the motor systems which enable it to alert the rest of the brain to possible danger.[3] The amygdala continually samples auditory events so as to detect those which are of emotional and motivational significance.

Initially, however, the auditory system also evolved so that organisms could turn and orient toward and localize unexpected sound sources. This was made possible by messages transmitted to the midbrain inferior tectum, which has also been called the **inferior colliculus**. This includes things that go bump in the night such as the settling of the house and the creaking of the floor, all of which can give rise to alarm reactions in the listener (in their amygdala). This is a sense humans and their ancestors have maintained for well over several million years, its main purpose being to maximize survival. In fact, the tectum (also called the colliculi, or little coolas [or butts]) of the midbrain (as well as the amygdala or the limbic system) was the main source of auditory analysis 200 million years before the evolution of neocortex, some 100 million years ago. This is a function it continues to perform among modern-day reptiles, amphibians, sharks, as well as mammals.

However, it was not until the appearance of the reptomammals, the therapsids, some 250 million years ago, that anything resembling an inner or even a middle ear first appeared.[4] It was probably at this time that sound first came to serve as a means of purposeful and complex communication. Nevertheless, somewhere between 70 to 100 million years ago, when the first true mammals appeared on the scene (creatures who evolved from the therapsids), vocalizing assumed an even greater importance in communication.[5] this in turn was made possible via further refinements within the inner ear and the development of auditory sensitive neocortex in the temporal lobe.

Auditory Transmission from the Cochlea to the Temporal Lobe

Within the cochlea of the inner ear are tiny hair cells which serve as sensory receptors. These cells give rise to axons which form the **cochlear division** of the **8th cranial nerve**, that is, the **auditory nerve**. This rope of fibers exits the inner ear and travels to and terminates in the **cochlear nucleus** which overlaps and is located immediately adjacent to the vestibular nucleus from which it evolved within the brain stem.

Among mammals the cochlear nucleus in turn projects auditory information to three different collections of nuclei. These are the **superior olivary complex**, the nucleus of the **lateral lemniscus**, and, as noted, the inferior colliculi/tectum (the superior portion of which is concerned exclusively with vision and which forms the optic lobe).[6]

A considerable degree of information analysis occurs in each of these areas before being relayed to yet other brain regions such as the amygdala (which extracts those features which are emotionally or motivationally significant), and to the **thalamus** where yet further analysis occurs. The thalamus (which sits above the hypothalamus at the center of the brain) is a very ancient sensory integration center which probably first evolved well over 450 million years ago.

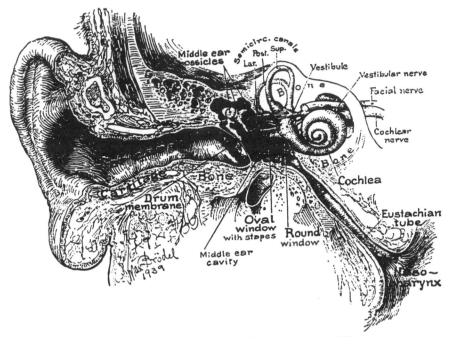

Figure 102. The human inner ear. From Max Brodel, 1939.

Among mammals, from the thalamus and amygdala, auditory signals are transmitted to the primary auditory receiving area located in the neocortex of the superior temporal lobe. This primary auditory area is referred to as **Heschl's gyrus**. Here auditory signals undergo extensive analysis and reanalysis and simple associations begin to be formed.[7] However, by time it has reached the neocortex, auditory signals have undergone extensive analysis by the thalamus, amygdala, and the other ancient structures mentioned above.

Unlike the primary visual and somesthetic areas located within the neocortex of the occipital and parietal lobes which receive signals from only one half of the body or visual space, the primary neocortical auditory region receives some input from both ears, and from both halves of auditory space. This is a consequence of the considerable **cross-talk** which occurs between different old cortical nuclei as information is relayed to and from various

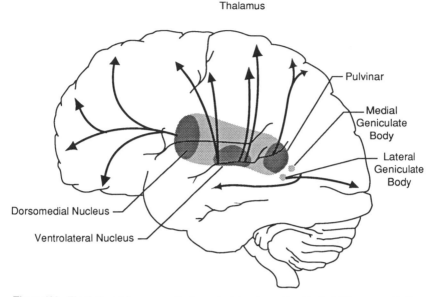

Figure 103. The thalamus is an exceedingly ancient structure, at least half a billion years old. It serves to integrate and analyze sensory signals received from the outside world and other brain regions, and to then transfer this information to select regions of the neocortex where further analysis takes place.

regions prior to transfer to the neocortex.[8] Predominantly, however, the right ear transmits to the left cerebral neocortex and vice versa.

Filtering, Feedback, and Temporal–Sequential Reorganization

The old cortical centers located in the midbrain and brain stem evolved long before the appearance of neocortex and have long been adapted and specialized for performing a considerable degree of information analysis.[9] This is evident from observing the behavior of reptiles, and other creatures, where auditory neocortex is either absent or minimally developed. Moreover, many of these old cortical nuclei also project back to each other such that each might *hear* and analyze the same sound repeatedly. In this manner the brain is able to heighten or diminish the amplitude of various sounds via feedback adjustment.[10] In fact, not only is feedback provided, but the actual

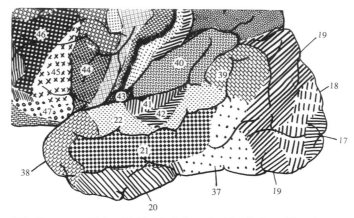

Figure 104. The temporal lobe of half of the left cerebral hemisphere. The primary auditory area is located in Broadman's area (41). Regions 42 and 22 correspond to the auditory association and Wernicke's area, and region 21 corresponds to the middle temporal lobe.

order of the sound elements perceived can be rearranged when they are played back. This same process continues at the level of the neocortex which has the advantage of being the recipient of signals that have already been highly processed and analyzed. It is here and in this manner that language-related sounds begin to be organized and recognized.

Sustained Auditory Activity. One of the main functions of the primary auditory neocortical receptive area appears to be the retention of sounds for brief time periods (up to a second) so that temporal and sequential features may be extracted and discrepancies in spatial location identified; that is, so that we can determine from where a sound may have originated.[11] This also allows comparisons to be made with sounds that were just previously received and those which are just arriving.

Moreover, via their sustained activity, these neurons are able to prolong (perhaps via a perseverating feedback loop with the thalamus) the duration of certain sounds so that they are more amenable to analysis. In this manner, even complex sounds can be broken down into components which are then separately analyzed. Hence, sounds can be perceived as sustained temporal sequences. In this manner, sound elements composed of consonants, vowels, and phonemes and morphemes can be more readily identified, particularly within the auditory neocortex of the left half of the brain.[12] Moreover, via feedback and sustained (vs. diminished) activity, the importance and even order of the sounds perceived can be changed, filtered, or heightened, an extremely important development in regard to the acquisition of language.[13]

Indeed, a single phoneme may be scattered over several neighboring units of sounds. A single sound segment may in turn carry several successive phonemes. Therefore, a considerable amount of reanalysis, reordering, or filtering of these signals is required so that comprehension can occur.[14] These processes, however, presumably occur both at the neocortical and old cortical levels. In this manner, a phoneme can be identified, extracted, analyzed, and placed in its proper category and temporal position.

For example, take three sound units, "t–k–a," which are transmitted from the thalamus to the superior temporal auditory receiving area. Via a feedback loop, this primary auditory area can send any one of these units back to the thalamus which again sends it back to the temporal lobe, thus amplifying the signal or rearranging the order, "t–a–k," or "k–a–t." A multitude of such interactions are in fact possible so that whole strings of sounds can be arranged in a certain order. Mutual feedback characterizes most other neocortical interactions as well, be it touch, audition, or vision.[15]

Via all these interactions sounds can be repeated, their order can be rearranged, and the amplitude of specific auditory signals can be enhanced, whereas others can be filtered out. It is in this manner that fine tuning of the nervous system occurs so that specific signals are attended to, perceived, processed, committed to memory, and so on. Indeed, a massive degree of filtering occurs throughout the brain not only in regard to sound, but visual and tactual information as well. Moreover, the same process occurs when organizing words for expression.

This ability to perform so many operations on individual sound units has in turn greatly contributed to the development of human speech and language. For example, the ability to hear human speech requires that temporal resolutions occur at great speed so as to sort through the overlapping and intermixed signals being perceived. This requires that these sounds are processed in parallel, or stored briefly and then replayed in a different order so that discrepancies due to overlap in sounds can be adjusted for,[16] and this is what occurs many times over via the interactions of the old cortex and the neocortex. Similarly, when speaking or thinking, sound units must also be arranged in a particular order so that what we say is comprehensible to others and so that we may understand our own verbal thoughts.

Hearing Sounds and Language

Humans are capable of uttering thousands of different sounds all of which are easily detected by the human ear. And yet, although our vocabulary is immense, human speech actually consists of about 12 to 60 units of sound, depending, for example, if one is speaking Hawaiian versus English. The English vocabulary consists of several hundred thousand words which are based on the combinations of just 45 different sounds.

Animals, too, however, are capable of a vast number of utterances. In fact, monkeys and apes employ between 20 to 25 units of sound, whereas a fox employs 36. However, these animals cannot string these sounds together so as to create a spoken language. This is because most animals tend to use only a few units of sound at one time which varies depending on their situation, for example, lost, afraid, playing.

Humans combine these sounds to make a huge number of words. In fact, employing only 13 sound units, humans are able to combine them to form 5 million word sounds.[17]

Phonemes

The smallest unit of sound is referred to as a phoneme. For example, *b* and *p* as in bet versus pet are phonemes. When phonemes are strung together as a unit they in turn comprise **morphemes**. Morphemes such as "cat" are composed of three phonemes, "k,a,t." Hence, phonemes must be arranged in a particular temporal order in order to form a morpheme.

Morphemes in turn make up the smallest unit of meaningful sounds such as those used to signal relationships such as "er" as in "he is old*er* than she." All languages have rules that govern the number of phonemes, their relationships to morphemes, and how morphemes may be combined so as to produce meaningful sounds.[18]

Each phoneme is composed of multiple frequencies which are in turn processed in great detail once they are transmitted to the superior temporal lobe. In fact, the primary auditory area is tonotopically organized, such that similar auditory frequencies are analyzed by cells which are found in the same location of the brain.[19] For example, as one moves across the temporal lobe, auditory frequencies perceived become progressively higher in one direction and lower when traveling in the opposite direction.

Temporal–Sequential and Linguistic Sensitivity

Although it is apparent that the auditory regions of both cerebral hemispheres are capable of discerning and extracting temporal–sequential rhythmic acoustics, the left temporal lobe contains a greater concentration of neurons specialized for this purpose since the left brain is clearly superior in this capacity.[20]

For example, the left hemisphere has been repeatedly shown to be specialized for sorting, separating, and extracting in a segmented fashion, the phonetic and temporal–sequential or linguistic–articulatory features of incoming auditory information so as to identity speech units. It is also more sensitive to rapidly changing acoustic cues be they verbal or nonverbal as compared to the right hemisphere. Moreover, via **dichotic listening tasks** (where sounds can be selectively transmitted to the right vs. left ear), the right ear (left temporal lobe) has been shown to be dominant for the perception of real words, word lists, numbers, backwards speech, Morse code, nonsense syllables, the transitional elements of speech, rhymes, single phonemes, consonants, and consonant–vowel syllables.[21]

Consonants and Vowels

In general, there are two large classes of speech sounds: consonants and vowels. Consonants by nature are brief and transitional and have identification boundaries which are sharply defined. These boundaries enable different consonants to be discerned accurately.[22] Vowels are more continuous in nature and in this regard, the right half of the brain plays an important role in their perception.

Consonants are more important in regard to left-hemisphere speech perception. This is because they are composed of segments of rapidly changing frequencies which include the duration, direction, and magnitude of sound segments interspersed with periods of silence. These transitions occur in 50 msec or less which in turn requires that the left half of the brain take responsibility for perceiving them.

In contrast, vowels consist of slowly changing or steady frequencies with transitions taking 350 or more msec. In this regard, vowels are more like natural environmental sounds which are more continuous in nature, even those which are brief such as a snap of a twig. They are also more likely to be processed and perceived by the right half of the brain, though the left cerebrum also plays a role in their perception.[23]

The differential involvement of the right and left half of the brain in processing consonants and vowels is a function of their neuroanatomical organization and the fact that the left hemisphere is specialized for dealing with sequential information. Moreover, the left brain is able to make fine temporal discriminations with intervals between sounds as small as 50 msec. However, the right brain needs 8 to 10 times longer and has difficulty discriminating the order of sounds if they are separated by less than 350 msec.

Hence, consonants are perceived in a wholly different manner from vowels. Vowels yield to nuclei involved in "continuous perception" and are processed by the right half of the brain. Consonants are more a function of "categorical perception" and are processed in the left half of the cerebrum. In fact, the left temporal lobe acts on both vowels and consonants, during the process of perception, so as to sort these signals into distinct patterns of segments via which these sounds become classified and categorized.

Nevertheless, be they processed by the right or left brain, both vowels and consonants are significant in speech perception. Vowels are particularly important when we consider their significant role in communicating emotional status and intent.

Spatial Localization, Attention, and Environmental Sounds

As noted, auditory neurons located in the neocortex of the temporal lobe receive input from both ears. Hence, the primary auditory area, in conjunction with the inferior auditory tectum, plays a significant role in orienting to and localizing the source of various sounds; for example, by comparing time and intensity differences in the neural input from each ear. A sound arising from one's right will more quickly reach and sound louder to the right ear as compared to the left.

Among mammals, a considerable number of auditory neurons respond or become highly excited in response to sounds from a particular location.[24] That is, just as certain parietal and motor neurons are responsible for only select regions of the body, many auditory neurons are selectively sensitive to certain locations, and to certain sounds including human screams and cries. Moreover, some of these neurons become excited only when the subject looks at the source of the sound, which is why some people feel they cannot hear as well when their eye glasses are off. Hence, these neurons act so that the source, identity, and location may be ascertained and fixated upon. Based on studies of humans who have sustained brain injuries, the right temporal lobe is more involved than the left in discerning location,[25] and in perceiving nonverbal sounds. As noted, certain cells in the auditory area are highly specialized and will respond only to certain meaningful vocalizations. In this regard they seem to be tuned to respond only to specific auditory parameters so as to identify and extract certain salient features. These are called feature detector cells, much like those in the inferior temporal lobe which respond only to faces or certain forms. For example, some cells will respond only to cries of alarm and others only to sounds suggestive of fear or indicating danger.

Although the left temporal lobe appears to be more involved in extracting certain linguistic features, the right temporal region is more adept at identifying and recognizing nonverbal environmental acoustics (e.g., wind, rain, animal noises), the prosodic–melodic nuances of speech, sounds which convey emotional meaning, as well as most aspects of music. Indeed, tumors in or electrical stimulation of the superior temporal gyrus, the right temporal lobe in particular, results in musical hallucinations. Frequently patients report that they hear the same melody over and over. In some instances, patients have reported the sound of singing voices and individual instruments.[26] Conversely, strokes or complete surgical destruction of the right temporal lobe significantly impairs the ability to name or recognize melodies and musical passages. Such injuries can also disrupt time sense, the perception of timbre and loudness, and tonal memory.

Indeed, the right temporal lobe's ability to perceive and recognize musical and environmental and sounds coupled with its sensitivity to the location of sound, no doubt provided great survival value to our early ancient ancestors. That is, in response to a specific sound (e.g., a creeping predator), one is immediately able to identify, locate, and fixate upon the source and thus take appropriate action. Of course, even modern humans rely upon the same mechanisms to escape being run over by cars when walking across streets or riding bicycles, or to ascertain and identify approaching individuals.

Cortical Deafness

In some instances, such as due to a middle cerebral artery stroke, the primary auditory receiving areas of the right or left cerebral hemisphere may be destroyed. This results in a disconnection syndrome such that sounds relayed from the thalamus, amygdala, or colliculi cannot be received or analyzed by the temporal lobe. In some cases, however, the strokes may be bilateral such that both auditory areas are injured. When this occurs, one can no longer hear sounds, and the patient is said to be suffering from **cortical deafness**.[27]

However, sounds continue to be processed by the thalamus, colliculus, and so on. Hence the ability to *hear* sounds per se, is retained. Since the sounds which are heard are not received neocortically and thus cannot be transmitted to the adjacent association areas, sounds become stripped of meaning. That is, meaning cannot be extracted or assigned by the neocortex, which is the seat or our language-dependent conscious mind. Rather, only differences in intensity are discernible. An individual so affected would not only lose the ability to hear sounds, but the sounds of speech.

Patients who suffer cortical deafness cannot respond to questions, lose the ability to discern the melody for music, cannot recognize speech or environmental sounds, and tend to experience the sounds they do hear as distorted and disagreeable, for example, buzzing and roaring, the banging of tin cans, and the like.

Individuals who are cortically deaf are not aphasic. They can read, write, speak, comprehend pantomime, and are fully aware of their deficit. Nevertheless, although not aphasic, per se, speech is sometimes noted to be hypophonic and contaminated by occasional literal paraphasias such that erroneous sound-order substitutions are produced.

More commonly, a destructive lesion may be limited to the primary auditory receiving area of just the right or left cerebral hemisphere. Patients with such lesions restricted to a single hemisphere are not considered cortically deaf. However, when the auditory receiving area of the left temporal lobe is destroyed, the patient suffers from a condition referred to as **pure word deafness**. If the lesion is in the right temporal receiving area, the disorder is described as an **auditory agnosia**,[28] that is, an inability to recognize or hear environmental sounds.

Pure Word Deafness

With a destructive lesion involving the left primary auditory receiving area, Wernicke's auditory association area becomes disconnected from almost all sources of acoustic input and patients are unable to recognize or perceive the sounds of language, be it sentences, single words, or even single letters. All other aspects of comprehension are preserved, including reading, writing, and expressive speech, and the ability to hear music and environmental–emotional sounds. This is because these cognitive capacities are dependent on the functional integrity of other brain areas.

Due to sparing of the right temporal region, the ability to recognize musical and environmental sounds is preserved. If someone were to knock on the door, this would be heard without difficulty. If instead they yelled, "is anybody home?" only an indiscernible noise would be perceived.

However, the ability to name or verbally describe these sounds is impaired—due to disconnection and the inability of the right hemisphere to talk. That is, these sounds, although recognized as sounds, cannot be transmitted to Wernicke's area which is located in the opposite half of the brain. However, they can describe what produced the sound, draw a picture of it, and so on.

Pure word deafness, of course, also occurs with bilateral lesions in which case environmental sound recognition is also affected. In these instances the patient is considered cortically deaf since the neocortex can no longer receive sound.

Pure word deafness, when due to a unilateral lesion of the left temporal lobe, is partly a consequence of an inability to extract temporal–sequential features from incoming sounds. Hence, linguistic messages cannot be recognized. Pure word deafness can be partly overcome if the patient is spoken to in an extremely slowed manner. The same is true of those with Wernicke's aphasia.

Auditory Emotional Agnosia

Mary was home sitting at her desk when she began feeling dizzy and then queasy and then a tremendous buzzing filled her left ear. For a brief moment, she thought she heard a weird crackling sound, and then it was gone. Returning to her work, she bent diligently over the papers when the dizzy, sick feeling and the buzzing began again, and then, just as suddenly again it was gone. She sighed and tried to concentrate but was bothered by the development of a tremendous headache.

Later, glancing at the clock, she furrowed her brow in irritation as she realized that the phone call she had been expecting had not come. However, as she pondered the significance of this, again came the buzzing, and then she noticed that the light on her answering machine was blinking. Somebody had left a message! How could that be? she wondered. She had been sitting by the phone for the last hour, and no one called. Pressing the button, she turned up the volume and listened intently to the message but couldn't make out the voice. Was somebody playing tricks on her? The voice claimed to be Dave, whose call she had been expecting, but it sure didn't sound like him. Even the "I love you" at the end sounded phony, like some kind of robot. Again her thoughts were interrupted by that stupid buzzing sound, and then, to her surprise, she watched as her answering machine switched on. Quickly turning up the volume she was shocked to hear a recorded voice that did not sound at all like hers, but like a female robot, and then, the robot voice of Dave came on the line, claiming to be angry that she wasn't home. But it didn't sound like Dave, and the voice didn't even sound angry, just loud. "What's wrong with me?" she said out loud, feeling ill, and then froze. That wasn't the sound of her voice!

An individual with cortical deafness also can suffer from a generalized auditory agnosia involving words and nonlinguistic sounds. Agnosia in these instances means, *to not know*, and those who suffer from agnosagnosia *do not know they do not know*.

Usually, an auditory agnosia for environmental and emotional sounds, but with preserved perception of language, occurs with lesions restricted to the right temporal lobe. In these instances, an individual loses the capability to correctly discern or even hear environmental sounds (e.g., birds singing, doors closing, keys jangling, telephones ringing) or to recognize emotional–prosodic speech and music. As such, speech may sound like a monotone, or melody may be distorted and displeasing. However, if the condition is severe, the person may not realize that she or he no longer hears sounds properly, simply because that part of the brain which would notice abnormalities in environmental and melodic–emotional sounds would be destroyed and unable to alert her or him to the problem. This is due to a disconnection and in this instances she or he suffers from an agnosagnosia, she or he doesn't know that she or he doesn't know, just as a wolf or a chimpanzee would not know that it doesn't know how to read or write as it does not have the brain areas which would alert them to this deficit.

As such, these problems are not likely to come to the attention of a physician unless accompanied by secondary emotional difficulties or if the stroke damages tissue beyond the right temporal lobe and disrupts motor functioning and the patient suffers a paresis. With a more restricted right temporal stroke, most individuals with this disorder, if they are completely agnosic, might not know that they have a problem and thus would not complain. If they or their families notice (for example, if a patient does not respond to a knock on the door), the likelihood is that the problem will be attributed to faulty hearing or even forgetfulness.

Mary had replayed the tape several times and had finally decided that something was definitely wrong with her machine. However, sometime later as she stood up to retrieve some aspirin for that terrible headache she was rudely shocked to see Dave walking into the room.

"Don't you believe in knocking?" she asked angrily. "What do you mean sneaking up on me like this?"

Dave stared at her with tightly pressed-down turned lips. It was clear from his expression that he was upset. "Sneaking up?" he replied loudly. "I pounded several times on your door and rang the bell."

"You did not!" she interrupted. "And what's wrong with your voice? You sound strange." Suddenly she felt frightened. Even her own voice didn't sound right.

"You're the one ignoring my calls and refusing to answer the door," he replied, giving her a funny look.

Mary stepped back feeling frightened. First the phone, and now this. She stared at him feeling increasingly upset and confused. Although it looked like Dave, he sure didn't sound like Dave, and Dave would never play tricks on her like this. And what about her own voice and the phone? What was going on? What was he up to? "Get out!" she suddenly demanded, as she reached for the phone and placed it next to her ear. Pulling the phone away, she momentarily forgot about Dave. What happened to the dial tone? What was that weird noise?

Because individuals suffering from auditory agnosia may also have difficulty discerning emotional–melodic nuances, it is likely that they will misperceive and fail to comprehend a variety of paralinguistic social–emotional messages. This includes difficulty discerning what others may be *implying*, or in appreciating emotional and contextual cues, including variables such as sincerity or mirthful intonation. Hence, a host of behavioral difficulties may arise as patients become upset and fearful, and as they realize that something is wrong. As such they may become paranoid, fear they are going crazy, or become convinced that loved ones have become replaced by impostors.

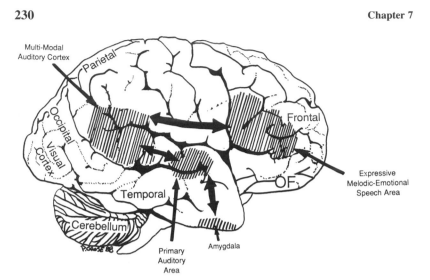

Figure 105. The melodic–emotional language axis of the right hemisphere.

For example, a patient suffering from a right temporal lobe injury and auditory emotional agnosia may complain that his wife no longer loves him, and that he knows this from the sound of her voice, which seems to him to be lacking in emotion. In fact, a patient may notice that the voices of friends and family, and even his own voice sounds in some manner different, which, when coupled with difficulty discerning nuances such as humor and friendliness may lead to the development of paranoia and what appears to be delusional thinking. Unless appropriately diagnosed, it is likely that the patient's problem will feed upon and reinforce itself and grow more severe.

It is important to note that rather than being completely agnosic or word deaf, patients may suffer from only partial deficits. In these instances they may seem to be hard of hearing, frequently misinterpret what is said to them, or slowly develop related emotional difficulties.

The Auditory Association Areas

Following the analysis performed in the primary auditory receiving area, auditory information is transmitted to the **middle temporal lobe** (which in the left hemisphere maintains neurons which store linguistic–auditory symbols, words, and names) and to the immediately adjacent auditory association area.[29] It is in this region where complex auditory associations are performed and where the comprehension of speech (Wernicke's area) or environmental–melodic, emotional sound occurs. This region also receives visual associations from the occipital lobe.

This middle temporal auditory and auditory–visual association neocortex is intimately interlinked with Wernicke's area which in turn has become coextensive with the angular gyrus and the inferior parietal lobule. It is in Wernicke's area where the units of sound are strung together in the form of words and sentences and are recognized as having specific meanings and as belonging to certain sample categories.[29] Hence, verbal comprehension takes place in Wernicke's and the middle temporal lobe area just as gestural comprehension takes place in the superior–inferior parietal lobe.

The middle temporal and Wernicke's areas, and the corresponding region in the right hemisphere are not merely auditory association areas as they also receive a convergence of fibers from the tactual and visual association cortices. They also receive input from the contralateral auditory association area via the **corpus callosum** and **anterior commissure** (large bundles of axonal nerve fibers which interconnect and link the two halves of the brain). Hence, a considerable number of multimodal linkages are made possible within the auditory association areas. Indeed, the richness of language and the ability to utilize a multitude of words and images to describe a single object or its use is very much dependent on these interconnections.

Receptive Aphasia

Caroline's grown children had at first thought that maybe she was becoming hard of hearing because she so frequently misunderstood what they were saying. Then they worried that she might be losing her memory or even becoming senile, because when they asked her a question or reminded her to keep an appointment or to do some task, she would sometimes just stare at them with a perplexed look on her face. Worse, sometimes she answered questions that they hadn't asked. Like when they wanted to know if Roger had come by, and she answered "fine, fine," as if they had asked how she was instead.

She seemed so reluctant to talk and when she did, it often didn't make any sense either. She had even quit reading the newspaper and had not opened any of the books she had been given on her birthday, which was very unusual since she loved to read.

Finally, since she seemed so depressed and confused and was increasingly acting like she did not understand anything that was being said to her, they brought her to their doctor, who ordered an MRI. It was then that they realized a tumor was growing in the superior temporal lobe of the left half of her brain.

When the left auditory association area is damaged, there results severe receptive aphasia, that is, **Wernicke's aphasia**. Individuals with Wernicke's aphasia, in addition to severe comprehension deficit, usually suffer abnormalities involving expressive speech, reading, writing, repeating, and word finding.[30] This is because Wernicke's receptive language area acts to retrieve, associate, and provide words, names, and the auditory equivalents of whatever is perceived, seen, or felt, as well as assisting in their temporal organization.

When a person with this disorder attempts to read a word, due to destruction of this tissue, she or he is able to see but unable to analyze the visual symbols linguistically and thus cannot comprehend them. She or he can no longer read.

Patients with severe forms of Wernicke's aphasia, in addition to other impairments, are necessarily "word deaf." As these individuals recover from their aphasia, word deafness is often the last symptom to disappear. Even when less severe, they are unable to perceive spoken words in their correct order, cannot identify the pattern of presentation, and become easily overwhelmed by the sounds of speech and must be spoken to very slowly.

Impaired Perception of Temporal Order

Frequently disturbances of linguistic comprehension are due to an impaired capacity to discern the individual units of speech and their temporal order.[31] Sounds must be separated into discrete interrelated linear units, or they will be perceived as a blur, or even as a foreign language. Hence, a patient with Wernicke's aphasia may *perceive* a spoken sentence such as the "pretty little kitty" as the "lipkitterlitty." They also may have difficulty establishing the boundaries of phonetic (confusing "love" for "glove") and semantic (cigarette for ashtray) auditory information.

Many receptive aphasics, be it due to brain tumors or stroke, can comprehend frequently used words but have increasing difficulty with those less commonly heard. Thus loss of comprehension is not an all-or-none phenomenon. They will usually have the most difficulty understanding relational or syntactical structures, including the use of verb tense, possessives, and prepositions. However, by speaking slowly and by emphasizing the pauses between each individual word, comprehension can be modestly improved. Nevertheless, in severe cases, comprehension is all but completely abolished, though the ability to perceive environmental sounds and emotional nuances will be preserved.

John scrunched his eyes closed for a brief moment to fight off the headache and the sick feeling in his stomach. Opening his eyes, he stared back into the newspaper and tried to read, but something was wrong. Closing his eyes again, he slowly put down the paper, which for some reason sounded awfully loud, so loud in fact that it startled him, causing him to almost knock over his coffee in the process. The clatter of the cup made his headache become worse, almost unbearable. Closing then opening his eyes he stared at the almost empty cup and then looked up and met the eyes of the heavyset waitress who was staring at him with a bemused expression. She held a coffee pot in her hand and was sliding some pie toward him.

"You did order pie, didn't you?" she asked.

John just stared at her, trying to figure out what she wanted and then glanced at the pie, frowned, and shook his head. The waitress shrugged and walked away. It was then that Roger, his partner, looked up from the paper and noticed that John looked angry. "You OK?" he asked.

John shook his head feeling irritable. "Piezgrazing and the waitress koasted black coffing I said to that drinker roze coffin . . ."

"What?" Roger asked, staring at him incredulously.

John glanced at the pie and then his empty coffee cup. "Coffing over on thuz the cup with pizelong she said but I'll be over if you want to pear peering, no way am I guzin, like the fat pig . . ."

Although comprehension has been lost, patients with damage to Wernicke's area are usually still capable of talking (due to preservation of Broca's area). However, they are completely unable to understand what they are saying. Moreover, since Wernicke's area has been destroyed, they do not know that what they are saying no longer makes sense; a condition referred to earlier as **agnosagnosia**—not knowing that one does not know.

Moreover, most of what they say is nonsensical and is not comprehensible to a listener, who instead may conclude that the person has suffered a nervous breakdown and is schizophrenic and psychotic.

The reason the speech of a person with Wernicke's receptive aphasia no longer makes sense is that Wernicke's area also acts to code linguistic stimuli for expression prior to its transmission to Broca's expressive speech area. Hence, expressive speech becomes severely abnormal, lacking in content, containing neologistic distortions (e.g., "the razgabin"), and/or characterized by nonsequiturs, literal (sound-substitution) errors, and verbal paraphasic (word-substitution) errors. Their speech is also characterized by a paucity of nouns and verbs, and the omission of pauses and sentence endings. This is because the correct words and linguistic concepts cannot be organized and provided by Wernicke's region, which has been destroyed. As such, what they say is often incomprehensible.

Presumably because the coding mechanisms involved in organizing what they are planning to say are the same mechanisms which decode what they *hear*, expressive as well as receptive speech becomes equally disrupted. This is why even those with normal brains are unable to talk and comprehend what others are simultaneously saying.

The spontaneous speech of Wernicke's aphasics is also often characterized by long, seemingly complex, grammatically correct sentences, the grammatical stamp being applied by the inferior parietal lobule and Broca's area. Sometimes their speech is in fact hyperfluent such that they speak at an increased rate and seem unable to bring sentences to an end with words being unintelligibly strung together. Hence, this disorder has also been referred to as **fluent aphasia** for although they can talk, they make absolutely no sense whatsoever. The reason a person can still speak is that Broca's area and the fiber pathway linking it to the inferior parietal lobule, the middle temporal lobe, and other areas where word sounds, names, phonemes, and morphemes are stored, are still intact. When severe, their speech deteriorates into jargon aphasia such that no meaningful communication can be made.

Among those with Wernicke's receptive aphasia, one gauge of comprehension can be based on the amount of normalcy in their language use. That is, if they can say or repeat only a few words normally, it is likely that they can only comprehend a few words as well. In addition, the ability to write may be preserved, although what is written is usually completely unintelligible consisting of jargon and neologistic distortions. Copying written material is possible although it is also often contaminated by errors.

Agnosagnosia

As noted, although the speech of a patient with Wernicke's aphasia is often abnormal or bizarre, with severe dysfunction these patients do not realize that what they say is meaningless. Moreover, they may fail to comprehend that what they hear is meaningless as well. Nor can you tell them since they are unable to comprehend. This is because when Wernicke's area is damaged, there is no other region left to analyze the linguistic components of speech and language. The rest of the brain cannot be alerted to the patient's disability. They don't know that they don't know, that they don't understand as these are not functions which the rest of the brain is concerned with. Not being informed otherwise, the rest of the brain assumes that what is being said is normal. Again, this is due to disconnection.

Emotional Awareness

Presumably, as a consequence of loss of comprehension, these patients may display euphoria, or in other cases, paranoia as there remains a nonlinguistic or emotional awareness that *something* is not right.[31] Emotional functioning and affective comprehension remains somewhat intact, though sometimes disrupted due to erroneously processed verbal input. That is, the patient's right hemisphere continues to respond to signals generated by others as well as the left half of the patient's own brain even though it is abnormal. Similarly, the ability to read and write emotional words (as compared to nonemotional or abstract words) is also somewhat preserved among aphasics. This is because the right hemisphere is intact and is dominant for all aspects of emotion.

Since these paralinguistic and emotional features of language are analyzed by the intact right cerebral hemisphere, sometimes the aphasic individual is able to grasp in general the meaning or intent of a speaker, although verbal comprehension is reduced. This in turn enables him or her to react in a somewhat appropriate fashion when spoken to. That is, he or she may be able to discern not only that a question is being asked, but that concern, anger, fear, and so on are being conveyed. Unfortunately, this also makes him or her appear to comprehend much more than he or she is capable of.

"Schizophrenia"

John was finally admitted to a psychiatric hospital after both his physician and the psychiatrist he had been referred to decided that he had suffered a "nervous breakdown." Roger still couldn't believe it. John was always the most emotionally stable individual he had ever known and then, out of the blue, he goes crazy and develops what those doctors call "paranoid schizophrenia."

Because these individuals display unusual speech, loss of comprehension, a failure to realize that they no longer comprehend or "make sense" when speaking, as well as paranoia and/or euphoria, they are at risk for being misdiagnosed as psychotic or suffering from a formal thought disorder, that is, "schizophrenia." Indeed, individuals with abnormal left-temporal-lobe functioning sometime behave and speak in a "schizophreniclike" manner.[32]

According to Benson,[33] "Those with Wernicke's aphasia often have no apparent physical or elementary neurological disability. Not infrequently, the individual who suddenly fails to comprehend spoken language and whose output is contaminated with jargon is diagnosed as psychotic. Patients with Wernicke's aphasia certainly inhabited some of the old lunatic asylums and probably are still being misplaced."

Conversely, it has frequently been reported that "schizophrenics" often display significant abnormalities involving speech processing, such that the semantic, temporal–sequential, and lexical aspects of speech organization and comprehension are disturbed and deviantly constructed. Significant similarities between schizophrenic discourse and aphasic abnormalities have also been reported.[34] In this regard it is often difficult to determine what a schizophrenic individual may be talking about. These same psychotic individuals have sometimes been known to complain that what they say often differs from what they intended to say. Temporal–sequential (i.e., syntactical) abnormalities have also been noted in their ability to reason. The classic example being: "I am a virgin; the Virgin Mary was a virgin; therefore I am the Virgin Mary."

Hence, there is some possibility that a significant relationship exists between abnormal left-hemisphere and left-temporal-lobe functioning and schizophrenic language, thought, and behavior.

The Language Axis

A large expanse of the cortex is involved in auditory perception, analysis, and production. This includes the amygdala and inferior, middle, and superior temporal lobe, the inferior parietal and inferior frontal areas as well as the thalamus.

The auditory area extends in a continuous beltlike fashion from the amygdala located deep in the inferior temporal lobe, to the primary and association (e.g., Wernicke's) area in the superior temporal lobe, toward the inferior parietal lobule, and via the **arcuate fasciculus** (a rope of axons interconnecting these areas) onward toward Broca's area in the frontal lobe.

Indeed, in the left hemisphere, this massive rope of interconnections forms a **language axis** such that the amygdala, Wernicke's area, the inferior parietal lobule, and Broca's area, together, are linked and able to mediate the perception and expression of most forms of language and speech via the extensive interconnections maintained.

This ropelike bundle of axons, the arcuate fasciculus, is a bidirectional fiber pathway. As noted, it runs not only from Wernicke's through to Broca's area but extends inferiorly deep into the temporal lobe where contact is established with the amygdala. In this manner, auditory input comes to be assigned emotional–motivational significance, whereas verbal output becomes emotionally–melodically *enriched*, a process that the four-layered

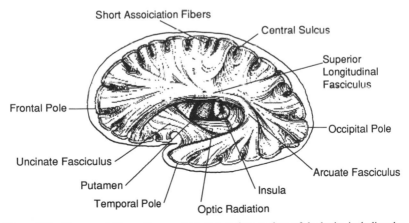

Figure 106. The axonal fiber pathways which link various regions of the brain, including the arcuate and superior longitudinal fasciculus which interlink with the language axis.

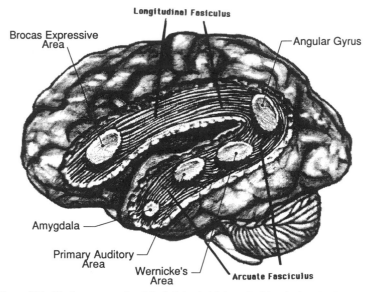

Figure 107. The language axis of the left (and right) cerebral hemisphere is linked together via a bundle of axons, the arcuate and longitudinal fasciculus, which interconnect the amygdala, Wernicke's area, the inferior parietal lobule, and Broca's area.

transitional limbic cortex, the cingulate gyrus, also contributes to. Within the right hemisphere, these interconnections which include the amygdala appear to be more extensively developed.[35]

It is via this linkage that when a sentence is read the information is transferred to the inferior parietal lobule and to Wernicke's area so that these two regions can in turn access the appropriate verbal labels for each letter, word, and so on. Moreover, through there interconnections, not just words, but concepts, categories, and associated attributes (e.g., color, form, size, shape, cost, desirability, use) and emotional significance of whatever is read, or heard, or felt can be accessed or determined as well. This information is then integrated within the language axis, associated, labeled, named, strung together, comprehended, provided some emotional coloration, and if necessary, talked about via transmission to Broca's area.

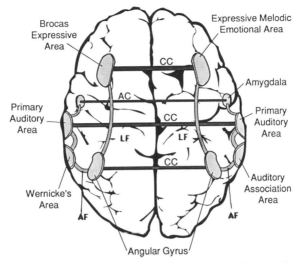

Figure 108. The language areas of the right and left cerebral hemispheres are interconnected and linked together via axonal fibers of the corpus callosum (CC) and anterior commissure (AC) which enable them to transfer information and coordinate their actions.

The Melodic–Intonational Axis

Language is not just a function of the left half of the brain, however, as the right half of the cerebrum makes significant contributions to the understanding and expression of its contextual, emotional, and melodic components. For example, it has been consistently demonstrated that the right temporal lobe (left ear) predominates in the perception of timbre, chords, tone, pitch, loudness, melody, intensity, prosody, and nonverbal, environmental, and emotional sounds.[36]

When the right temporal lobe is damaged, there results a disruption in the ability to perceive, recognize, or recall tones, loudness, timbre, and melody, and one's sense of humor may be abolished as well.[37] Similarly, the ability to recognize even familiar melodies and the capacity to obtain pleasure while listening to music is abolished or significantly reduced; this is a condition referred to as *amusia*. In addition, lesions involving the right temporal–parietal area have been reported to significantly impair the ability to perceive and identify environmental sounds, comprehend or produce appropriate verbal prosody, emotional speech, or to repeat emotional statements. Indeed, when presented with neutral sentences spoken in an emotional manner, right temporal–parietal damage has been reported to disrupt the perception and comprehension of humor, context, and emotional prosody regardless of its being positive or negative in content.[38]

Hence, the right temporal–parietal area is involved in the perception, identification, and comprehension of environmental and musical sounds and various forms of melodic and emotional auditory stimuli. This region then probably acts to prepare this information for expression via transfer to the right frontal lobe.

Just as Broca's area in the left frontal region is responsible for the production of words and sentences, a similar region in the right frontal lobe is dominant for the expression of emotional–melodic and even environmental sounds. Hence, it appears that an **emotional–melodic–intonational axis**, somewhat similar to the language axis of the left half of the brain in anatomical design, is maintained within the right hemisphere.[39]

When the posterior portion of the melodic–emotional axis is damaged, that is, the right temporal–parietal area, the ability to comprehend or repeat melodic–emotional vocalizations is disrupted. Such patients are thus agnosic for nonlinguistic sounds. With right frontal convexity damage, speech becomes bland, atonal, and monotone or the tones produced become abnormal.

Limbic Language

Maternal Behavior and Infant-Separation Cries

Fish and many other creatures who swim the shining sea are capable of producing sounds. Marine mammals, such as the dolphin and a variety of whales, in fact produce numerous informative and highly meaningful sounds. However, these, like most other mammals, first developed and evolved upon dry land and only later returned to live within the deep. Nevertheless, creatures who roamed the planet or swam beneath the waves, some 300 or more million years ago, also relied on sound as a means of communication, although the brain structures and external hearing apparatus they possessed were probably quite rudimentary, and not much more developed than that seen in modern-day frogs and lizards.[40] As such, they were attuned to hear low-level vibrations and sounds, such as croaking, tails being thumped on the ground, and a few distress calls and sounds of contentedness, at least among the amphibia.

When reptiles began to differentiate and evolve into the reptomammals, and then, many millions of years later, when reptiles again diverged and the first tiny dinosaurs began to roam the Earth, several major biological alterations occurred. These include changes in skeletal structure, thermo-regulation, reproduction, and skull and brain size and organization, all of which coincided with the development of audio–vocal communication.[41]

For example, most reptiles show little or no maternal care, seldom vocalize, and are capable of hearing at best only low-frequency vibrations. Moreover, infants generally must hide from their parents, and other reptiles,

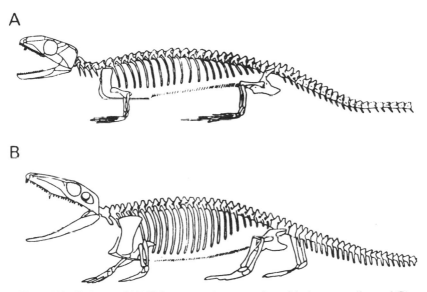

Figure 109. Skeletons of (A) Hylonomous, the most ancient of the known reptiles, and (B) a therapsid, a primitive, mammallike reptile, Varanosaurus. Note the increase in the length of the temporal cavity of the skull, and the relocation of the legs from alongside the body (where the chest walls and lungs would be constricted) in the reptile, to beneath the body. Hence, now, when running, instead of having to stop every few feet in order to take a breath (and thus probably being consumed by a predator in the process), one could run for long distances and breathe simultaneously. From P. Maclean. *The Evolution of the Triune Brain*. New York: Plenum Press, 1990.

in order to avoid being cannibalized. One might presume that the first reptiles were not in any manner more advanced than their predecessors in this regard.

In contrast, many of the reptomammalian therapsids, and then later, many of the various dinosaurs, lived in packs or social groups, and presumably cared for and guarded their young for extended time periods lasting up until the juvenile stage.[42] Hence, one of the hallmarks of this evolutionary transitional stage, some 250 million years ago, was the first evidence of maternal feelings and what would become the family.

Among the reptomammals and the dinosaurs, significant alterations also occurred within their brain, including the development of brain cells that were capable of producing highly meaningful social and maternally related sounds and yet other regions which would inhibit the tendency to consume one's young.

Among social, terrestrial vertebrates, the production of sound is very important in regard to infant care, for if an infant becomes lost, separated, or in danger, a mother would have no way of quickly knowing this by smell alone. Such information would have to be conveyed via a cry of distress or a sound indicative of separation fear and anxiety. It would be the production of these sounds which would cause a mother to come running to the rescue.

The Cingulate Gyrus

As noted, all such creatures, including reptiles and fish, possess a limbic system, consisting of an amygdala, hippocampus, hypothalamus, and septal nuclei. It is these structures, the amygdala in particular, which are important in maternal care and infant bonding, and which are most highly developed in mammals and humans. It is the presence of these particular limbic nuclei which enable a group of fish to congregate together, that is, to school, or for reptiles to form territories which include an alpha female, several subfemales, and a few juveniles.[43]

Such creatures, however, do not care for their young and do not produce complex meaningful sounds as they are lacking more recently acquired brain tissue. These more advanced tissues which they lack include not only the neocortex, but a cortical mass which is intermediate between neocortex and old cortex, and which is the most recently developed region of the limbic system. This new, or rather, transitional limbic–cortical tissue is the cingulate gyrus. It is likely that the expansion which occurred within the reptomammalian brain, some 250 million years ago, was due to the evolution of this transitional, four-layered cingulate limbic cortex.[44] It is only rather late in evolution, about 100 million years ago, that the first thin sheets of neocortex would begin to spread across and cover the old limbic brain and the cingulate gyrus.

The cingulate cortex is intrinsically linked to the hippocampus and the amygdala and is absent in reptiles and fully present only in mammals. Moreover, the cingulate in humans maintains direct interconnections with the left and right frontal regions, which, as noted, are responsible for the production of speech and emotional sounds. More importantly, when the cingulate cortex is electrically stimulated, the separation cry, similar if not identical to that produced by an infant, is elicited,[45] and it is only this area which produces it. It thus appears that the cingulate, in conjunction with the amygdala and other limbic tissue, may well be the responsible agents in

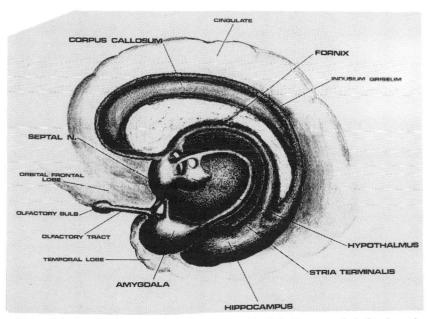

Figure 110. The limbic system, including the transitional limbic–neocortical, four-layered cingulate gyrus, as well as the amygdala, septal nucleus, hippocampus, and hypothalamus.

regard to infant care and the initial production of what would become language. I have referred to this elsewhere as "limbic language."[46]

Limbic Language: Amygdala and Cingulate Contributions

Phylogenetically and ontogenetically, the original impetus to vocalize springs forth from roots buried within the depths of the ancient limbic lobes (e.g., amygdala, hypothalamus, septum, cingulate).[47] Although nonhumans do not have the capacity to speak, they still vocalize, and these vocalizations are primarily limbic in origin being evoked in situations involving sexual arousal, terror, anger, flight, helplessness, and separation from the primary caretaker when young. The first vocalizations of human infants are similarly emotional in origin and are limbically mediated.

Although cries and vocalizations indicative of rage or pleasure have been elicited via hypothalamic stimulation, of all limbic nuclei the amygdala is the most vocally active, followed only by the cingulate.[48] In humans and animals a wide range of emotional sounds has been evoked through amygdala activation, including those indicative of pleasure, sadness, happiness, and anger.

Conversely, in humans, destruction limited to the amygdala, the right amygdala in particular, has abolished the ability to sing, convey melodic information, or to enunciate properly via vocal inflection. Similar disturbances occur with right-hemisphere damage. Indeed, when the right temporal region (including the amygdala) has been grossly damaged or surgically removed, the ability to perceive, process, or even vocally reproduce most aspects of musical and emotional auditory input is significantly curtailed. Moreover, the person loses much of his or her social interest and ceases to express affection, love, or emotions.

Similarly with destruction of the anterior cingulate, there results a loss of fear or socially appropriate behavior and the melodic qualities of the voice become altered. Humans will often become initially mute and socially unresponsive, and when they speak, their vocal inflectional patterns and the emotional sounds they produce sound abnormal.[49] Animals, such as monkeys who have suffered cingulate destruction, will also become mute, will cease to groom or show acts of affection, and will treat their fellow monkeys as if they were inanimate objects. For example, they may walk upon and over them as if they were part of the floor or some obstacle rather than a fellow being.[50] That is, their social behavior seems "reptilian." Maternal behavior is also abolished, and the majority of infants soon die from lack of care.

The amygdala and the cingulate gyrus, however, appear to influence language in a different manner. As noted, the amygdala is buried within the depths of and maintains rich interconnections with all regions of the temporal lobe, including the neocortex and fiber pathways which link Wernicke's receptive speech area with Broca's expressive speech area, that is, the arcuate fasciculus.

Via its interconnections with the Wernicke's area in the left temporal lobe, and the sound-receiving areas in the right temporal lobe, the amygdala is able to discern as well as apply emotional coloration to all that is heard. It is also able to add emotional tone to all that might be said. This is why when the right amygdala has been destroyed or surgically removed, the ability to sing as well as to properly intonate is altered.

In this regard, the amygdala should be considered part of the melodic–intonational axis of the right hemisphere, and part of the language axis of the left hemisphere, as it not only responds to and analyzes environmental sounds and emotional vocalizations but imparts emotional significance to auditory input and output processed and expressed at the level of the neocortex.

The cingulate, via its rich interconnections with the amygdala, the neocortex, and auditory association area, adds yet another level of motivational and emotional significance to all that is heard, including possible feelings of maternal concern. Through its connections with the more evolutionarily advanced neocortical regions of the left and right frontal lobes, and thus with Broca's expressive speech area and the melodic–emotional speech area in the right half of the brain, the cingulate is able to directly influence the emotional and melodic tone of all that is said. Hence, the cingulate gyrus, like the amygdala, is an important part of the language and melodic–emotional axes of the right and left half of the brain. These two limbic nuclei provide the motive source from which spoken language derives its origins.

Grammar and Auditory Closure

Despite claims regarding "universal grammars" and "deep structures," it is apparent that many people when conversing together, speak in a decidedly nongrammatical manner with many pauses, repetitions, incomplete sentences, irrelevant words, and so on. However, this does not prevent comprehension since the structure of the nervous system enables us to perceptually alter word orders so they make sense, and even fill in words which are left out.

In one experiment reviewed by Peter Farb in the book *Word Play*,[51] listeners were played a tape of a sentence in which a single syllable (gis) from one word "le*gis*latures," had been deleted and filled in with static (i.e., le . . . latures). However, no one could detect it and instead filled in the missing sound (gis) so that they heard "legislatures."

In another experiment, when the word "tress" was played on a loop of tape 120 times per minute ("tresstresstress . . ."), subjects reported hearing words such as dress, florists, purse, Joyce, and stress. In other words, they organized these into meaningful speech sounds which were then coded and perceived as words.

The ability to engage in gap filling, sequencing, and to impose temporal order on incoming (supposedly grammatical speech) is important because human speech is not always fluent as many words are spoken in fragmentary form and sentences are usually four words or less. Much of it also consists of pauses and hesitations, "uh" or "err" sounds, stutters, repetitions, and stereotyped utterances, "you know," "like."

Hence, a considerable amount of reorganization as well as filling in and deletion must occur before comprehension, and this requires that these signals be rearranged or filtered in accordance with the temporal–sequential rules imposed by the structure and interaction of Wernicke's area, the inferior parietal lobe, and the nervous system (what Noam Chomsky referred to as "deep structure")[52] so that they may be understood.

Consciously, however, most people fail to realize that this filling in and reorganization has even occurred, unless directly confronted by someone who claims that she or he said something she or he believes she or he didn't. Nevertheless, this filling in and process of reorganization greatly enhances comprehension and communication.

Universal Grammars

Regardless of culture, race, environment, geographical location, parental verbal skills, or attention, children the world over go through the same steps at the same ages in learning language.[53] Unlike reading and writing, the ability to talk and understand speech is innate and requires no formal training. One is born with the ability to talk, as well as the ability to see, hear, feel, and so on. However, one must receive considerable training in reading, spelling, and mathematics as these abilities are acquired only with some difficulty and much effort. On the other hand, just as one must be exposed to light or one will lose the ability to see, one must be exposed to language or one will lose the ability to talk or understand human speech.

In his book *Syntactic Structures*, Noam Chomsky argues that all human beings are endowed with an innate ability to acquire language as they are born able to speak in the same fashion, albeit according to the tongue of their culture, environment, and parents.[54] They possess all the rules which govern how language is spoken, and they process and express language in accordance with these innate temporal–sequential motoric rules which we know as grammar.

Because they possess this structure, which in turn is imposed by the structure of our nervous system, children are able to learn language even when what they hear falls outside this structure and is filled with errors. That is, they tend to produce grammatically correct speech sequences, even when those around them fail to do so. They disregard errors because they are not processed by their nervous system which acts to either impose order even where there is none, or to alter or delete the message altogether. It is because we all possess these same inferior-parietal-lobe "deep structures" that speakers are also able to realize generally when a sentence is spoken in a grammatically incorrect fashion.

It is also because apes, monkeys, dogs, and cats do not possess these deep structures or an inferior parietal lobe that they are unable to speak or produce grammatically complex signs or vocalizations. Hence, their vocal production does not become punctuated by temporal–sequential gestures imposed upon auditory input or output by the language axis.

Language Acquisition: Fine Tuning the Auditory System

By the time we reach adulthood, we have learned to attend to certain sounds and to ignore the rest. Some sounds we almost never hear because we learned a long time ago that they are irrelevant or meaningless. Or we lose the ability to hear these sounds because of actual physical changes which occur within the auditory system, such as deterioration and deafness.

Initially, however, beginning at birth and continuing throughout life, there is a much broader range of generalized auditory sensitivity. It is this generalized sensitivity that enables children to rapidly and more efficiently learn a foreign tongue, a capacity that decreases as they age. It is due, in part, to these same changes that generational conflicts regarding what constitutes "music" frequently arise.

Nevertheless, since much of what is heard is irrelevant and is not employed in the language the child is exposed to, the neurons involved in mediating their perception drop out and die from disuse, which further restricts the range of sensitivity. This further aids the fine-tuning process so that, for example, one's native tongue can be learned.

For example, no two languages have the same set of phonemes. It is because of this that to some cultures, certain English words, such as pet and bet, sound exactly alike as they are unable to distinguish between or recognize these different sound units. Via this fine-tuning process, only those phonemes essential to one's native tongue are attended to.

Language differs not only in regard to the number of phonemes, but the number which are devoted to vowels versus consonants and so on. Some Arabic dialects have 28 consonants and 6 vowels. By contrast, the English language consists of 45 phonemes which include 21 consonants, 9 vowels, 3 semivowels (y, w, r), 4 stresses, 4 pitches, 1 juncture (pause between words), and 3 terminal contours which are used to end sentences.

It is from these 45 phonemes that all the sounds are derived which make up the infinity of utterances that comprise the English language. However, in learning to attend selectively to these 45 phonemes, as well as to specific consonants and vowels, required that the nervous system become fine tuned to perceiving them while ignoring others. In consequence, those cells which are unused die.

Children are able to learn their own as well as foreign languages with much greater ease than adults because initially infants maintain a sensitivity to a universal set of phonetic categories. Because of this, they are predisposed to hearing and perceiving speech and anything speechlike regardless of the language employed.

These sensitivities are either enhanced or diminished during the course of the first few years of life so that those speech sounds which the child most commonly hears become accentuated and more greatly attended to such that a sharpening of distinctions occurs. This generalized sensitivity thus declines as a function of acquiring a particular language and the loss of nerve cells not employed. The nervous system becomes fine tuned so that familiar languagelike sounds become processed and ordered in the manner dictated by the nervous system; that is, the universal grammatical rules common to all languages. The ease of learning a second language thus also diminishes. Fine tuning is also a function of experience which in turn exerts tremendous influence on nervous-system development and cell death. Hence, by the time most people reach adulthood, they have long learned to categorize most of their new experiences into the categories and channels that have been relied upon for decades.

Nevertheless, in consequence of this filtering, even sounds that arise naturally within one's environment can be altered, rearranged, suppressed, and thus erased. By fine tuning the auditory system so as to learn culturally significant sounds and so that language can be acquired occurs at a sacrifice. It occurs at the expense of one's natural awareness of one's environment and its orchestra of symphonic sounds. In other words, the fundamental characteristics of reality are subject to language-based alterations.

Language and Reality

According to Edward Sapir: "Human beings are very much at the mercy of the particular language which has become the medium of their society . . . the real world is to a large extent built up on the language habits of the group. No two languages are ever sufficiently similar to be considered as representing the same social reality."[55]

According to Benjamin Whorf, "language . . . is not merely a reproducing instrument for voicing ideas but rather is itself the shaper of ideas. . . . We dissect nature along lines laid down by language." However, Whorf believed that it was not just the words we used but grammar which has acted to shape our perceptions and our thoughts.[56]

A grammatically imposed structure forces perceptions to conform to the mold which gives them not only shape, but direction and order. Moreover, distinctions imposed by temporal order, as well as by necessity, not only result in the creation of linguistically based categories but labels and verbal associations which enable them to be described. Again, however, they are described in accordance with the rules and vocabulary of language.

Eskimos possess an extensive and detailed vocabulary which enables them to make fine verbal distinctions between different types of snow, ice, and prey, such as seals. To a man born and raised in Kentucky, *snow* is *snow*, and all seals may look the same. But if that same man from Kentucky has been raised around horses all his life, he may in turn employ a rich and detailed vocabulary so as to describe them, for example, appaloosa, paint, pony, stallion, and so on. However, to the Eskimo a *horse* may be just a *horse*, and these other names may mean nothing to him. All horses look the same.

Moreover, both the Eskimo and the Kentuckian may be completely bewildered by the thousands of Arabic words associated with camels, the 20 or more terms for rice used by different Asiatic communities, or the 17 words the Masai of Africa use to describe cattle.

Nevertheless, these are not just words and names, for each cultural group is also able to see, feel, taste, or smell these distinctions as well. An Eskimo sees a horse, but a breeder may see a living work of art whose hair, coloring, markings, stature, tone, height, and so on speak volumes as to its character, future, and genetic endowment. Indeed, many ranchers are able to differentiate and recognize their cows as individuals, many of whom look almost identical to a city slicker or Burger King patron.

Through language, one can teach another individual to attend to and to make the same distinctions and to create the same categories. In this way, those who share the same language and cultural group, learn to see and talk about the world in the same way, whereas those speaking a different dialect may in fact perceive a different reality.

For example, when A. F. Chamberlain visited the Kootenai and Mohawk Indians during the late 1800s, he noted that they even heard animal and bird sounds differently from him.[57] For example, when listening to some owls hooting, he noted that to him it sounded like "tu–whit–tu–whit–tu–whit," whereas the Indians heard "Katskakitl." However, once he became accustomed to their language and began to use it, he soon developed the ability to hear sounds differently once he began to listen with his "Indian ears." When listening to a whippoorwill, he noted that instead of saying whip–poor–will, it was saying "kwa–kor–yeuh."

Observations such as these thus strongly suggest that if one were to change languages one might change one's perceptions and even one's thoughts and attitudes. Consider, for example, the results from an experiment reported by Farb. Bilingual Japanese-born women married to American servicemen were asked to answer the same question in English and in Japanese.[58] The following responses were typical:

"When my wishes conflict with my family's . . ."
". . . it is a time of great unhappiness." (Japanese)
". . . I do what I want." (English)
"Real friends should . . ."
". . . help each other." (Japanese)
". . . be very frank." (English)

Obviously, language does not shape all our attitudes and perceptions. Moreover, language is often a consequence of these differential perceptions, which in turn requires the invention of new linguistic labels so as to describe these new experiences. Speech, of course, is also filtered through the personality of the speaker and the listener and is influenced by their attitudes, feelings, beliefs, prejudices, and so on, all of which can affect what is said, how it is said, and how it is perceived and interpreted.

On the other hand, regardless of the nature of the experience, be it visual, olfactory, or sexual, language can influence our perceptions and experiences and our ability to derive enjoyment from them, for example, by labeling them *cool*, *hip*, *sexy*, *bad*, or *sinful*, and by instilling guilt or pride. In this manner, language serves not only to label and filter reality, but also affects our ability to enjoy it.

8

Self-Deception and the Origin of Language and Thought

A multitude of neuronal structures and fiber pathways are involved in the formulation, expression, and comprehension of speech and verbal thought. These include Broca's area in the left frontal convexity, which mediates expressive speech and programs the external facial–oral musculature; Wernicke's receptive speech and auditory association area (located in the left superior temporal lobule); and the angular gyrus of the left inferior parietal lobe, which promotes the formation of multimodal associations, classifications, and categories.[1] These regions, coupled with the thalamus, amygdala, and cingulate gyrus of the limbic system, and the multiple interactions which link them, comprise the **language axis** of the left hemisphere.[2]

When listening to someone speak, this information is first received in the primary auditory cortex and then is transferred to Wernicke's area as well as the middle temporal lobe. It is within Wernicke's area that the temporal–sequential, semantic, and related linguistic features are stabilized, extracted, analyzed, and comprehended.

The **middle temporal lobe** maintains extensive interconnections with the superior and inferior temporal lobes and the auditory association area. As noted, the left-middle-temporal region plays a particularly important role in the discrimination and organization of speech and other auditory sounds, word finding, naming, the maintenance of temporal order, as well as auditory memory. Hence, when activated, it draws associations from memory and also acts to supply names and verbal labels to that which is seen or heard. Moreover, since it is interconnected with the rest of the language axis via the arcuate fasciculus, the left middle temporal lobe also acts to supply information not only to Wernicke's area and the memory centers, but to the inferior parietal lobule and Broca's area.[3]

Via these interactions, if a question is asked, the semantically correct and suitably chosen reply is organized and transferred from Wernicke's region and the inferior parietal lobe, via the arcuate fasciculus to Broca's area which then programs the speech musculature and neocortical motor areas so that the reply can be expressed.

Broca's Area

Broca's area is located within the left frontal lobe and is responsible for the expression of speech. This area is named after Paul Broca, who over 100 years ago delineated the symptoms associated with damage to this area. Immediately adjacent to Broca's area is the portion of the primary neocortical motor area which subserves control over the oral–facial musculature. Directly above Broca's area is Exner's writing area, which controls and programs the hand when engaged in the process of writing.

Various forms of linguistic information are transferred from the temporal–parietal portion of the language axis and converge upon Broca's area. It is here where they receive their final sequential (syntactical, grammatical) imprint so as to become organized and expressed as temporally ordered motoric linguistic articulations, that is, speech. These impulses are then transferred to the primary frontal motor areas which in turn control the oral musculature. If the reply is to be written, the information is transferred to Exner's area, which acts to program the hand. Hence, verbal communication and the expression of thought in linguistic form is made possible.

Expressive Aphasia

If individuals were to sustain massive damage to the left-frontal convexity, their ability to speak would be dramatically curtailed. Often, immediately following a stroke in this region, individuals are almost completely mute and suffer a paralysis of the right upper extremity. This disorder is called Broca's (or expressive) aphasia. Comprehension, however, is generally intact.

Although symptoms differ, depending on the severity of the lesion, in general, individuals with Broca's aphasia are very limited in their ability to speak. In severe cases, speech may be restricted to a few stereotyped phrases and expressions, such as "Jesus Christ" or to single words such as "fine," "yes," and "no," which are produced with much effort. Even if capable of making longer two-word statements, much of what they say is poorly articulated or mumbled such that only a word or two may be intelligible. However, this allows them to make one-word answers in response to questions. Nevertheless, speech is almost always agrammatical (i.e., the production of some correct sounds and words but in the wrong order), and contaminated by verbal paraphasias, that is, "orrible" for "auto," and the erroneous substitution of semantically related words, for example, mother for father.[4]

Broca's expressive aphasia is also characterized by the omission of relational words such as those which tie language together, that is, the prepositions, modifiers, articles, and conjunctions. Their ability to repeat what is said to them, although grossly deficient, is usually not as severely reduced as is conversational speech.

The Inferior Parietal Lobule

The primary sensory receiving areas for vision, audition, and somesthesis are located in the occipital, temporal, and parietal lobes, respectively. Adjacent to each primary zone is a secondary-association neocortical region where higher-level information processing occurs and where complex associations are formed. Wernicke's region is one such zone, as is the middle temporal and the superior parietal lobe.

The inferior parietal lobule (which includes the angular and supramarginal gyri) is located at the junction where the three secondary association areas meet and overlap. In this regard, the inferior parietal region receives converging higher-order information from each sensory modality and in fact makes possible the formation of multiple associations based on the assimilation of this divergent sensory input.[5] One can thus feel an object while blindfolded and know what it would look like and be able to name it as well.

Through its involvement in the construction of cross-modal associations, this region acts so as to increase the capacity for abstraction, categorization, differentiation, and the verbal as well as visual labeling of sensorimotor experience. One is thus able to classify a single stimulus or event in multiple ways. In part this is made possible because the inferior parietal lobule is the recipient of the simple and complex associations already performed in the primary and association cortices via the 10 billion axonal interconnections that occur in this region.

Anomia or Word-Finding Difficulty

Not surprisingly, a common disturbance resulting from even mild abnormalities involving the left inferior parietal area is "word-finding difficulty," or what some people refer to as "tip of the tongue," when they have trouble coming up with or remembering a particular word. Every individual with aphasia has some degree of word-finding difficulty, that is **anomia** (or **dysnomia** when mild).[6]

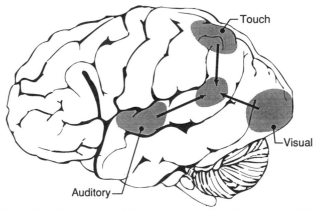

Figure 111. All sensory modalities converge on the inferior parietal lobe.

Anomia is very common and can occur with fluent output, good comprehension, and ability to repeat, and in the absence of paraphasias or other aphasic abnormalities. In fact, it can occur with damage anywhere within the left half of the brain. Although many *normal* individuals who have trouble finding a word sometimes experience it as on the "tip of the tongue," anomia is a much more pervasive abnormality and a patient so affected may have considerable difficulty naming objects, describing pictures, and so on, even when hints are provided.

However, in some cases, depending on where the damage and thus the disconnection occurs, persons may not be able to name an item when it is shown to them but can name it if they touch it, or it is described to them out loud. This would occur with damage disconnecting the inferior parietal lobe from the visual cortex. Since the axonal fibers leading from Wernicke's area and the tactual association areas have not been injured, the person can name it if he or she touches it or if it is described or makes a sound (e.g., running

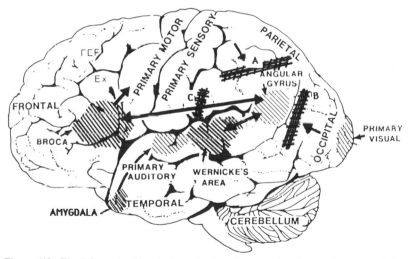

Figure 112. The left cerebral hemisphere. Lesions disconnecting the angular gyrus of the inferior parietal lobe (IPL) from the (A) remainder of the parietal lobe result in an inability to name or recognize what is held in the hand. However, objects can be named when looked at or if it makes a sound. (B) Disconnection of the IPL from visual cortex results in inability to name objects or letters that he sees. However, he can name the same object if he touched it or if it makes a sound. (C) Disconnection of the IPL from Broca's area results in conduction aphasia.

the fingers over the teeth of a comb). If the lesion disconnected the inferior parietal lobe from the rest of the parietal lobe, they would be able to name an object if they saw or heard it, but not by touch alone.

Frequently, with large lesions and thus severe word-finding difficulty, the patient suffers difficulty regardless of the sensory modality of presentation. If provided with hints or even the initial letter, there is little improvement. Moreover, once the word is supplied, the patient may again experience the same problem almost immediately. In general, the ability to generate nouns is more affected than verbs.

In cases where the inferior parietal region has been severely injured, while conversing, the patient may be so plagued with word-finding difficulty that speech becomes "empty" and is characterized by many hesitations and long pauses as the patient searches for words. This condition is sometimes referred to as **anomic aphasia**.[7] These same patients may erroneously substitute phrases or words for the ones that cannot be found. They may call a "spoon" a "stirrer," a "pencil" a "writer," or various objects "the whatchmacallit." This can lead to circumlocution as they tend to talk around the word they are after: "Get me the uh, the uh thing over on the uh, on the top over there . . ." Children are often plagued by such difficulties, which they grow out of, which is the result of immaturity within the inferior parietal lobe and the late establishment of the necessary axon fiber connections with other brain areas.

Conduction Aphasia.   Normally, once the linguistic organizational process has been completed by the inferior parietal lobule and the rest of the language axis, the resulting words and sentences are transmitted via the arcuate fasciculus nerve-fiber pathways to Broca's area so that expression can occur either through speech or writing. However, if for some reason the resulting verbal organization is emotionally suppressed, or if a single word is prevented from being placed in the temporal–sequential hierarchy, the person responds with a "tip-of-the-tongue" experience.

Sometimes, however, the person knows what the word is and just can't say it, which is due to a failure of conduction. When the person knows what the word or sentence he or she wants to say is, but cannot say it or write it, then he or she is suffering from conduction aphasia, which is what occurs when the interconnections between Broca's area and the inferior parietal lobule (i.e., the arcuate fasciculus) have been severed. Although the inferior

parietal and Broca's areas are intact and functional, they have been disconnected. The line is down.

Stimulus Anchors and the Train of Thought: Assimilation and Association within the Inferior Parietal Lobe

The angular gyrus and the inferior parietal lobule of which it is part,
makes possible the assimilation of complex associations which have been
constructed elsewhere.[8] This includes the assimilation of their tactual,
visual, verbal, and other sensory labels so that multiple classifications,
categorizations, and descriptions are possible. It also acts to integrate and
arrange them according to preestablished gestural temporal sequences and
the requirements of what needs to be communicated.

When activated, this area acts to transfer from other neocortical and old
cortical regions whatever associations are needed. Moreover, via its interconnections with Wernicke's area, it is able to associate auditory/verbal
labels with other sensory experiences such that we can describe things as
"loud, green, sweet, angry, hard, unpleasant," and so on. This is particularly important in regard to reading and naming as already described. For
instance, when a word is read, the pattern of visual input is transmitted from
the various visual areas in the occipital lobe to the angular gyrus which then
performs an auditory visual-matching function. That is, it calls for and
integrates the auditory equivalent of what is viewed or read so that we know
what animals, objects, words, and letters look as well as what their name
sounds like. If this area were damaged, reading ability would be lost.

In most instances in which the inferior parietal area of the brain is
activated via internal or external sources of stimulation, multiple trains of
inquiry are initiated via the numerous interconnections this area maintains.
Impressions, memories, ideas, and feelings which are in any manner
associated with the initial stimulus probe are aroused in response.[9]

If a student is asked: "What did you do in school today?" a number of
verbal and memory associations are aroused and integrated within the
language axis, all of which are related in some manner to each element of the
eliciting stimulus. Finally, in the process of associational linkage, those
associations with the strongest stimulus value and which most closely match
each element of the question in terms of internal and external appropriate-

ness and thus with the highest probability of being the most relevant rapidly take a place in a hierarchical and sequential, grammatical arrangement that is being organized in a form suitable for expressing a reply.

To return to the question regarding "school," each speech segment and sound unit like a magnet draws associations accordingly. All aroused forms of mental imagery, verbal associations, and so on which are activated and accepted are then arranged, individually matched, and group matched such that all associations that match all sources of relevant input with sufficient value of probability then act as templates of excitation that stimulate and attract other relevant ideas and associations. These in turn are assimilated and associated or are subsequently deactivated due to their low probability in contrast to the association already organized. Moreover, because the strength and value of closely linked associations change in correspondence to the developing sequential hierarchy, previously aroused and assimilated material may subsequently come to have a now-lower value of probability or appropriateness within the matrix of overall activity and may be deactivated.[10]

"What, pray tell, is furry, small, loves milk, and makes the sound Meoww?" At the level of the neocortex, each word, "furry," "small," "milk," and "meoww," acts to trigger associations (e.g., "furry = coat–animal-. . .," "milk = liquid–white–cow-. . ."). The grammatical linkage of these words also acts to trigger certain associations (e.g., "furry–milk–meoww = animal–cat-. . .") while deactivating others (e.g., "cow"). Following the analysis and comprehension of these sounds and words in Wernicke's area, the angular gyrus, and the middle temporal lobe, the inferior parietal lobule continues to call forth associations so that a reply to the question can be generated.

So that the animal can be named, the inferior parietal lobe via its interactions with the temporal lobe, activates the necessary phonemic elements (e.g., "k–a–t"), and then transfers this information to Broca's area and the question is answered: "Cat." If instead the individuals replies "tak" this would indicate a problem in organizing the correct phonemic elements once they were activated.

The final product of this hierarchical, highly grammatical parallel arrangement of mutually determining associational linkages is the train of thought or a stream of auditory associations in the form of words and sentences. However, before this occurs, these verbal associations must receive a final temporal–sequential grammatical stamp which is a consequence of the organization imposed on this material as it passes from Broca's area to the oral–speech musculature.

Self-Deception, Confabulation, Gap Filling, and Denial

Assimilation of input from diverse sources is a major feature of the language axis in general. However, if due to an injury the language axis is functionally intact but isolated from a particular source of information about which the patient is questioned, he or she may then suffer word-finding difficulty.[11]

In some cases following massive lesions of a brain area with which it normally communicates, the language axis sometimes begins to invent an answer or reply to questions based on the information available such as in cases of denial of blindness (following massive injuries to the visual neocortex) or of left extremity paralysis (due to massive right-cerebral injuries involving the motor and parietal neocortex). They confabulate. To be informed about the left leg or left arm, the language axis must be able to communicate with the neocortical area which is responsible for perceiving and analyzing information about the extremities. When no message is received, due to destruction of the neocortical receiving area, and when the language axis is not informed that no messages are being received (because the brain area which would alert them is no longer functioning), the language zones instead rely on some other source even when that source provides erroneous input.[12] Substitute material is assimilated and expressed, and corrections cannot be made (due to loss of input from the relevant knowledge source) and the patient begins to **confabulate**.[13]

Lying versus Confabulating

A long time ago, I discovered when anyone dared tell me: "I never lie," that much of what followed bordered on the pathological and delusional. Such individuals frequently turned out to be chronic liars.

People lie for any number of reasons, and sometimes they even lie to themselves and then believe their own fabrications, contradictory evidence notwithstanding. According to Jean-Paul Sartre, when people attempt to hide the truth from themselves, they are in "bad faith."[14] Elsewhere I have discussed this phenomenon in terms of "self-deception."[15]

Self-deception implies a knowing and a not knowing, or simultaneously telling a lie and believing it. This most commonly occurs when someone says or engages in some act which she or he later finds embarrassing and unacceptable to the conscious self-image and thus conjures up any number of believable explanations so as to appease friends, lovers, the public, or his or her own guilty conscience.

Knowing Yet Not Knowing: Seduction and Self-Deception

As noted, the brain is organized in such a way that the limbic system and the right and left half of the brain are specialized to perceive and respond to different forms of information. As such, some impulses, feelings, desires, or aspects of language cannot be shared or comprehended by different regions of the brain. This gives rise to situations where the limbic system or one half of the brain may initiate certain actions in response to certain desires, much to the surprise or chagrin of the remainder of the cerebrum. However, when the language-dependent half of the brain is called upon to explain these actions or feelings which have originated elsewhere, it may respond by filling in the gaps in its knowledge by confabulating. However, sometimes one half of the brain may choose to remain ignorant as well.

Christy had a conservative background and well-meaning, upper-middle-class, semireligious parents who had impressed upon her the need to maintain certain values and morals. Christy in her first semester at college was living in the dorm.

One day a handsome, muscular, sexy upperclassman, who she has seen about campus, invites her out. She consents. They have dinner, stop at a local nightclub, dance close together, and have a drink. So far it has been an exciting, romantic evening.

On the way back to the dorm he invites her to stop by his place to "check out the new CD player and stereo equipment" he has just purchased. She knows, or rather, is *aware* of the possibility of his intentions and what he may really be up to. Indeed, the look in his eyes, his tone of voice, and body language have made his real interests completely clear to the right half of her brain and limbic system long before they even began dinner. However, insofar as the language axis of her left brain is concerned, this guy just wants to show her his stereo.

Due to her right cerebral, emotional, sexually laced nonverbal *aware-*

ness of his intentions including the manner in which she may later respond, her left brain begins to invent and think up reasons why it is OK for her to drop by as it does not have access to these feelings. This awareness is not something she can consciously acknowledge. She rationalizes her decision. He seems real nice. They had a real good time. He has not acted inappropriately. Why not?

Once at his home he makes certain advances; sitting close to her, looking in her eyes, he tells her she is "really beautiful." The right half of her cerebrum responding to his tone, body language, and so forth is now almost certain of his desires and what he means by his words. Moreover, her own limbic system is entertaining similar feelings. Moreover, she is not only nonverbally, emotionally aware of what might occur next, she is also aware that she may have to make a decision regarding what does or does not happen. However, she does not want to "think about that" (she is a nice girl). This social–emotional awareness remains confined to the limbic system and right half of her brain, and the language axis of the left half of her brain, being disconnected from these sources of input, draws certain conclusions based on only partial information and makes decisions accordingly.

Due to her conscious self-image, this information is suppressed and ignored. The left cerebrum does not gain access to what the right brain is fully aware of. The language axis thinks: This guy is only being nice.

Slowly he places his hand on her knee. The act risks changing the situation by calling for an acknowledgment and an immediate decision as to its meaning. To leave his hand there is to consent, to acquiesce to his desires, and to consciously, verbally acknowledge them willingly. Yet to withdraw his hand is not only a recognition of what it implies, but a refusal. Her aim is to postpone her decision, to stall conscious, verbal recognition. She has "no idea" as to what he is up to. She leaves his hand there because she "doesn't notice it."

And yet, as they talk, he has moved closer to her, his hand inching its way past her knee. But she is concentrating on what he is saying, the curve of his lips, the white of his teeth, her reply to his questions; this is what she is verbally conscious of and to what she addresses herself. She reacts as a personality that does not know that a hand lies there upon her leg. She is being seduced.

She is also being self-deceived. She has restricted her consciousness and her verbal thoughts to only selective aspects of what is occurring. She is failing to acknowledge consciously what she is fully aware of nonverbally.

Nevertheless, she was aware that his actions were sexual and had been socially sexually aware of everything that might occur since the moment he asked her out. Indeed, it is because she entertained similar secret limbic desires that she consented in the first place.

However, if someone were to have told her this guy would bring her home and make possible sexual advances, she would have turned down the date. After all, her conscious, verbal self-image is that of a "good girl," a "nice girl." Although she is not a "prude," as she has always told her friends, she will have sex only with a man that she loves. The possibility of "sex" was so far from her (conscious) mind that it never even occurred to her that this guy would make certain advances.

So why did she go out with him? Why is she not taking his hand and indicating that his behavior is not acceptable: "Please stop." One would have to assume that her limbic system was "turned on," she was sexually attracted to him, and the possibility of his making sexual advances was not only acceptable but welcome; at least on an unconscious limbic level. However, what is acceptable at an unconscious level may not be acceptable to her conscious mind and verbally reinforced self-image. Unconsciously she is in fact yielding to his and her limbic desires. Consciously, at the level of the language axis of the left cerebral hemisphere, she still does not know what is going on. His behavior seems harmless.

If we accept Jean-Paul Sartre's position,[16] there is a possible unity in a situation such as this, for her actions necessitate a recognition of the intentions of her male friend as well as a disavowal. Otherwise she would have asked: "What is your hand doing on my thigh? Please remove it." Even if unable to verbalize her conscious disapproval (as some young women presumably become too intimidated to speak up), she would have made some attempt to back off or disengage herself if that is not what she wanted. However, consciously she is verbally disconnected from what is really occurring and from what she is fully aware of, so she does nothing except yield to her own limbic lusts.

This is the beauty of self-deception; it allows a person to engage in a certain behavior while simultaneously preserving his or her conscious self-image. It preserves the possibility of an event's occurring or a desire being acted upon while simultaneously behaving as if the event or action is not happening.

Within the realm of the limbic system or the right half of one's brain a person may be fully aware of what she is doing or intends to do, while consciously, linguistically, she may behave as if nothing of the sort were going to occur. This allows a person to deceive herself into thinking she is not responsible for her own behavior or for what happens.

He takes her hand, leans close, and kisses Christy upon the lips. The kiss takes her by "surprise." He snakes his other arm around her and kisses her again. She does not know what to do. She is confused. And yet, her limbic system is proclaiming: "I want it now!"; her right brain is saying: "Go for it!"; and the language axis of the left half of the brain in all innocence is saying: "What's going on?" He begins kissing and touching her passionately, and she feels herself yielding as her limbic system begins to overwhelm her conscious inhibitions. Her limbic system, in effect, has taken over, and the left half of her brain, essentially, has gone "off line," and is now a passive observer to behavior over which it has no control. She is swept away by passion.

Indeed, it often happens that people behave in a certain manner, or say certain things that are shocking even to themselves. They don't know "what came over" them. They may see themselves as moral, just, kind, loving, and for some unknown reason act in a cruel, selfish, sexual, or spontaneous manner and yet retain the conviction that what they have done or said is not a true representation of how they really feel or who they really are. Often, however, once the limbic system has been satisfied and the left brain regains dominance, it may then begin to impose and cause feelings of guilt on itself.

Their language axis may then begin to confabulate explanations so as to defend their conscious self from the verbal truth or to minimize these self-induced feelings of unnecessary guilt. Such behavior, they reason, has an explanation outside themselves, being only a rare and momentary lapse, is justified by certain mitigating circumstances, or is due to the horrible shocking behavior of someone else: "They pushed me too far that time!"

And, if Christy was "taken advantage of" by "that rat," "that scoundrel," "that womanizer," then she is of course not responsible for what happened. She "had no idea as to what he was up to," or that "he was that kind of guy." And, besides she "had too much to drink" (although she in fact had only one) and "was too tired to fight him off" (although she never tried) and before she "knew it, it was too late," and so on. Her conscious self-image remains untarnished. She is free of conscious guilt. As pointed out by Sartre, she has deceived herself. And yet, she knows and is aware of the truth. It is precisely because she knows that she invents innumerable excuses for her behavior.

Confabulation

A less common but more extreme form of "lying," or self-deception, is confabulation. Unlike self-deception, in which the individual resists or is psychologically unable to gain access to his or her true store of knowledge, the confabulator is often neurologically unable to gain access to information and is consequently unable to recognize the erroneous nature or absurdity of his or her statements, even in the face of seemingly apparent contradictory evidence. Moreover, rather than relinquishing an incorrect belief when confronted by evidence to the contrary, confabulators sometimes make further erroneous extrapolations which partially incorporate some aspects of what they have been told or confronted with.

For example, in cases where individuals suffer from blindness or gross visual disturbances due to injuries in the visual cortex, some patients will continue to claim that they see even when they bump into objects and trip over furniture.[17] When questioned about their obvious visual impairment, they will deny that they are blind and instead claim that it is too dark in the room, or that they forgot their glasses, or that the floor is slippery, or that something must be wrong with their legs. Apparently, they maintain these claims because the areas of the brain that normally would alert them to their blindness (i.e., visual cortex) are no longer functioning.

In these instances, delusions and confabulatory responses occur as a result of an attempt by the language axis to fill the gaps in the information received from the visual cortex. It responds by calling forth associations and ideas which are in some manner related to the fragments available and that have something to do with seeing.[18] Hearing the doctors comment about their seeming blindness but receiving no internal confirmation of being blind (because they no longer have a visual cortex), the language axis instead begins to conjure up explanations which can explain or fill in the gap in the information received; that is, "it must be dark" or "I need my glasses." In this regard, confabulatory–delusional statements can contain some accurate elements around which erroneous, albeit related, ideas and associations are anchored.

The language axis of the left half of the brain is thus very dependent upon its interconnections not only with diverse regions of the brain but with the functional integrity of the nuclei which comprise it. Indeed, all of linguistic comprehension and understanding is dependent on the language axis, for if these regions are injured, comprehension suffers. However, because it is an interdependent functional unit, it tends to comprehend only what is available to it.

The language axis cannot comprehend an absence or a vacuum but instead requires yet other regions of the brain to inform it. Hence, if an individual loses his or her sight, it is the visual cortex of his brain which comprehends this and informs the language axis. If the visual cortex were destroyed and the person were deprived of sight, the language axis would remain in the dark as to his or her condition and would even deny being blind. This is because the language axis cannot see. Only the visual cortex can see, and only the visual cortex knows when it cannot see, and in this case the visual cortex has been destroyed. A similar process occurs with Wernicke's aphasia. Since the visual cortex cannot be informed as to his condition, neither can the rest of the brain. Instead the language axis believes only what is available to it and then claims to see.

Confabulation and Denial of the Body

Confabulation is most clearly demonstrated in cases where individuals suffer profound memory loss, or severe right-cerebral parietal injuries which erase half the body image. Confabulation associated with right-parietal and right-frontal injuries often seems to be the most bizarre and transparent, in that often the patient will make up lies about what seems most obvious such as denying the existence of the left half of his or her body.[19] This is due to the destruction of the body image maintained by the parietal lobe.

Frequently, patients with right parietal lesions, when confronted with their left hands and arms or immobile limbs may (at least initially) deny that they belong to them or swear there is nothing wrong, since the body image has been erased. In some cases, however, patients may perceive the left half of their body but refer to it using ego-alien language, such as "my little sister," "my better half," "my friend Tommy," "my brother-in-law," "spirits," etc. For example, Gerstmann[20] describes a patient with left-sided hemiplegia who "did not realize and on being questioned denied that she was paralyzed on the left side of the body, did not recognize her left limbs as her own, ignored them as if they had not existed, and entertained confabulatory and delusional ideas in regard to her left extremities. She said another person was in bed with her, a little Negro girl, whose arm had slipped into the patient's sleeve." Another declared (speaking of her left limbs), "That's an old man. He stays in bed all the time."

One such patient engaged in peculiar erotic behavior with his "absent" left limbs which he believed belonged to a woman. A patient described by

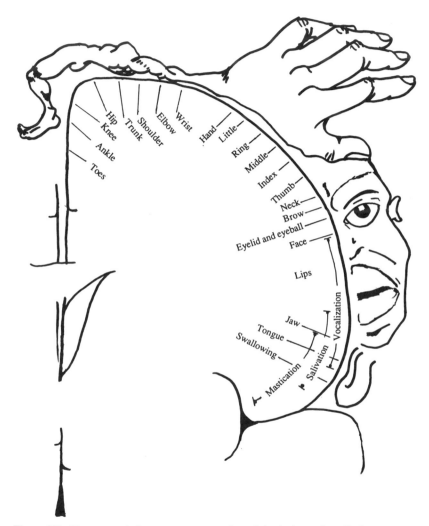

Figure 113. The neocortical sensory representation of the body surface. Body parts are distorted in proportion to their sensory or motoric importance.

Bisiach and Berti,[21] "would become perplexed and silent whenever the conversation touched upon the left half of his body; even attempts to evoke memories of it were unsuccessful." Moreover, although "acknowledging that all people have a right and a left side, he could not apply the notion to himself. He would affirm that a woman was lying on his left side; he would utter witty remarks about this and sometimes caress his left arm."

Some patients may develop a dislike for their left limbs, try to throw them away, become agitated when they are referred to, entertain persecutory delusions regarding them, and even complain of strange people sleeping in their beds due to their experience of bumping into their left limbs during the night. One patient complained that the *person* tried to push her out of the bed and then insisted that if it happened again she would sue the hospital. Another complained about "a hospital that makes people sleep together." A female patient expressed not only anger but concern lest her husband should find out; she was convinced it was a man in her bed.

Disconnection, Confabulation, and Gap Filling

In some respects it seems quite puzzling that individuals with right-parietal injuries may deny what is visually apparent and what they should easily be able to remember, that is, the presence of the left half of their body. It is possible, however, that when the language-dominant left hemisphere denies ownership of the left extremity, it is in fact telling the truth. That is, the left arm belongs to the right, *not* the left hemisphere, which in turn maintains a neuronal sensory image of the opposite half of the body.

Moreover, as discussed in detail elsewhere,[22] memories are sometimes unilaterally stored only in the right or only in the left half of the brain.[23] The left half of the brain predominantly stores verbal, numerical, and related forms of information within its memory banks, whereas the right half of the cerebrum is more likely to store visual, emotional, musical, imaginal, facial, and spatial information including knowledge of the left half of the body. Since each half of the brain is responsible for processing certain forms of information, certain memories and the events upon which they are based are processed and stored only in the right versus left half of the brain. Hence, just as the visual cortex cannot talk and does not receive verbal information, and the auditory cortex cannot see and does not process visual information, certain actions, events, thoughts, feelings and so on are processed and stored only in one versus the other half of the brain because this same information would be less meaningful or even meaningless to the opposite cerebral hemisphere.

In this regard, the left half of the brain may in fact have no memory regarding the left half of the body because those memories have been stored in the right brain, which in the cases described above has been severely injured. The left brain, which controls the ability to talk, cannot gain access to those memories, and thus what it says does not strike it as preposterous. Of course, all this seems preposterous, and yet, patients will deny what is obvious, that is, the existence or ownership of the left half of their body.

In some respects, denial is the reverse of an amputee's experience of a phantom limb. With a phantom limb the neocortical body image is retained and continues to be referred to and experienced. In neglect and denial, the body image has been deleted because of destruction of the parietal neocortex which maintains this image, and hence it no longer exists.

Thus, there is no "conscious" attempt on the part of the confabulator to deceive or lie, nor is the lie evoked in an attempt to please the interviewer or to avoid embarrassment, for they believe their own erroneous statements, which are often presented with "rocklike certitude." Confabulations are motivated and organized no differently than is factual information, with the single exception that the facts are not available to the language axis which attempts to organize and express whatever is available to it, even erroneous and preposterous substitutes.[24]

Inevitably, in order for an individual to confabulate, erroneous information must become integrated in some fashion so that the confabulated response can be expressed. When the parietal lobes are compromised, there results a disconnection. Information received in the language axis is incomplete and riddled with gaps. Still, the patient is not aware of these gaps, for if he or she were like a normal individual, he or she would plead ignorance.

Nevertheless, the larger the gap the more bizarre the confabulation may become. This is because the language axis attempts to fill the gaps in the incomplete information received with associations and ideas which are related in some manner to the fragments available. When the number of fragments available is reduced, the accuracy of whatever is reported similarly declines. It is around these fragmentary elements that related, albeit erroneous ideas and associations are anchored.

Consider again the question, "What is small, furry, drinks milk, and makes the sound Meoww?" If instead the person were asked, "what is small . . . drinks milk . . . ?" any number of associations would be aroused and might be expressed accordingly, for example, "a baby." The more information that is available, the greater is the likelihood that one will come up with correct associations and dismiss those which are irrelevant. When one is presented with gaps, the ideas and associations one comes up with are likely to be correspondingly erroneous.

However, one need not suffer a brain injury in order to confabulate. Confabulation is a common feature of everyday life as we go about attempting to figure out and explain or understand our behavior. For example, in a study performed in a supermarket many years ago, two experimenters, R. E. Nisbett and T. D. Wilson, laid out four identical nylon stockings from left to right on a table and invited shoppers to indicate their preference. The majority of shoppers picked the nylon on their far right, even though it was in all respects identical to the other three. Although the experimenters repeatedly changed the order, most shoppers still picked whatever stocking happened to be on their right.

When asked the reasons for their preference, all gave perfectly logical, rational, and sensible explanations. "The color." "The texture." "The shading." "It seems more durable." "More elastic." "Softer." "Smoother." And so on. However, not a single shopper came up with the real reason, that the stocking was on their right. Nevertheless, all were perfectly satisfied that the explanation they provided was the correct one.

Most people, at least those who are right handed, have a right-sided response bias. They are more likely to gesture with their right hand (versus the left) when talking, and they are more likely spontaneously to activate the right half of the body when asked to make a movement. Many advertisers are aware of this response bias and will usually, but not always, put "Brand X" to the left and give their own product the favored right-sided exposure. This is especially evident in television commercials. However, as is apparent, most people, having no knowledge of this bias, instead fill the gap in their knowledge with what they consider reasonable explanations.

However, self-deception and confabulation are not always due to information gaps but to emotional interference. Some feelings, thoughts, ideas, and actions are unacceptable or unavailable to the left half of the brain or the conscious self-image and are correspondingly suppressed and denied. When this occurs, the language axis is again presented with incomplete information around which erroneous associations are anchored and reported as fact when in truth they are fiction. However, the words chosen still have some bearing on what is hidden, which explains the presence of slips of the tongue or the efficiency of Carl Jung's famous Word Association test at teasing out emotional disturbances and complexes.

In addition, not all forms of confabulation are due to gaps in the information received in the language axis. Some forms of brain injury result in the flooding of the association and language areas with tangential and irrelevant information, much of which is amplified completely out of proportion to more salient details. Consequently, both salient and irrelevant, highly arousing and fanciful information is expressed indiscriminately as the normal filtering process is disrupted. Conditions such as these are most likely to occur when the frontal lobes have been compromised.[25]

The frontal lobes serve as the senior executive of the brain and personality and serve to inhibit irrelevant impulses and to direct and control information processing in the rest of the brain. It is via the frontal lobes that long-term planning skills and the concern for consequences are maintained.

Flooding of the Language Axis

Injuries causing left frontal damage can result in speech arrest or significant reductions in verbal fluency, that is, Broca's aphasia. Right frontal damage has frequently been observed to result in speech release, extreme verbosity, and overtalkativeness (i.e., "motor mouth"), tangentiality, and in the extreme, confabulation.[26] When secondary to right frontal lobe damage, confabulation seems to be due to extreme impulsiveness and disinhibition, difficulties monitoring responses, withholding answers, or suppressing the flow of tangential and irrelevant ideas such that the language axis of the left hemisphere becomes overwhelmed and flooded by irrelevant associations. In some ways, right frontal lobe damage can create problems similar to the initial effects of excessive alcohol intake. Hence the saying, "I'd rather have a bottle in front of me, than a frontal lobotomy."

In some cases the content of their speech may be bizarre and fantastical, as loosely associated ideas become organized and anchored around fragments of current experience. For example, one 24-year-old individual who received a gunshot wound, which resulted in destruction of the right frontal lobe, attributed his hospitalization to a plot by the government to steal his ideas. When it was pointed out that he had undergone surgery for removal of bone fragments and the bullet, he pointed to his head and replied, "that's how they are stealing my ideas."

Another patient, formerly a janitor, who suffered a large right frontal blood clot (which required evacuation) soon began claiming to be the owner of the business where he had formerly worked. He also alternatively claimed to be a congressman and to be fabulously wealthy. When asked about his work as a janitor, he reported that as a congressman he had been working undercover for the C.I.A. Interestingly, this patient, several months later, stated that he realized what he was saying was probably not true. "And yet I feel it and believe it though I know it's not right."

Tangential and Circumlocutious Speech

With bilateral or right frontal lobe damage, tangentiality is frequently observed in speech. For example, when a patient with severe right orbital damage was asked by Blumer and Benson if his injury affected his thinking,[27] he replied, "Yeah—it's affected the way I think—It's affected my senses—the only things I can taste are sugar and salt—I can't detect a pungent odor—ha ha—to tell you the truth it's a blessing this way."

When I asked one frontal patient what he received for Christmas, he replied, "I got a record player and a sweater." (Looking down at his boots) "I also like boots, westerns, popcorn, peanuts, and pretzels." Another right frontal patient, when asked in what manner an orange and a banana were alike, replied, "Fruit. Fruitcakes—ha ha—tooty fruity." When next asked how a lion and a dog were alike, he responded, "They both like fruit—ha ha. No. That's not right. They like trees—fruit trees. Lions climb trees, and dogs chase cats up trees, and they both have a bark."

Tangentiality is in some manner related to impulsiveness and the flooding of the language axis with closely related, albeit nonsensical associations which it then organizes and expresses. That is, the person becomes very impulsive and says whatever pops into his or her mind (or rather, into their language axis).

In contrast, patients with circumlocutious speech often have disturbances involving the left cerebral hemisphere and frequently suffer from word-finding difficulty and sometimes receptive or expressive aphasia. They continuously experience difficulty expressing a particular idea or describing some need as they have trouble finding the correct words. Thus, they seem to talk around the central point and only through successive approximations are they able to convey what they mean to say.

Patients with right frontal damage and thus confabulatory or tangential speech lose the point altogether. Instead, words or statements trigger other words or statements which are related only in regard to sound (e.g., like a clang association) or some obscure and ever-shifting semantic category. Speech may be rushed or pressured, and the patient may seem to be free associating as he or she jumps from topic to topic. Again, this is due to the importance of the frontal lobes in regulating the activity in the remainder of the cerebrum. In fact, much of the brain functions in regard to inhibition. It is this inhibition which makes possible the filtering out of irrelevant sensory experiences including fleeting limbic desires. It is via the frontal lobes that one is able to suppress these influences and put off one's desires until later. The frontal lobe thus acts as a massive inhibitor, and when injured (or intoxicated) the brain begins to process and express information indiscriminately; the person begins to act on his or her desires and says whatever occurs to him or her. He or she acts and speaks without thinking.

Isolation of the Language Axis

The formation of spoken language and verbal thoughts are a result and a reaction to internal and external stimulus inputs. It is a result of an attempt to make linguistic sense of a wide variety of information. However, what is understood and comprehended is an interpretation, the results of previous perceptual filtering, and the resulting organization of associations and the manner in which they were organized.

In many ways what we comprehend linguistically is based on an interpretation which may be related only loosely to the original signal, impulse, need, question, or desire. Moreover, if the question, ideas, or impulse is emotionally significant, unpleasant, or if it is not fully amenable to linguistic labeling, then the language axis, having to select from a flood of probable but not completely relevant or valid associations, will organize and express that which seems to be the most likely or which has the greatest strength of association. If what is received is riddled with gaps, and if disconnected or isolated from a needed source of input, the language axis makes up a response based on the fragments received and the person engages in self-deception.

It sometimes happens, however, that the language axis comes to be completely disconnected and isolated from almost all sources of information except that which is made directly available to it by hearing. This occurs when all neocortical tissue surrounding the language axis has been destroyed. Since it is still intact but now functionally isolated from all other neocortical areas, almost all relevant associations are unavailable to it. All of experience becomes one big gap. This condition is referred to here as **isolation of the language axis (or speech area)**.[28]

Isolation of the speech area is a condition in which the cortical border zones surrounding the language axis have been destroyed due to occlusion of the tiny tertiary blood vessels which supply these regions. That is, the language axis of the left hemisphere becomes completely disconnected from surrounding cortical tissue but presumably remains an intact functional unit. This is in contrast to **global aphasia**, in which the three major zones of language have been destroyed. In global aphasia, comprehension, speech output, reading, writing, and virtually all aspects of language functioning are abolished as the language areas have been completely destroyed.[29]

Unlike global aphasia, when the language axis has been isolated, the rest of the cerebrum is unable to communicate with the language zones which are still intact and functional. It is the rest of the brain which has been injured. As such, individuals are unable to describe verbally what they see, feel, touch, or desire. Moreover, because the language axis cannot communicate with the rest of the brain, linguistic comprehension is largely abolished. That is, communication among Wernicke's area, Broca's area, and the inferior parietal lobule is maintained, but associations from other brain regions cannot reach them. Though able to talk, the patient has nothing to say. Moreover, he or she may be able to hear but is unable to linguistically understand what he or she perceives. However, such patients are capable of generating automaticlike responses to well-known phrases, prayers, or songs.

In an interesting case described by Geschwind, Quadfasel, and Segarra,[30] a 22-year-old woman with massive destruction of cortical tissue due to gas asphyxiation was found to have a preserved language axis. It was noted once the patient regained "consciousness" that "she sang songs and repeated statements made by the physicians. However, she would follow no commands, resisted passive movements of her extremities, and would become markedly agitated and sometimes injured hospital personnel." In all other regards, however, she was completely without comprehension or the ability to communicate. The patient's spontaneous speech was limited to a few stereotyped phrases, such as "Hi daddy," "So can daddy," "mother," or "dirty bastard."

She never uttered a sentence of propositional speech over the nine years of observation. She never asked for anything and showed no evidence of having comprehended anything said to her. Occasionally, however, when the examiner said, "ask me no questions," she would reply "I'll tell you no lies," or when told, "close your eyes" she might say "go to sleep." When asked, "Is this a rose?" she might say, "roses are red, violets are blue, sugar is sweet and so are you": To the word "coffee" she sometimes said, "I love coffee, I love tea, I love the girls and the girls love me."

An even more striking phenomenon was observed early in the patient's illness. She would sing along with songs or musical commercials sung over the radio or would recite prayers along with the priest during religious broadcasts. If a record of a familiar song was played the patient would sing along with it. If the record was stopped she would continue singing correctly both words and music for another few lines and then stop. If the examiner kept humming the tune the patient would continue singing the words to the end. New songs were played to her and it was found she could learn these as evidenced by her ability to sing a few lines correctly after the record had been stopped. Furthermore, she could sing two different sets of words to the same melody. For example, she could sing "Let me call you sweetheart" with the conventional words, but also learned the parody beginning with "Let me call you rummy." Her articulation of the sounds and her production of melody were correct although she might sometimes substitute the words "dirty bastard" for some of the syllables.

In this case, and a few others like it, it is likely that when comprehension and learning ability remain, such as in regard to singing and praying, it is largely a product of the undamaged portions of the limbic and right half of the brain.

The Origin of Thought

Left Hemisphere Thinking

Thinking, like spoken language, is clearly a form of communication, and one form of thought, that is, linguistic thought (thinking in words), is a form of inner language; an organized hierarchy or train of associations which are silently heard or which appear before an observer in the "mind's eye."[31]

The train of thought, such as when thinking in words, often serves as a means of explaining something to the thinker, as in thinking (or figuring) out a problem. Verbal thoughts usually appear as a linked progression of words, or an associative advance with an initial or leading idea which is then followed by a series of related words and ideas.[32] However, verbal thinking can also be a means of questioning, deducing, planning and forming goals, as well as explaining, be it an idea, action, thing-in-the-world, or upcoming event so that it may be understood, communicated, and possibly acted upon or prevented.[33]

It is through linguistic thought that the left half of the brain and the conscious mind with which it is associated is able to better understand and manipulate the world. Of course, thinking also occurs in images, feelings, musical ideas, and may in fact be a mixture of associations which are sometimes visual, sometimes verbal, or even simultaneously both coupled with an underlying or predominant tone of emotion which directs the entire process.

Verbal thinking often serves the function of rehearsal, description, or as a means of explaining things. Persons may listen to their thoughts, or observe the images as they dance upon the stage of the mind's eye as they figure out a problem, or rehearse what they intend to say or do. As such, one is both a listener and a speaker producing the verbal thoughts, or an observer and a producer, creating the dancing images for an audience of one.[34]

Yet the need to communicate and explain things to oneself seems paradoxical. Similarly, the act of producing and the act of observing also seem somewhat redundant in that the audience, actor, producer, and director are one and the same, seemingly coexisting as a single unity in the privacy of one's head. A functional duality is thus implied in the production and reception of some aspects of verbal and visual thought.

It might be asked, "Who is explaining what to whom?" "Who is the producer and who is the audience?" Since the "I" that I am explains and communicates these ideas, feelings, and thoughts to the "I" that I am, does this not imply a split in psychic functioning and mental processing? That is, since "I," the thinker, serves as both the explainee and the explainer, producer and audience, shouldn't I know what I am going to think before I think it since the source is within me?

The answer is yes and no, and in this regard there is no paradox since we have two cerebral hemispheres that are specialized to deal with different information, as well as a limbic system which also has a mind of its own.

Sometimes thoughts produced in one region of the brain and the mind serve to explain conclusions, deductions, ideas, feelings, fears, hopes, desires, and so on, originating in yet a different region; a realm of the mind to which the questioning, observing, or explaining aspects of the psyche do not have complete access or understanding.[35]

This also implies that some forms of verbal thought are often based on information and knowledge that is already in existence but in a tacit, nonlinguistic, imaginal, emotional, sensory, or nonorganized form. That is, this information seems to first exist in a form that cannot be recognized or understood by the language axis or the conscious mind. For the left brain and the language axis to gain an understanding of this nonverbal source of information requires that it be translated into and presented in a linear temporal sequence of language-related ideas and images.

That is, the only way this information can be understood by the left hemisphere is if it is interpreted and transformed and put into words. The only way it can be transformed and comprehended linguistically is by thinking about it in a verbal manner. Thus, linguistic thought often serves to explain and communicate something that we are already aware of non-linguistically and unconsciously, to the language-dependent conscious mind. Linguistic thought is the silent verbal language utilized by the left half of the brain, and it is comprehended via the aid of Wernicke's area and angular gyrus which acts to apply whatever verbal labels seem appropriate.

Thinking, when in words, is associated with the conscious mind and the language axis (Broca's and Wernicke's areas, the angular gyrus) of the left cerebral hemisphere. In this regard, thought often serves as a means of organizing, interpreting, and explaining impulses which arise in the non-linguistic portions of the nervous system so that the language-dependent regions may achieve understanding.

Verbal thinking thus seems to act as a conscious inner language which organizes and assigns verbal labels and grammatical order to our "not-thought-out" ideas, which are not fully conscious, into an organization which may then be understood consciously. As noted, however, there are also many other forms of "thought," for people think in pictures, images, feelings, numbers, visions, musical notation, and so on.

Egocentric Speech and the Internalization of Thought

Insofar as thinking is verbal, it is clearly associated with the functional integrity of the left half of the brain and often serves as a means of covert self-explanation. That is, our thoughts occur within the privacy of our head and are meant for no one other than ourself. Although we may listen to our thoughts, we usually listen alone; unless, of course, one is thinking "out loud." However, even when we think out loud, or talk to ourselves, our thoughts are still usually meant for our ears alone, no one else's.

Thinking and "talking to ourself" are ontogenetically linked. Before children are able to think in words, they must be able to *hear* them, in order to know what the words sound like. Because of this, in part, people first learn to think out loud as children. It is only over the course of the first 7 years of life that verbalized thoughts become progressively internalized as the private dialogue that we all experience as our train of thought.[36] This internalization in turn, corresponds to the increasing maturation and development of the corpus callosum,[37] which subserves information transfer between the right and left cerebral hemispheres. That is, during the first few years of life, information transfer is quite limited and only progressively increases as the child ages.[38]

As we've noted, not only are the different lobes of the brain specialized in regard to the type of information they can receive and process, but the two halves of the brain are differentially organized as well. The left half of the brain is more concerned with language and the sequencing of information, and the right half with emotional, musical, visual–spatial, environmental, and body-image information. Because of this, just as the primary auditory area cannot see or feel tactual sensations, and the primary tactual area is deaf to sound, some forms of information can only be processed in the right versus the left half of the brain. In consequence, sometimes one half of the brain has little or no idea or understanding of what is occurring in the opposite half of the cerebrum since it cannot process certain types of information. It is due to this differential organization of the right and left halves of the brain, as well as the limbic system, that "Christy" was able to engage in sex and self-deception while sparing her self-image.

Although the brain is divided into a right and left cerebral hemisphere, much communication is made possible via the corpus callosum, a major rope of axonal nerve fibers which interconnect them. However, like certain other regions of the brain, the corpus callosum, as noted, also takes many years to mature.

Hence, given the degree of corpus callosum immaturity during childhood, children have some difficulty accessing and sharing information between the two halves of the brain, much more so than adults who are also limited. One consequence of this immaturity is that the right and left half of the brain are much more likely to store their own unique memories which not only are not shared, due to these limitations in information transfer, but which may become permanently hidden from the opposite half of the brain. As such, certain memories, although seemingly hidden or forgotten by one half of the brain, may in turn plague the other half of the cerebrum later in life.[39]

Another consequence of this immaturity of the callosum is lying and confabulation as the language axis of the left half of the brain attempts to make sense of what the right half of the brain may be up to. In very young children, however, the left half of the brain will pose its explanations by "thinking out loud." This is because thinking internally within the privacy of one's own head is slower to develop.

Children first begin to think out loud and verbalize their thoughts around the age of 2–3, a form of communication that has been referred to as "egocentric speech."[40] Prior to this period, although children may think in terms of visual images, songs, feelings, desires, emotions, rhymes, and an occasional word, and may frequently engage in daydreaming, they do not yet think in connected words and sentences. That is, left hemisphere verbal thinking has not yet developed.

Egocentric speech is a form of thinking out loud and talking to one-self, and it usually takes the form of commenting on and explaining one's actions. Children engage in this when they play alone and when they are in groups. At its peak, almost 40% of all language production in children is egocentric.

When a child is engaging in egocentric speech, however, she does not appear concerned with the listening needs of those who might be near as her words are meant for her ears alone. Moreover, the child who is engaged in her egocentric monologue will seem oblivious if someone were to respond to her explanatory monologue by asking a question or making a comment. In fact, often several young children can be observed playing together all of whom are engrossed in their own monologues and none of whom is listening to the other.

When egocentric speech first makes its appearance (around age 2–3) and children begin to comment on their actions, the monologue will be initiated only after a behavior has been completed. For example, children may be drawing a picture and once it is finished they will tell themselves what they have done and explain the drawing. As children age, the egocentric monologue begins to occur at an earlier point in the action. They may explain their creation when they are only half done and still drawing. Finally, around age 6–7 and as children mature, they will announce what they will draw and then draw it.

Essentially, as children age they appear to receive advance warning of their intentions until finally the information is available before rather than after they act. First, children comment on an action after it occurs, then at an older age while the action is still occurring, and then finally children have reached an age where they describe what they are going to do before they perform it. It is soon after reaching this stage, at about age 7, that egocentric speech becomes almost completely internalized as verbal thought.[41]

The Left Hemisphere, Egocentric Speech, and Corpus Callosum

That egocentric speech initially appears only after the action has occurred indicates that the language axis and the left hemisphere mental system are responding to impulses and actions initiated outside their immediate realm of experience and understanding. It seems the left brain in the initial production of egocentric speech is attempting to organize the results of its experience (the observation of its behavior) into a meaningful verbalized description or explanation, which it then linguistically communicates to itself, out loud, but only after the action has occurred. However, insofar as it can only describe what has happened after it occurs, this indicates that the impulse to act originated elsewhere.[42]

Predominantly, egocentric speech is a function of the left hemisphere's attempt to verbally label and make sense of behavior initiated by the limbic system or right half of the brain.[43] Nevertheless, because of limits in communication between the right and left half of the brain, which is more pronounced in children, the left hemisphere utilizes language to explain to itself the behavior in which it observes itself engaged and which is being produced by the right hemisphere or limbic system. Later in life, in response to certain actions and extremes, it may simply proclaim, "I don't know what came over me."

When the efficiency of corpus callosum transmission begins to approximate that of an adult (which is still limited), the child can now explain or comment on his or her actions before they occur. This does not mean the explanation is correct, however.

Hence, as the corpus callosum matures, the left hemisphere is now able to gain access to this information internally rather than via external observation. In consequence, it begins to create its explanations internally as well. It begins to think thoughts as well as speak them.

Essentially, as the child ages, the production of egocentric speech appears to diminish. However, it does not disappear, it just goes underground, and eventually becomes completely internalized as verbal thought.

Delusional Playmates and Egocentric Speech

Young children not only produce egocentric speech, in which they comment upon, explain, and describe their own behavior, but sometimes describe their behavior to others (such as their parents) using ego-alien language. When a child performs a "bad" behavior and is then confronted by an adult, he or she may blame the actions on someone else, or he or she might conjure up some fantastic imaginary explanation. In this regard, if the left speaking half of a child's brain denies responsibility for something he or she has obviously done, it may, in fact, be telling the truth, and it may be as upset as his mother or father. That is, her left brain didn't do it. It was her right brain or limbic system all along. However, blaming others or responding to questions by making up explanations is not egocentric speech, it is confabulation.

It is also not at all unusual for a child to develop imaginary friends with whom he or she shares secrets, plays games, or upon whom he or she may place the blame for some untoward incident. These same "imaginary" friends sometimes urge them to commit certain acts or explain to them the actions and motives of others, and why mommy or daddy may have acted in a certain way.

Not all children, of course, develop elaborate imaginary friends. However, all (or almost all) children develop egocentric speech and at times employ ego-alien descriptions when confronted with their own disagreeable behavior. Largely, much of this is secondary to corpus callosum immaturity and certain actions being initiated by the limbic and right half of the brain much to the surprise of the left. However, among this same age group (up to age 7 and even 10), the parietal lobes are also quite immature, the inferior parietal area in particular.

As seen with adults with destruction of the parietal lobe, they sometimes uncontrollably comment upon their actions when given commands (e.g., "Now I'm waving goodbye"), whereas in other instances they may claim that the person performing the action is someone other than themselves, or they may completely deny that the left half of their body is their own and claim it belongs to another.

Although an injured parietal lobe is not the same as an immature parietal lobe (particularly in that immaturity is bilateral and damage is usually unilateral), there remains a curious similarity in the behavior of these adults and children at least insofar as their imaginary friends and ego-alien explanations.

Perhaps this is because the ego and self is first identified with the body whereas the image of the body is maintained by the parietal lobe. When one parietal lobe (due to damage or immaturity) is unable to communicate with the other half of the brain, that brain half is unable to recognize a continuity of self. Consequently, behaviors initiated by the opposite (usually right) half of the brain, or the half of the body controlled by the right brain are recognized by the speaking half of the cerebrum only from a disconnected (i.e., alien) perspective. When the speaking (i.e., left) hemisphere is questioned about certain behaviors, or about its left limbs, the child or brain-injured adult may claim the action was initiated by or the left limbs belong to someone else, such as "an old man," "my brother-in-law," or an imaginary friend.

Unconscious Influences on Language and Thought

The manner in which we speak, the words we employ, and the fashion in which we employ them are not just a function of the culture or geographic location in which we are reared. Rather, the production of speech, language, and thought can be influenced by any number of variables ranging from a simple touch, a certain fragrance, a pleasing melody, or brain damage, emotional dysfunction, or embarrassment. Just as a smile is not just a smile but can hide as much as it reveals, the words we use are influenced by a number of important variables, the associations available to the language axis, the gaps in the information received, the nature of the question, the person who is asking, and the words and intonation used in asking them.

The organization and expression of thought and language are also influenced by the status of those whom we are interacting with, as well as the context, social appropriateness, and our own feelings about what is being said. Indeed, emotional factors may sometimes prevent certain associations from being transmitted to the language axis and may instead cause completely irrelevant or only tangential associations to be available. Moreover, personality and emotional variables may cause us to say certain things without thinking first or to hesitate and say nothing at all.

Nevertheless, what is not said is sometimes just as informative as what is said. By carefully listening to the words as well as intonation and other speech variables that persons employ when conversing, we will discover that they unwittingly provide us with a wealth of data regarding their own unconscious motivations, their hidden thoughts, and even the structure of their mind and brain.

9

Sexual, Social, Emotional, and Unconscious Speech

. . . But is it too early to ask the commissioners, are you beginning, of course, the commission's been, what, in effect seventy-seven days or something like that. Seventy-seven days traveled to many stages, uh states, which I think is very important, because I think it's important when that report comes in it has a national concept to it, that it isn't regional in any sense. So that report, we can't ask the commissioners. . . .

GEORGE HERBERT WALKER BUSH

The manner in which we listen to and create the sounds of language are tremendously influenced by the neurological hardware that supports these functions as well as the environment in which we are raised. Not all brains are created equal, however, and the environment in which various people are reared can differ drastically, not only for those brought up in different parts of the world, but for those raised on the same block.

Take for instance the agrammatical and often nonsensical speech of former President Bush. This is a problem that plagued his listeners for most of his life, and which reflects a fundamental difference in the manner in which his brain is organized for language. Mr. Bush, who is left-handed, is different from most of us not only because he is rich, the son of a senator, highly intelligent, and former President of the United States, but because his brain is organized differently as well. In fact, many individuals who are left-handed have language represented in the right instead of the left half of the brain.

285

Even those who possess brains that are organized similarly to the norm often repeat themselves, insert or delete pauses inappropriately, speak too loudly or too softly, hesitate, stammer, employ jargon or nonsense words, suffer slips of the tongue, or are plagued by word-finding difficulty. Although some linguists may believe that these problems merely reflect a difference in linguistic style which is supposedly a consequence of if you are born in Greece or New York, they may in fact reflect structural abnormalities in the brain, or more likely, structural differences in one's personality and emotional state. What someone says, and how they say it often is a function of what they feel as well as what they are trying to hide from others and from themselves.

Linguistic Styles and the Personality of Speech

When human beings interact, they often employ stereotyped phrases and words which are often seemingly devoid of meaningful content or information but which nevertheless serve important ritualized social and emotional functions, for example, "How are you?" "Have a nice day." These seemingly meaningless patterns of speech are often employed as a means of maintaining contact and social harmony, but little else. They serve as a form of linguistic glue, or social cement, and act to hold relationships together.

Because such speech patterns serve social–emotional functions, their production, not surprisingly, is generally under the guidance of the limbic system and the right-brain language system. Frequently individuals who have become aphasic following a stroke are still able to express these stereotyped phrases and may be able to say "hello," "fine," and so on.

Consequently, one might suppose that the right half of the brain may be able to comprehend these frequently used phrases as well. Indeed, this is the case. Familiarity sometimes breeds comprehension. The right half of the human brain is able to comprehend a few words which are simple, concrete, social and emotional, and repeatedly employed.

Similarly, frequently used social phrases such as "good boy . . . go for a walk . . . ride in the car . . ." or "want a cookie . . ." are also easily comprehended and understood by apes and canines, even though they, like the right half of the human brain, are linguistically quite limited.

Dogs, chimps, human beings, and right brains in general, however, need and expect to be greeted either physically or verbally and learn and look forward to these stereotyped forms of verbal interactions. When the expected and the familiar are not forthcoming, their feelings are likely to be hurt.

Stroking

Stroking, a concept discussed in detail by Eric Berne,[2] is a general term for specific forms of interpersonal and social interaction. Although the term seems to denote physical contact, that is, a physical stroke or touching, it can be nonphysical, verbal, or conveyed via gesture, facial expression, or tone of voice.

In general, a single stroke corresponds to a single unit of social interaction. When individuals exchange and restrict their interactions to a single stroke, their relationship is similarly one-dimensional. Nevertheless, like a touch, a smile, or a wave, a single-stroke relationship promotes mental health and maintains social harmony.

Let's consider Bob, who is an engineer at a local computer company and Jack, who works in marketing. Although they have never held a conversation and neither remembers the name of the other, they nevertheless nod or say hello as they pass each other in the hall each morning, giving each other one stroke each. As they are not friends but see each other every day, the single stroke is appropriate and not only maintains harmony but reduces potential tension.

Our closest genetic relatives, chimpanzees, engage in a similar form of stroking, with the exception that they may touch hands or grin at one another (such as a grin of appeasement). If in the presence of a higher-ranking chimp and depending on his mood and whatever may have occurred between them previously, a lower ranking individual instead of touching may respectfully present his or her rump as if seeking to be mounted. If, however, these chimps continually ignored each other, social relations would soon deteriorate and probably become hostile if not aggressive.

Similarly, if one day Bob ignores Jack as they pass in the morning, it is likely that Jack will assume that Bob is being hostile, and he may even take offense at this slight. Jack feels entitled to one stroke and when it is not forthcoming, he feels deprived and in need; this is a frequent cause of tension and social animosity.

Jason works in the same department with Ralph, and they sometimes discuss projects together. Because they are more familiar with one another, they typically exchange two or three strokes each morning. "Morning. . . . What's new? Nothing much. Sure is hot today. I'll say. Well, see ya. Yeah, later."

The transaction between Ralph and Jason is obviously ritualized, just as it is between Jack and Bob, as neither exchange is really designed for the purposes of information exchange and true communication. However, each party has a general notion of the number of strokes that need to be exchanged to maintain the homeostatic social structure. If Jason merely said "morning" and walked away, Ralph would probably think little of it, realizing that the necessary remaining strokes would be exchanged later or the next day. However, if several days passed and Jason continued to withhold strokes, Ralph would realize that something was terribly amiss.

On the other hand, in these unidimensional relationships, if perchance one member suddenly provides more strokes than is usual, this, too, can be a source of tension as it is a disruption of an established ritual. Probably since the dawn of human consciousness it has been taboo to break with custom and ritual as it is likely to anger the spirits or bring forth the wrath of God. When this occurs, one is left vulnerable to the unknown which in turn can be a source of anxiety or confusion. Thus if for some reason Jack suddenly began offering and seeking more than his customary one to two strokes as they walked by each morning, Bob might in turn feel some degree of discomfort.

On one particular morning instead of exchanging their customary number of strokes as they pass in the hall, Jack disrupts the ritual by continuing the conversation.

Jack: Hey, how ya doing?

Bob: Fine, and you?

Jack: Eh, OK. Think it will ever stop raining?

Bob (turning to go): Sure doesn't seem like it.

Jack: Well, I'm sure getting tired of this. The traffic sure was crazy this morning, too. It's like they have all the traffic lights purposefully desynchronized.

Bob (glancing at his watch): Yeah, it sure seems like it sometimes.

Jack: Its just one red light then another. Anyway, catch that Forty-niners game last night?

Bob: No, been kind of busy *(glances at his watch again)*.

Jack: Hell of a game. They'll probably make the playoffs this year if they ever straighten out that quarterback situation.

Bob: Maybe. Anyway, I got to get this memo off. See ya later.

Jack: Yeah, later.
Bob (to himself): What in the world is with him? I wonder what the hell he wants?

As pointed out by Berne, in transactional terms all that is owed between Jack and Bob is one or two strokes each. Anything more or less is unusual and demonstrates a disturbance in the harmony of their relationship. By breaking with ritual, the consequences may interfere with, change, or enhance their ability to successfully interrelate; that is, they may become better friends. Or, one party may reduce his strokes and begin to avoid the other.

Stroking, however, insofar as it involves power, ritual, aggression, and social status, is often of more concern to males than females.[3] Females are far more likely to use these types of strokes to maintain and promote interpersonal harmony.[4] Females frequently provide them not only at the beginning of a conversation, but periodically throughout as well. In contrast, most men will provide the strokes they think appropriate and end the conversation at that, or they may not vocalize them at all and either nod, ignore each other, or just get down to business.

Nevertheless, in terms of stroking, some people consistently fail to deliver the appropriate number of strokes, either providing too many or too few, and this puts others off. It throws them off balance, requires them to think or analyze the situation, and makes them feel uncomfortable and uneasy. As such they, too, may begin to avoid the stroker.

Individuals who fail to provide the correct number of strokes are often viewed as being odd, stuck up, sexist, socially dense, or are considered to have poor social skills or some type of emotional disturbance. Sometimes, in fact, individuals who respond in this fashion are suffering an abnormality in the neural organization of the right brain or limbic system such as an imbalance between the amygdala and septal nuclei.[5] Frequently, as noted, this is due to conditions of emotional deprivation during early childhood and lack of maternal care, as well as a function of a disturbance in their family life and inappropriate upbringing (e.g., being ignored, abused, over-protected, etc.).

Of course, sometimes people provide more strokes than necessary because of something that has upset them, and they are looking for an outlet or an opportunity to acquire strokes because someone else has ceased to provide them. Sometimes people are just lonely, or overly friendly, which in turn may be due to loneliness. However, sometimes this extrafriendliness results in loneliness as others are not desirous of so many strokes and may reject them, or worse, complain of harassment. Or, conversely, if people fail to deliver sufficient strokes they may be viewed as aloof, weird, stuck up, and so on and then come to be avoided, which also can be a source of confusion to the avoided party.

However, some changing situations sometimes require that different strokes be applied to the same folks at different times. In one situation the interaction may require one stroke, whereas in yet a different social situation, three or more strokes are required, although the same person is being dealt with.

Hence, part of the secret for good social success is recognizing what are the appropriate strokes that are to be exchanged. In general, however, it is better to be generous with one's strokes and to smile. Smiling actually makes the smiler feel good.

Indirectness

When conversations move beyond strokes they may in turn be contaminated by repetitions, circumlocutions, projections, as well as our secret hopes, needs, and desires. This is true regardless of the motives for conversing, for example, exchanging information, giving directions, or cooing words of love or sorrow. Even the words we choose may be influenced by emotional conflicts, in which case we may experience slips of the tongue, word-finding difficulty, or find we must hold our tongue.[6] For instance, the secret pedophile who in complaining about his wife's love making notes that he has tried "flowers, candy, candles, and even incest."

Often we must avoid certain words and topics which are either taboo or which may be approached only indirectly; for example, obtaining a raise, getting a date, obtaining sex.

Some people are chronically indirect which in turn represents not just their conversational style (as if we were talking to a computer or black box) but their personality and underlying emotional state. It is these same personal and emotional factors which sometimes give rise to hesitations, repetitions, perseverations, and the avoidance of certain subjects.

In our culture people are often indirect in their conversations for a variety of reasons.[7] They may be sparing the feelings of their listener, trying to hide the truth, or attempting to engage in an act of seduction. Depending on the reactions of the listener one might be emboldened to say more or to skip to a different subject altogether. This also enables the listener to exert some control. For example, in its most common and subtle manifestations, indirectness is often used as a strategy for verbal seduction. This also enables the woman the option of not understanding what is being said so that it may be ignored. Or she can respond thus showing her appreciation for the implied double entendre and her willingness for the man to continue or even use more direct tactics.

Moreover, being indirect allows a person to retract what she was implying by claiming that it is not what she meant should her listener take offense. Being indirect allows one to take a sample of potential reactions so as to guide future responses.

Many people often hint as to what they want and even in close relationships they may refuse to be blunt with their requests. Some people feel that to be direct is to be demanding, or they claim to fear that it puts the listener in the awkward position of having to say *yes* or *no* to a direct request, as if there were something wrong with being straightforward. Certainly, being direct could be very straining on certain relationships, unless, of course, one party prefers bluntness or straight talk. In general, however, being indirect can represent respect, caution, fear of rejection, insecurity, ambivalence, passive aggressiveness, or any combination thereof. On the other hand, bluntness can also be cruel and insensitive.

Often we must be indirect and thus dishonest not just to spare the feelings of others, but to successfully maneuver through social space. When Carol called and asked Bob to go to a dinner party which he did not want to attend, rather than gracefully declining, he first acted enthusiastic and then after a moment of feigned reflection, he lied by sorrowfully informing her that he had a big date planned for that very evening and he already had the tickets. She was completely understanding.

However, when she called her friend Nancy, Nancy declined because she said she hated formal affairs. At this Carol was offended, for it did not seem to be a very good reason and was in fact a frank rejection.

However, in truth, Nancy declined not because she hated affairs but because of her own insecurity. She indirectly expressed her insecurity by claiming to hate something that she didn't so as to avoid the possibility of rejection by those who attend such functions. Moreover, she was sure that Carol had only called and asked her to be polite and was convinced that Carol did not want her around her fancy friends. Nevertheless, she would have liked to have gone, and one of her other reasons for declining was that she indirectly wanted to be asked a second time for the sake of reassurance. When Carol simply accepted her responses and did not act more interested, Nancy felt offended and rejected and took this as confirmation that her friend was not really interested in having her come over.

Unfortunately, Nancy is trapped in a fog of self-deceptions and the deceptive allure of language. Her words and explanations and the thoughts she thinks in their support in fact not only result in her experiencing what she fears but prevent her from seeing the reality of her own, unknown self. She is completely out of touch and is completely under the magic spell of language and its power to hide and conceal as much as it reveals. She has not just responded in an indirect manner; she has engaged in self-deception.

Melody and Personality

The words an individual chooses or avoids, their hesitations and pauses, and even alterations in pitch and melody convey volumes of information regarding what a speaker wishes you to hear, what one wants you to infer, how strongly one believes one's own statements, if one is asking a question or expressing doubt, and even as to what one does not want you to know. For example, alterations in pitch and melody may indicate uncertainty, anxiety, fear, contempt, sarcasm, a desperate need for approval or that a simple question is being asked, in which case pitch will go up. However, via pitch alone multiple features may be expressed and deduced simultaneously.

Alterations in pitch and melody provide contextual cues and indicate meaning as well as one's emotional involvement and attitude about what they are saying and who they are saying it to.[8] For example, the tone and melody of voice will tell if a person is rushed or wishes to stick around and chat. In this way persons can say they are rushed but interested to hear what you have to say by their tone alone. These variations also provide a listener with presumably unintentional details about the speaker's degree of self-confidence, desires, and emotional state.

Sex Differences, Approval Seeking, and Qualifiers

When dealing with men whom they are not intimately involved with, women in general tend to be more tentative as well as more interpersonal when asking questions and making statements, such as "That movie really was just so delightful, wasn't it?" or "Why don't we all go out tonight, OK?" Whereas a man might say, "That movie sure was good." "Let's go have a couple drinks."[9]

In this regard, women are also more likely to employ qualifiers such as "kind of," "maybe," "a little," whereas a male is more likely to say, "yes," "no," "never," "always." Of course, the use of these qualifiers also makes the speech of females sound friendlier and, when coupled with their tonal control and range, more enthusiastic. These characteristics are also expressed via their more frequent use of intensifiers such as "quite," "very," "really," "so," as well as their tendency to incorporate others in their questions. Moreover, many of these sex differences are apparent in children as young as 3 to 4 years of age.[10]

However, some men also tend to use these questioning speech patterns, "It's sure hot today. Isn't it?" "That was some game, wasn't it?" "She sure knows how to dress, doesn't she?" Frequently, in these instances it could be assumed that perhaps the speaker is not merely making a statement, or really asking a question. Rather, one might suspect he is asking for verification and validation of his thoughts and feelings. Such phrasing suggests that one is seeking approval.

Sex Differences in Pitch and Intonation

In regard to intonation and pitch, women tend to employ five to six different variations and to utilize the higher registers when conversing. They are also more likely to employ glissando or sliding effects between stressed syllables, and they tend to talk faster as well. Men tend to be more monotone, employing two to three variations on average, most of which hover around the lower registers.[11] Even when trying to emphasize a point, males are less likely to employ melodic extremes but instead tend to speak louder. Perhaps this is why men are perceived as more likely to bellow, roar, or growl, whereas females will shriek, squeal, and purr.

Unfortunately, when mixed-sex pairs are conversing, men who use the lower register are often perceived as lecturing, boring, unfriendly, or overly authoritative, at least by women. Women tend to employ the upper registers, which are more likely to express friendliness and receptivity, and even sexiness, even when that is not the intent.

Because some females more than men tend to employ extremes and greater shifts when employing pitch they are sometimes seen as more emotional, hysterical, or overdramatizing what they are trying to convey, that is, to a man. This in turn is a function of the manner in which the melodic–intonational speech centers of the right brain are organized in men versus women. Moreover, men are likely to perceive these emotional or friendly overtures, such as smiling, as sexual, and then act on it, much to the surprise of the female who truly was only being friendly. However, these differences in perception are not a function of "linguistic style" but sex differences in the organization of the brain and mind.

For example, just as there is an auditory association area in the left temporal lobe which analyzes language, the association area in the right temporal region analyzes melodic and emotional sounds including pitch. Women, it turns out, are more sensitive to these cues, and their right temporal lobe (and limbic system) is far superior and sensitive as compared to that of men in discerning and extracting these paralinguistic variables and in correctly identifying their meaning. In consequence, they not only employ these emotional variables more in speech, but expect to hear them as well from others including men.[12]

On the other hand, since men are not so well endowed, although they produce and perceive pitch and melodic variations, they are not as adept or as sensitive at employing or interpreting them and are not as accurate in discerning their meaning. Instead, they tend to talk louder or softer.

These sex differences in language use and sensitivity gives rise to considerable difficulties as men and women often misinterpret these cues when employed by members of the opposite sex. Many women complain that men seem overly crude or do not seem as emotionally invested or willing to share their feelings, as men tend to underutilize these prosodic speech variables.[13] In contrast, a man may view the woman as unprofessional, emotionally unbalanced, or overly emotional simply because he is overwhelmed by these intonational, contextual cues when she provides them. He is not as well equipped as women to process this information.

Not surprisingly, these cognitive and other sex differences in the organization of the brain[14] can give rise to a considerable degree of misunderstanding. Moreover, because of this reduced sensitivity (at least as compared to women) when men use emotional speech, such as in swearing or in discussing sex, they tend to utilize a greater degree of intensity, or rather, crudeness, especially insofar as some women are concerned. Being comparatively less sensitive, males use a greater degree of intensity to make an impact because they themselves require it. Hence, they naturally assume that their listeners are similarly limited in their abilities to appreciate such nuances, when in fact women are much more sensitive. Sometimes, in consequence, what is normal banter between males is viewed as insensitivity and even verbal sexual harassment by a female recipient, who is more sensitive to all such social and emotional displays.

It is also, at least in part, because the female right cerebral hemisphere is organized in a superior fashion in regard to social and emotional language functions that women are more resistant to the effects of brain injuries and developmental disorders when it comes to speech.[15] That is, given similar-sized left-cerebral injuries, males are more likely to become aphasic, will show greater language impairment, and will experience less recovery as compared to females. Females have more right cerebral space devoted to emotional language functions such that when the left hemisphere is injured, they can continue to perceive and express social–emotional paralinguistic nuances which in turn enables them to continue to comprehend, or at least to appear to comprehend, even in cases where in fact comprehension of the denotative aspects of speech is more or less lost. Consider, for example, the aphasic woman who would smile and respond, "fine" to questions such as "how are you today?" or "It's raining outside?" simply because the melody used was identical in both.

In any case, we find that not only are there sex differences in the employment of pitch, intonation, and melody, but females also tend to employ more emotionally laden and personal terms and phrases as well as adjectives of adoration; for example, "hope," "cute," "precious," "wish," "feel," "love," "adorable," "sweet," "dear," such as in "Oh, she is just such a cute, precious, sweet, little dear."[16]

Sex Differences in Speech Production: Women Talk More Than Men

Women tend in general to be more grammatical, fluent, and articulate than men, a characteristic that is apparent beginning in early childhood.[17] They also tend to speak faster and are less likely to employ foul language, or empty space fillers such as "uh," "hmm," "umm" when speaking, although such interjections are often employed to indicate they are listening.[18]

It is also in part because women have more brain space devoted to language that they in fact talk more than men and are more likely to start spontaneous conversations with people, usually other women, they don't even know.

Over the last few years, some linguists and social psychologists have argued that men talk more than women, and this is true in some situations. However, in many others it is clear that females are much more talkative and much more willing to share information with friends, family, their husbands, or even total strangers. Go to any nightclub, sporting event, fair, circus, opera, symphony, or college campus and linger near the male versus the female restrooms, or the lines forming outside, and it is immediately apparent that numerous conversations are being held simultaneously among the women including many who do not know one another, whereas a pall of silence perhaps broken only by an isolated word or two characterizes the men. Amble over to where the pay phones are situated at such events, and we find that the majority are in use by females.

Certainly men like to talk, and sociolinguistic research has shown that in many formal or power-oriented situations such as at work or in a classroom with a female instructor, males tend not only to talk more but to interrupt others more frequently.[19] Many females are more reticent to speak up when in the presence of a strange male or group of males who the female does not know well, or in very power-oriented situations (as some law and medical-school classes tend to be), or in formal business meetings with an excess of males.[20] However, in a classroom with a male instructor, the willingness of the female students to participate dramatically increases and more often than not drowns out whatever contributions are made by male classmates, although the males continue to interrupt them. Moreover, once outside the class, females are more likely to congregate together as they leave and to engage in animated conversation about any number of subjects, whereas the men are less talkative and more likely to restrict their conversation to what occurred in class, or sports, politics, or sex, much of which is communicated in few words and short sentences.

Do women talk more than men? Based on my experience with a variety of populations in a variety of situations, it seems the answer is generally yes. Women in general tend to ask more questions, and are more articulate, socially and emotionally more adept, and forthcoming in the bargain.[21] This is why it is a common complaint among many women that their husbands and boyfriends do not share their feelings and seem more interested in watching sports, reading the paper, or staring at the TV than engaging in meaningful conversation. Conversely, this is why many men complain that their wives or girlfriends share information with friends and relatives that they believe should not only be kept secret, but which borders on betrayal as a breach of his trust. Men are much more reluctant to share feelings and personal details, whereas women find the failure to do so is unusual and maybe even annoying. Of course, there are numerous exceptions as some men, such as political commentators, sports announcers, and politicians, can go on and on and on.

Lecturing versus Sharing

For some men, one of the most valuable uses of speech is not for the purposes of describing and expressing their needs and inner feelings, but for concealing them. Indeed, many men prefer to keep their personal thoughts to themselves because, being so power-oriented and knowing other men are similarly inclined, they know that to reveal their innermost feelings puts them at risk. Other men might tease, laugh, or use the information against them, so as to establish dominance over them.

However, there are also many other contributing factors in regard to this reticence, including the presence or absence of certain hormones, evolutionary influences, sex differences in the structure of the limbic system, as well as social–cultural pressures. Some cultures so value a man who can hold his tongue that by time most reach 40 they are almost always silent, such as the Paliyans of Southern India.

Many men not only are less talkative but attempt to purposefully control what and how much they say. They not only seem reticent to speak but act as if they are actively censoring or inhibiting their thoughts, even among close friends. In contrast, many women can become highly animated and uninhibited in their speech production, particularly when among women.

In part this later difference may not only be due to differences in the amount of brain space devoted to language. The anterior commissure which connects the two amygdalas is much larger in the female than the male. This presumably allows for more emotional information to travel from one half of their brain to the other, at least insofar as emotional language is concerned. Moreover, the language centers of the left brain appear to have greater access to the language centers in the right brain in women versus men because their corpus callosum is larger as well. Hence, linguistically women appear not only more vocal but to be more emotionally in touch. However, they are also more likely to be subject to and influenced by these unconscious emotional concerns as well.

Tone and Intensity

The ability to communicate and to convey meaning is often reflected via changes in the melody, intonation, volume, and intensity of a person's voice. For example, becoming softer suggests the person is almost done and has run out of things to say. Becoming louder indicates he thinks what he is saying is important for you to hear, so please don't interrupt. The expression and perception of pauses and hesitations often communicates similar information to the listener.

Like melody, intonation, intensity, and pitch, loudness and the use of pauses and hesitations are affected not only by the linguistic community in which one is raised but how one's own family members spoke when one was growing up. Similarly, one's personality and emotional state are also significant contributors in the use of pauses and amplitude as well as one's feelings regarding his or her listeners. Intensity, volume, loudness, and amplitude, like other aspects of social and emotional speech, are greatly influenced and under the dominant control of the right half of the brain.[22]

Robert always became frustrated when talking with his mother because she tended to whisper, a tendency she had possessed as long as he could remember. In contrast, his father often seemed to shout, even when he was feeling emotionally neutral and was not upset. Robert, however, thought he had taken the middle ground in his vocal intensity. Nevertheless, he found it upsetting to talk with his mother because he had continually to ask her to repeat herself to which she would often state "never mind," or chastise him for shouting at her.

Again, although this could be seen as a stylistic difference in communicating, Robert could never shake the sense that his mother was being "passive-aggressive" and was purposefully whispering so as to control and strike out at her listeners. Her frequent refusal to repeat herself only confirmed the diagnosis insofar as Robert was concerned.

In fact, his mother was behaving passive-aggressively, a style that began soon after she met his father and which became only worse with time. His mother spoke quietly so as to set an example for her husband and his relatives who struck her as uncouth barbarians the way they always shouted at one another. Moreover, as her husband was a very forceful and loud individual, she found that by whispering she could often force him to converse with her on her own terms. That is, she used whispering to control him. Moreover, she had long learned that by whispering, she was often able to take center stage because this acted to make others think she was saying something secret and thus important of notice.

Intensity or loudness generally indicates the importance the speaker attributes to his or her words. Although one might whisper so as to keep important information confidential, generally raising our voice indicates we have something important to say or that we are not finished saying it yet, or that we are upset or are demanding a response. However, whispering often causes others to pay close attention as well for it sometimes alerts them to the fact that something secret and important is being discussed, maybe something about them. In general, however, becoming softer often indicates a person has no more to say or he or she feels embarrassed by what he or she is about to say. It can also be a sign of respect or determined by the situation in which one speaks.

In any case, intensity and amplitude, like melody, generally reflect the mood state and attitude of the speaker's right cerebral hemisphere and limbic system. Indeed, sometimes these nuances in fact betray feelings and emotions the party would rather keep hidden, or which he or she does not even know he or she possesses. One can inadvertently and thus unconsciously draw attention to certain words or phrases by the intensity one uses to convey them.

Similarly, one's own emotional state and right- and limbic-brain mental system will greatly color one's perceptions. Sometimes this prevents us from catching the drift of what another person is saying. Or, the listener may impose his or her own secret hopes, desires, or fears on other persons' statements.

Many relationships and conversations are influenced by mood, self-esteem, and similar misunderstandings. Although it would make life simple if factors such as indirectness, intensity, melody, and so on were in fact merely representative of "conversational style," it just isn't so, as they do not comprise the whole picture. Indeed, the words we use, our pauses, hesitations, tendencies to repeat ourselves, often have nothing to do with "conversational style" and everything to do with the modeling of our parents, unconscious emotional problems, and even the manner in which we were raised.[23]

Personality, Complexes, and Unconscious Speech Production

Communication between humans is multifaceted, being expressed via gesture, odor, melodic tone, facial expression, touch, and so on. Conversational style is also affected by numerous variables, including the manner in which one is raised, one's conscious and unconscious self-concept, context, the words used by others, their listening needs, and what is required in order to respond to what has just been said.

Sometimes when someone utters certain seemingly neutral words she or he may trigger an emotional reaction in her or his listener which in turn influences the words he or she may employ in response. Sometimes neutral words will also be misinterpreted by a listener depending on his or her self-image or the image he or she has of the person who is speaking. In fact, these underlying emotional states and beliefs can affect even what we read.

For example, while an undergraduate student, I one day showed up at school carrying a book titled "**word** power" (I was attempting to improve my vocabulary). Nevertheless, although the title was in large print and easy to read, many of my friends and acquaintances, in noting the book, joked about me trying to take over the "**world**." That is, these individuals perceived the title as "**world** power." I can only assume this perception was influenced by their erroneous presumptions regarding my personality and goals in life.

Emotionally significant words are also frequently misinterpreted or are perceptually deleted so that a listener in fact does not hear them, depending on his or her self-concept, personality, and underlying emotional state. This is also why certain words that are invested with emotion are replaced by euphemisms which essentially are indirectly meant to mean what the taboo word had implied. Thus, a lady might perspire or menstruate, whereas a waitress might sweat or bleed.[24]

Oddly enough, like *sweat* and *menstruate*, many taboo words refer to the body or body functions such as defecating. However, the only purpose served in applying euphemisms in these instances is to escape guilt or prevent embarrassment, as if it were abnormal to relieve oneself, or worst of all, engage in sexual intercourse, which is why many people say "sleep with," "hit the hay," "make love," and so on.

Being related to the body and invested with emotion, what are known as "dirty words" fall within the linguistic repertoire of the right hemisphere and limbic system. It is the language axis of the left half of the brain which invents euphemisms so that it may engage in a bit of self-deception and confabulation as well as wrest control over certain acts and functions which it has learned to associated with embarrassment. This, however, is a consequence of the process of socialization. In many societies, parents put considerable effort into forcing a child to learn to control his body and to put body waste in some receptacle, and in consequence instill guilt and shame. So that the left half of the brain can escape the shame it has been taught to feel and since it deals in words, it invents new words in place of the bad words so as to protect itself from feeling guilt or embarrassed.

It is interesting to note, however, that over the course of recent history, many euphemisms invented to replace nasty words have in turn eventually become equally undesirable and in fact identical with the words they were meant to replace. For example, the baddest of bad words, the *F-word*, at one time was a euphemism whose original meaning may have been "to knock" or "to pound repeatedly." The *F-word* was originally an indirect and essentially neutral replacement for what long ago were consider the baddest of bad words—jape, sard, and swive—all of which directly referred to the sex act.

Thus, such as with the F-word, euphemisms often become replaced by euphemisms in a never-ending repetitive cycle of left hemisphere induced linguistic deceptions, all so it can pretend that the word or act being referred to does not in fact exist. I would presume it possible that at some point in the future the original banished word could be reintroduced as a euphemism in which case one might employ "swive" or "sard" when discussing the sex act. In any case, the euphemism, when employed in most conversations, is used as a protective defense, for humor, in acts of seduction, and when in the midst of polite company.

However, depending on one's personality and emotional upbringing, bad words can be so highly disturbing and offensive that in consequence one might in fact fail to hear such language and will in fact filter it out when uttered in his or her presence. In experiments performed almost 40 years ago, student volunteers were told they were participating in a perceptual experiment where they were merely to report the words flashed on a screen.[25] When words such as "dog, cat, chair, cloth," and so on were viewed, students were able to accurately report them with very short viewing times, usually less than 150 msec. However, when words such as "bitch, fuck, fart, bastard," and so on were presented, viewing times for recognition not only doubled, but the majority of subjects claimed to have seen a nonemotional neutral word, "branch, friar, Fred, basket," and the like. Unfortunately these experimenters did not explore why these particular euphemisms were invented, or what the association, say between "fart" and "Fred," might be.

Experiences such as these are not limited to the laboratory but can characterize conversations held between friends, business associates, and lovers. People fail to remember hearing certain information or instructions, claim others made statements they swear they did not say, or will deny having made certain statements or used certain expressions. For example, in 1992, Governor Cuomo, of New York, while being interviewed on television, once referred to former Vice President Quayle as "Danny the cabin boy." The next day when he was confronted, he adamantly claimed that no such words passed his lips, although his words were in fact preserved on videotape. He was, in effect, confabulating. Married couples have similar experiences.

Interestingly, although it is well known that that which is emotionally or motivationally significant is better retained in memory than the mundane or emotionally neutral, how one reacted and what one might have said when in the throes of passion, often is not. Somehow the verbal memory centers fail to store these details. In fact, things said in the heat of the moment are not only forgotten (at least by the language-dominant half of the brain) but when confronted later the accused may actively deny having produced such language or having said such things. Again, this is a form of confabulation and is due to a purposeful disconnection between the language axis and the limbic system and the right half of the brain.

The Unconscious and Word Associations

Sometimes people think about saying something, say it, and then later fail to realize they gave expression to their thoughts. Sometimes the underlying feelings and motives which triggered their speech are in fact unconscious, and they say things which hint at what is unsaid and secret. Turmoil within the unconscious can give rise to outbursts, slips of the tongue,[26] and other unfortunate statements which a person consciously might not realize he or she said. For example, the politician who says: "A vote for me is a vote for dishonesty." Insofar as these influences are unconscious they are mediated by the limbic system or the right half of the brain. Although these structures do not control or produce the denotative aspects of speech per se, they influence the choice of words indirectly during stages of emotional arousal.

Almost 100 years ago, Carl Gustav Jung began exploring these unconscious influences on speech production and perception.[27] One of the methods he helped to develop and subsequently employed included the use of the word association test, a version of which was used by Galen and experimentally explored by Wilhelm Wundt during the 1800s. In a word association test, an examiner presents a list of words, and the subject is supposed to answer as quickly as possible, saying the first thing which pops into his or her mind.

Jung, in fact, developed a list of 100 words based on a formula constructed after many years of experimentation. He chose these stimulus words based on his beliefs and experiences regarding the unconscious and the existence of "complexes." According to Jung, complexes "slip the wrong word into one's mouth, they make us forget the name of the person one is about to introduce, they tickle the throat just when the softest passage is being played on the piano at a concert, they make the tiptoeing latecomer trip over a chair with a resounding crash. They bid us congratulate the mourners at a burial instead of condoling them and they are the actors in our dreams whom we confront so powerlessly "[28]

Jung developed and used the association test because he believed that words represent condensed actions, situations, and things. Hence, when requiring someone to respond to a certain single word he is in fact being presented with many options and questions such as "How do you behave toward it? What do you think of it? What would you do in this situation? How do you feel about it?" Words act on and elicit reactions from people much in the same manner as do much more direct and in depth personal and emotional questions.

Jung felt this test was best with those who were not overly educated or in love with words for they tend to produce linguistically deep-rooted associations which are often unrelated to emotional disturbances, whereas the less well-educated form highly revealing associations. He also found that people who are lacking in creativity or a deep emotional inner life tend to provide definitions, whereas some intelligent people with an inferiority complex tend to give flowery, overdrawn responses such as in response to a kiss: perception of friendship; and anxiety: heart anguish.

These responses, he felt, usually betrayed inferiority complexes. Similarly, those who display excessive emotional expressions are sometimes trying to overcome or compensate for an emotional deficiency of some sort.

Sometimes, a flood of words is produced which indicates how easily associations are formed and can be attached to the stimulus word when it touches on and activates an underlying and hidden unconscious complex and emotional disturbance. Consider Bob, who breaks up with Lisa and then wherever he goes he sees cars, colors, clothes, hairstyles which remind him of her. His feelings attach themselves almost indiscriminately to even neutral stimuli in their attempt to make their presence known.

In addition, by examining a person's mean reaction time, his hesitations, pauses, memory lapses, repetitions and perseverations, as well as his facial expressions, body movement, gestures, coughing, stammering, laughing, or movements of the body, hands or feet, and so on, Jung was also able to make certain deductions regarding his patient's personality and any emotional trauma affecting him. For example, in one case, Jung gave the test to a patient who showed disturbances to the words *knife, lance, beat, pointed,* and *bottle.* Later this fellow admitted to Jung that while drunk he got into a fight and used a knife in self-defense and had killed a man many years before.

In general, the longer a person waits before responding to a word such that reaction time is prolonged is of particular importance in revealing some underlying emotional disturbance. As all patients do not respond the same, Jung would determine each individual's mean response time by averaging the time to react to all 100 items per subject. Those reactions that exceeded the average time were of the most significance.

Jung also took note of unusual and suspicious phrasing of the reaction, particularly if they were in the form of sentences. Some subjects would also respond to certain words as if they were a question, or they would forget or misinterpret and claim to have heard a word that was not on the test, sometimes providing highly potent associations in the process. In other cases, patients would simply draw a blank and fail to respond altogether.

According to Jung, when the stimulus word is immediately forgotten or there is a failure to respond, this is indicative that the word is very upsetting and close to the heart of the complex. Hence, by hesitating or forgetting, it is almost as if the potential response was immediately sucked down into the complex and disappeared into the depth of the unconscious. However, sometimes the stimulus word causes a delayed reaction such that the next word presented elicits the unusual response or omission. For example, if the word "mother" was upsetting, the emotional contagion associated might instead attach itself to the next word on the list. Indeed, a word that triggers a complex may in turn result in contamination of the next several stimulus words so that each of them are affected in some manner although unrelated to the complex.

Jung also utilized a third method for discovering the complex and that is the "reproduction method," in which the subject is asked to recall the specific words he had associated with the original stimulus word. For example, if he said "death" to the word "mother," later Jung would ask him to recall the word he had used. If it was forgotten, this was further evidence of the existence of an underlying complex. For example, if his mother was in fact alive, the underlying disturbance may be an unconscious wish for his mother's death.

Jung also noted that sometimes the same response was given to several different words, which he referred to as *perseverations*. According to Jung, perseverations, where the same response word is repeated to various stimuli also shows the strength of the complex to attach itself almost indiscriminately as it seeks to make itself known, like the forlorn lover who cannot keep thinking about his lost love. Perseverations and repetitions also appear in regular speech which sometimes, too, indicates a complex, desire, guilty conscience, or some other psychic disturbance. Nevertheless, as I have discussed in detail elsewhere, the need to repeat is a fundamental aspect of learning and often may have nothing to do with hidden emotional needs or disturbances.

In real life, sometimes a complex or emotional disturbance will become associated with certain words which, although seemingly neutral and unrelated, are in fact linked to an underlying disturbance. Moreover, sometimes a person will draw attention to an underlying emotional problem by using or repeating certain words or phrases which, when linked together, hint at its existence.

For example, Peter and Jessica had been dating for several months and were thinking about getting married. One evening, however, she called him rather late and although he had not asked about her whereabouts, she informed him that she and her sister had shopped all day at a particular mall. Although she didn't mention it, he recognized that the mall was located in a town just a few miles away, Sunnyvale. Two days later she spontaneously mentioned that she was going to play bingo (an activity she had never before engaged in) with a friend and absently commented that the bingo parlor was in Sunnyvale. She mentioned the mall again a few days later and in the same breath told him that she and her sister were going to visit a new friend of her sister's. Later she mentioned that her sister had found a new boyfriend. Several days later, she indicated she was thinking about joining a health spa which he realized, without her saying it, was in Sunnyvale.

Later, after they broke up, Peter discovered that she had been seeing another man who, you guessed it, lived in Sunnyvale. In this case, although she never gave a direct hint as to what she was up to, she had nevertheless communicated to her boyfriend that something significant was occurring in her life, in the city of Sunnyvale, and that somebody was having an affair. What was significant to her had managed to intrude upon and contaminate her speech, although it was her desire to keep her true actions secret.

During a long and lengthy discussion with a married colleague, I noted that she used the words "guilt," "angry," "sex," "affair," "lonely," and "shame" several times, sometimes out of context. When I finally pointed this out and began to question her she reluctantly admitted that she was contemplating having an affair. However, I suspected that she had already done so based on what else I heard. In either case, it was also clear that she felt guilty about her intentions or her actions.

In general, often the complex or source of emotional discomfort remains hidden due to the actions of the conscious mind and the language axis of the left half of the brain, or the unconscious aspects of the ego which are under the jurisdiction of the frontal lobes and the right half of the brain. By their very nature, complexes and the unconscious, insofar as their contents are nonlinguistic, are part of the mental system that we associate with the right half of the brain and limbic system.

In that the complex is actively prevented from achieving verbal representation by the left half of the brain, frequently all attempts to explore the complex are generally rebuffed. In these cases the person may actively deny or attempt to mislead so that it will not be discovered unless it repeatedly intrudes on consciousness and affects behavior. In these instances, one might resort to the use of "free associations" such as was pioneered by Freud, so as to get at the heart of the disturbance.

On the other hand, by carefully listening to others when they speak, their repeated use of certain words or phrases, as well as gesture, facial expression, and the melody and volume they employ, one can often make quite astute guesses as to what a person wants to say, but cannot, and what he or she prefers to keep hidden. Similarly, a person's reactions to what others say, just as in Jung's word association test, can convey volumes as to his or her underlying thoughts and feelings. Indeed, what a person says, and even what he or she fails to say, including hesitations, are a virtual treasure trove of information.

Pauses and Hesitations

People often hesitate to make certain statements or hesitate in reaction to certain words or questions asked by others.[29] This in turn indicates that the stimulus is very important, personal, or highly significant. This is why Jung placed great store in reaction time in revealing the presence of an emotional disturbance.

Hesitations are also meaningful in everyday speech. Individuals may hesitate because of embarrassment, confusion, or because something is emotionally or even intellectually significant to them. When we hear things important to us this is often accompanied by physiological changes in the body such as preparedness to take some action, which in turn is reflected by a moment's hesitation. We are still geared to physically respond to emergencies. When the hesitation is prolonged, then more effort is being put into creating and preparing a defense.

Most individuals indicate they are done speaking by pausing or ceasing to speak. This gives a second conversant a chance to speak her or his mind. Sometimes people pause before they are done which may permit a second party to jump in. Many people also hesitate in order to avoid making definite statements or to put the burden on another.

If a listener jumps in, this is determined not only by the urgency with which he or she wishes to speak but by his or her own expectations as to what pauses imply. For example, an individual who frequently inserts long pauses as he gathers his thoughts, may not only have his conversation interrupted but may forever wait for his chance to speak as individuals not so inclined may resume talking or see this as their opportunity to get their three cents in.

For example, Jack is a vice president of his software company and while at work his employees are used to his long pauses as he gathers his thoughts and wait patiently for him to signal them via gesture or subtle changes in posture or facial expression when they may speak. Jack knows this frustrates some of his employees. However, he basks in his power and the respect they must demonstrate. His use of the pause in these situations in fact is a reflection of that power as well as the underlying feeling of insecurity he must feel that causes him to flaunt it.

When Jack returns home he feels that his beautiful, young wife completely dominates him and does not respect him. She not only constantly interrupts but complains that he is a lousy conversationalist who either has nothing to say or who will not share his feelings or his thoughts. In contrast, Jack feels he cannot get a word in edgewise and has complained to her as such. However, he wife states that whenever she says anything he simply stares at her so she just starts in again for he apparently has nothing to say.

The problem is that Jack uses long pauses and his wife uses short pauses. However, Jack also feels unsure of himself with his wife. Hence, his pauses at home reflect insecurity, whereas at work they reflect his careful consideration of facts, or at least so he consciously believes. Although the need to be careful in organizing one's thoughts may also reflect insecurity at work, insecurity rules him at home as he has become convinced that his wife has no interest in anything he has to say. This in turn erodes his sense of love and security with her, and makes him hesitate that much more, which erodes her respect for him. This is probably why he flaunts his power at work, so as to make up for those missing strokes at home.

Fortunately, Jack wisely sought counseling for his difficulties and realized that his use of pauses was not some reflexive style of communicating that he had inherited but was a manifestation of his personality and fears. After a few sessions of assertiveness training, he was able to overcome this difficulty, much to the delight of his wife.

Parental Influences on Speech Production

The psyche of the child is tremendously influenced and molded by his or her parents. Children mimic and model what they observe, including acts of verbal abuse or physical violence. These influences extend well beyond whatever direct influences parents purposefully or inadvertently exert and includes their facial expression, touch, tone of voice, and gestures, and the manner in which the parents treated each other.[30]

Children model, mimic, and incorporate what they see, hear, feel, and even infer, and these influences can last a lifetime. Even the superficial aspects of the intellect and the mind are subject to these forces, as well as other important functions such as wishes, hopes, intentions, and emotions.

As pointed out by Jung, one can even incorporate and model the emotional inner life and turmoil of a parent such that a young boy can act like an old man, bitter and disappointed with no hope for the future, or a young girl can give up and accept the fate of an unschooled mother who is abused and battered and then live her life accordingly.

> What passes over from the mother to the child? [Jung asks.[31]] It is not the good and pious precepts, nor is it any other inculcation of pedagogic truths that have a molding influence upon the character of the developing child, but what most influences him is the peculiarly affective state which is totally unknown to his parents and educators. The concealed discord between his parents, the secret worry, the repressed hidden wishes, all these produce in the individual a certain affective state with its objective signs which slowly but surely, though unconsciously, works its way into the child's mind, producing therein the same conditions and hence the same reactions to external stimuli. . . . The father and the mother impress deeply into the child's mind . . . which is as soft and plastic as wax . . . the seal of their personality.

Parents not only shape personality and pass on complexes and personal shortcomings but greatly influence the style and content of what their children think and how they use language. The manner in which a child speaks in turn is a reflection of how the child and the parent feels. As such, children sometimes incorporate and feel as their parents which is reflected in their choice and use of words. In fact, when Jung gave these word association tests to the same members of dysfunctional families, many of the associations produced by family members were the same.

For example, Jung described a family that consisted of an alcoholic father, his 45-year-old wife, and their 16-year-old daughter. In the word association test, Jung noted that the responses of the daughter were extremely similar to that of her mother, and 30% were in fact identical. In other words, she suffers from many of the same complexes and emotional upheavals. According to Jung, given the close similarities in their reactions, such a girl comes out into the world much like her 45-year-old mother with the same kinds of thoughts and ideas and will search for a "man who is alcoholic and marry him; and if by chance he should not be one, she will make him into one."[32]

Due to the incorporation of these parental patterns of behaving and thinking (which elsewhere I have discussed as a complex of associations called the unconscious parent), upon reaching adulthood, many people may remain attached unconsciously to one or both of their parents (or perhaps to an overbearing older sibling) and thus always seek that parent's love, or someone who matches up to their parent ideal even if it is negative and harmful.[33] If perchance they meet someone who cannot meet these expectations, they quickly become dissatisfied and end the relationship.

Parental power guides the child when he or she is small and continues to exert formidable influence after the child has reached adulthood. However, these influences, like other aspects of emotional experience, remain under the guiding domain of the limbic system and right half of the brain, and if not expressed via direct actions, may be conveyed through right-hemisphere and limbic influences on gesture, facial expression, and spoken language.

10

The Universal Language
Music, Limbic, and Right-Hemisphere Speech

Origins of Speech

Human speech, like a musical composition, consists of many elements, some of which are under the controlling influence of the left half of the brain, and others being mediated by the right brain or limbic system. Speech consists of more than vocabulary and grammar, but emotion, melody, pitch, and other paralinguistic nuances. What we recognize as human speech is the fusion of these different elements so that what we say is grammatical, emotional, melodic, and so on.

The successful fusion of these different elements probably did not arise until the appearance of our *Homo sapiens sapiens* ancestors, the Cro-Magnon, about 130,000 years ago in North and South Africa. The Neanderthals of Europe and the Middle East, who essentially appeared during the same time frame as the Cro-Magnon, had throats and mouths that would not support the production of complex speech patterns.[1] These differences, which include bone structure, height, cranial capacity, the length of the arms and legs, and even life span, may in turn be a function of the Neanderthal and Cro-Magnon races descending from different branches of the human family tree. Nevertheless, for the next 60,000 years the Neanderthals shared the planet with the Cro-Magnon people until dying off and possibly being exterminated by the superiorly endowed Cro-Magnon, who began invading Eurasia some 40,000 years ago.

Be it modern humans, Cro-Magnon, or their ancestors, speech has always been firmly moored by its limbic and melodic roots to the emotional core of our being.[2] Similarly, among human infants, the first sounds produced are also limbic in origin and are expressive of mood, emotion, and homeostatic needs.[3] These first vocalizations are typically characterized by variations imposed on pitch, amplitude, and the melodic qualities of the voice. It is only with the development of a second babbling stage, at around 3 to 4 months of age, that the left half of the brain begins to mature and to impose temporal sequences and syllabication on these right brain–limbic intonational contours and prosodic vocalizations.[4] As such, left-hemisphere speech comes to be superimposed over right hemisphere and limbic speech.

Obviously, the creation of modern speech involves more than stamping units and sequential patterns on the prosodic and melodic inflections of the right cerebrum so as to produce what we recognize as grammar and vocabulary. Not just sounds, but words, too, have their origins.

It is likely that long before human beings held their first conversation and long before Eve held that fateful discussion with the serpent that the sounds of speech consisted predominantly of barks, grunts, screams, moans, coos, whimpers, clicks of the tongue, as well as imitative sounds such as those of the animal world. In fact, the ability to "ape" and mimic animal and environmental sounds could well have led to the formation of the first denotative words. If someone makes the sound of a wolf or a tiger, the idea of a wolf or tiger is thus conjured forth in the mind of friend, foe, and stranger alike, whereas a scream or a cry is more likely to indicate simply an emotion, which in itself is highly communicative.

Not surprisingly, the names of at least a few animals are derived from the sounds they make, and in many instances their cries and calls are similarly labeled. This includes, for example, the cuckoo and the whippoor-will, and words such as "buzz," "purr," and "chirp," which are essentially the verbal reproduction of an animal sound. Of course, this only accounts for a fraction of the words in our vocabulary and is reflective of our amazing ability to mimic and intimate. Moreover, such as in the case of the whippoor-will, the presence of language often shapes the sounds we hear and mimic.

Limbic Language

There are many aspects of speech and spoken language which are unrelated to mimicry but appear instead to be emotional in origin and related to and similar to the sounds produced by many animals. These emotional sounds constitute the foundations of what we have referred to as right-hemisphere and limbic speech. However, in regard to its temporal sequential and grammatical functions, spoken language in many ways falls within a completely different category which is separate from the agrammatical sounds produced by beasts and insects.

Nevertheless, the very origins of spoken language are emotional in origin and are a product of certain nuclei within the limbic system, the amygdala, and the cingulate gyrus in particular.[5]

The Innate Languages of the Limbic System

It has been argued in detail elsewhere that phylogenetically and ontogenetically, the original impetus to vocalize springs forth from roots buried within the depths of the ancient limbic lobes (e.g., amygdala, hypothalamus, septum, cingulate gyrus). For example, although nonhumans do not have the capacity to speak, they still vocalize, and these vocalizations are primarily limbic and emotional in origin, being evoked in situations involving sexual arousal, terror, anger, flight, helplessness, and separation from the primary caretaker when young.

The first vocalizations of human infants are similarly emotional in origin and limbically mediated, consisting predominantly of sounds indicative of pleasure and displeasure. Indeed, these sounds and cries are produced soon after birth, indicating they are innate and are produced even by infants who are born deaf and blind.

Similarly, apes and monkeys reared in isolation or with surgically muted mothers utilize complex and appropriate calls in order to convey a wealth of information, including the presence of danger, and will respond to these same calls with appropriate reactions, even when they had never before been heard. It is important to note, however, that the production of these sounds is not completely involuntary for learning and even cultural-geographic factors can influence their production. That is, the infants' initial emotional sound production appears to convey generalized meanings which are context-dependent and which gradually become shaped and tied to specific mood states or events.

For example, as demonstrated by Cheney and Seyfarth, vervet monkeys employ three distinct calls so as to signal the presence of eagles, snakes, and leopards. Experienced adults respond to these calls by looking up, looking down, or climbing up a tree, depending on which call is produced, even when played from a tape recorder. However, infants reared in isolation merely respond with generalized alarm and are as likely to look up as look down as climb a tree. Squirrel monkeys reared in isolation also respond with fear and anxiety in response to warning "yapping" calls when they are first heard, and will produce an appropriate "yapping" cry when they are first exposed to a potential predator.

Hence, it is apparent that many of these emotional sounds are not learned. Moreover, due to the tremendous immaturity of the neocortex during infancy, it is also apparent that these initial emotional sounds are produced by the limbic system. In fact, the semi-independence of the limbic system from the neocortex in the production of these cries and calls has also been demonstrated in adult humans as well as among nonhuman primates with neocortical injuries involving areas known to be responsible for other aspects of language production.

For example, destruction of the left frontal region in the brain of a squirrel monkey has little or no effect on the vocalization rate or the acoustic structure of their calls. Lesions involving the superior temporal lobe and the primary auditory area of the squirrel and rhesus monkey and the macaque significantly affect their ability to differentiate and discriminate between complex sounds and species-specific calls. However, the capacity to detect and recognize them is not affected. Interestingly, left temporal destruction is more effective in disrupting discrimination than right-sided injuries in these primates. Again, however, the ability to recognize or to detect the call is not affected.

Limbic Localization of Sound Production. Emotional cries and warning calls have been produced via electrode stimulation of wide areas of the limbic system, and these same areas often become activated in response to certain emotional sounds. Indeed, the limbic system is more vocal than any other part of the brain. In contrast, motor areas are not associated with sound production or perception.

Nevertheless, the type of cry elicited, in general, depends upon which limbic nuclei have been activated. For example, portions of the septal nuclei, hippocampus, medial hypothalamus, and the periaqueductal gray have been repeatedly shown to be generally involved in the generation of negative and unpleasant mood states, whereas other limbic tissues, including the lateral hypothalamus and amygdala, and portions of the septal nuclei are associated with pleasurable feelings. Not surprisingly, areas associated with pleasurable sensations often give rise, when sufficiently stimulated, to pleasurable calls, whereas those linked to negative mood states will trigger cries of alarm and shrieking.

However, as pointed out by Jurgens, the threshold for vocalization is higher than that for eliciting avoidance or positive approach behaviors or feelings states. Moreover, in the more recently evolved, four-layered transitional limbic cortex, the cingulate gyrus, completely different emotional calls can be elicited from electrodes which are immediately adjacent, and the calls do not always correlate with the mood state. This suggests considerable flexibility within the cingulate, which also appears to have the capability of producing emotional sounds that are not reflective of mood. This suggests a high degree of voluntary control within the cingulate.

Sound Production in the Amygdala. Although cries and vocalizations indicative of rage or pleasure have been elicited from a variety of limbic nuclei, the amygdala is the most vocally active, followed only by the cingulate. In humans and animals, a wide range of emotional sounds have been evoked through amygdala activation, including those indicative of pleasure, sadness, happiness, and anger. Moreover, the amygdala is tightly linked with other auditory areas (e.g., the tectum, medial geniculate of the thalamus, vestibular nuclei) so as to trigger startle reactions, such as in response to transient sounds such as those typically made by predators, prey, or potential mates. Indeed, from an evolutionary perspective it was probably the vestibular nucleus, the amygdala, the lateral geniculate nucleus of the thalamus, and a very primitive, vibration sensitive inferior tectum where the first neuronal rudiments of "hearing" and the analysis of sound took place.

Be it lower mammals or humans, the auditory system remains tightly linked with the limbic system as well as with the motor systems which enable it to alert the rest of the brain to possible danger and to then take appropriate action. Indeed, the amygdala, for example, continually samples auditory (as well as visual and tactual) events so as to detect those which are of emotional and motivational significance and when activated can trigger intense emotional expressions, including flight or fight reactions accompanied by appropriate vocalizations.

Conversely, in humans, destruction limited to the amygdala, the right amygdala in particular, has abolished the ability to sing, convey melodic information, or to enunciate properly via vocal inflection and can result in great changes in pitch and in the timbre of speech. Even the capacity to perceive and respond appropriately to social-emotional cues is abolished, and the animal ceases to show startle or emotional reactions to dangerous or friendly overtures.

Maternal Behavior and the Evolution of Infant Separation Cries

Fish and many other creatures who swim the shining sea are capable of producing sounds. Marine mammals, such as the dolphin and a variety of whales, in fact produce numerous informative and highly meaningful sounds. However, these, like most other mammals, first developed and evolved upon dry land and only later returned to live within the deep. Nevertheless, creatures who roamed the planet or swam beneath the waves some 300 or more million years ago, such as reptiles, also relied on sound as a means of communication, although the brain structures and external hearing apparatus they possessed were probably quite rudimentary, and not much more developed than that seen in modern-day frogs and lizards. As such, they were attuned to hear low-level vibrations and sounds, such as croaking, tails being thumped on the ground, and a few distress calls and sounds of contentedness. Moreover, most but certainly not all reptiles show little or no maternal care, rarely vocalize, and are capable of hearing at best only low-frequency vibrations. Infants generally must hide from their parents, and other reptiles, in order to avoid being cannibalized. One might presume that the first reptiles were not in any manner more advanced than their forebears in this regard.

When reptiles began to differentiate and evolve into the reptomammals (the therapsids) some 250 million years ago, and then many millions of years later, when reptiles again diverged and the first tiny dinosaurs began to roam the Earth, several major biological alterations occurred, including the development of teats for suckling young and alterations in brain size and organization, all of which coincided with tremendous advances in the capacity to engage in audio-vocal communication and to nurse their young. Indeed, it was not until the appearance of the therapsids that mammary glands came into being and the first rudiments of an inner and even a middle ear first developed.

It was probably at this time, some 250 million years ago, that sound first came to serve as a means of purposeful and complex communication, such as occurs not only between potential mates or predator and prey, but also between mother and infant. This ability in turn was probably made possible by the amygdala as well as through the evolution of a four-layered transitional neocortex, the cingulate gyrus. Indeed, it is the limbic system and the interactions of limbic nuclei such as the amygdala and the cingulate gyrus which stimulate the desire to communicate and form attachments, social groups, and eventually the family.

Indeed, the reptomammalian therapsids, and then later many of the various dinosaurs (who diverged from a different line of reptiles, the thecodonts), lived in packs or social groups, and presumably cared for and guarded their young for extended time periods lasting until the juvenile stage. As noted, the first appearance of rudimentary nipples coincided with therapsid development. Hence, one of the hallmarks of this evolutionary transitional stage, some 250 million years ago, was the first evidence of maternal feelings and what would become the family.

Mother–Infant Vocalization. Among social terrestrial vertebrates the production of sound is very important in regard to infant care, for if an infant becomes lost, separated, or in danger, a mother would have no way of quickly knowing this by smell alone. Such information would have to be conveyed via a cry of distress or a sound indicative of separation fear and anxiety. It would be the production of these sounds which would cause a mother to come running to the rescue. Conversely, vocalizations produced by the mother would also enable an infant to continually orient and find its way back if perchance it got lost or separated. Hence, the first forms of limbic social-emotional communication were probably produced in a maternal context.

Indeed, considerable vocalizing typically occurs between human and nonhuman mammalian mothers and their infants, and the infants of many species, including primates, will often sing along or produce sounds in accompaniment to those produced by their mothers. In fact, among primates, females are more likely to vocalize and utter alarm calls when they are near their infants versus nonkin, and vice versa, and adult males are more likely to call or cry when in the presence of their mother or an adult female versus an adult male. Similarly, infant primates will loudly protest when separated from their mother so long as she is in view and will quickly cease to vocalize when isolated. It thus appears that the purpose of these vocalizations is to elicit a response from the mother.

Hence, the production of emotional sounds appears to be limbically linked and associated with maternal-infant care, and with interactions with an adult female. In fact, human females in general tend to vocalize more than males and their speech tends to be perceived as friendlier and more social.

The Cingulate Gyrus.   As noted, most creatures, including sharks, amphibians, reptiles, and fish, possess a limbic system, consisting of an amygdala, hippocampus, hypothalamus, and septal nuclei. It is these limbic nuclei, the amygdala (as well as the later-appearing and more advanced mammalian cingulate gyrus), which are important in maternal care and infant bonding and attachment, and which are most highly developed in mammals and humans. It is the presence of these former limbic nuclei which enable a group of fish to congregate together (i.e., to school), or for reptiles to form territories which include an alpha female, several subfemales, and a few juveniles. Such creatures, however, although sometimes showing parental investment, generally do not care for their young and do not produce complex meaningful sounds, presumably because they are lacking a more recently acquired cingulate cortex.

Among humans and lower mammals, destruction of the anterior cingulate results in a loss of fear, lack of maternal responsiveness, and severe alterations in socially appropriate behavior. Humans will often become initially mute and socially unresponsive, and when they speak, their vocal melodic-inflectional patterns and the emotional sounds they produce sound abnormal. Animals, such as monkeys who have suffered cingulate destruction, will also become mute, will cease to groom or show acts of affection, and will treat their fellow monkeys as if they were inanimate objects. For example, they may walk upon and over them as if they were part of the floor or some obstacle rather than a fellow being. Hence, behavior becomes somewhat reptilian. Maternal behavior is also abolished and the majority of infants soon die from lack of care.

The cingulate cortex is intrinsically linked to the amygdala and is absent in reptiles and fully present only in mammals. Moreover, the cingulate in humans maintains direct interconnections with the left and right frontal lobes, neocortical regions which, among humans, are responsible for the production of speech and the emotional sounds which accompany propositional language.

Hence, the cingulate probably contributes to the setting of thresholds for vocalization, including modulating some of the prosodic and melodic features which characterize different speech patterns (e.g., happiness versus sadness).[6] More important, when the cingulate cortex is electrically stimulated, the separation cry, similar to that produced by an infant, is elicited. It thus appears that the cingulate, in conjunction with the amygdala (which are intimately linked) and other limbic tissue, may well be the responsible agents in regard to infant care and the initial production of what would become language. This has been referred to elsewhere as "limbic language" and "limbic speech." Moreover, although language has come to be hierarchically subserved and expressed by the evolutionary advanced neocortical tissues of the human brain, the limbic system has retained its imminence in the mediation, production, and comprehension of emotional-social sounds, including sex differences in their production.

Emotional Speech and Expressive Aphasia. As noted, if an individual were to sustain massive damage to the left frontal convexity, their ability to speak would be dramatically curtailed. However, although almost all aspects of expressive, grammatical speech can be greatly reduced, many patients with Broca's expressive aphasia are capable of singing, cursing, making emotional statements, and even praying. In fact, they are able to sing words they cannot say.

Destruction of the neocortical frontal regions corresponding to Broca's area in nonhuman primates also leaves the capacity to produce emotional sounds and cries intact, whereas limbic destruction alters or abolishes these capabilities. Similarly, among humans melodic-intonational disturbances in speech production have been noted with deep left frontal or deep right frontal lesions that appear to involve the anterior cingulate. Such patients sound as if they are speaking with an accent.

Emotional Comprehension in Wernicke's Receptive Aphasia. When the left auditory association area is damaged, there results severe receptive aphasia (i.e., Wernicke's aphasia). Although comprehension has been lost, patients with damage to Wernicke's area are usually still capable of talking (due to preservation of Broca's area). However, they are unable to understand what they are saying.

The reason the speech of a person with Wernicke's receptive aphasia no longer makes sense is that Wernicke's area also acts to code linguistic stimuli for expression prior to its transmission to Broca's expressive speech area. As such, what they say is often incomprehensible. Nevertheless, prosody and emotional tone often remain fairly normal.

Moreover, although comprehension for vocabulary and denotative speech is lost, in some cases patients may display euphoria, or in other cases, paranoia as there remains a nonlinguistic or emotional awareness that *something* is not right. That is, the patient's limbic system and right hemisphere, being undamaged, continue to respond to signals generated by others. Similarly, the ability to read, write, or respond to emotional words (as compared to nonemotional or abstract words) is also somewhat preserved among aphasics, and even those who have undergone left hemispherectomy. Again, this is because the right hemisphere and limbic system are intact and remain dominant for all aspects of emotion be they positive or negative in content.

Since these paralinguistic and emotional features of language are analyzed by the limbic system and intact right cerebral hemisphere, sometimes the aphasic individual is able to grasp in general the meaning or intent of a speaker, although verbal comprehension is reduced. This in turn enables him to react in a somewhat appropriate fashion when spoken to. That is, he may be able to discern not only that a question is being asked, but also that concern, anger, fear, and so on are being conveyed. Often our pets are able to determine what we mean and how we feel by analyzing similar melodic-emotional nuances. In fact, many individuals with severe receptive (Wernicke's) aphasia can understand and respond appropriately to *emotional* commands and questions (e.g., "Say 'shit' " or "Do you wet your bed?").

Hierarchical and Cultural Factors.   Limbic speech is not bound up with thinking, the expression of thought, or conscious reflection. Although communicative, limbic speech occurs essentially independent of thought as it is predominantly emotional and concerned with the immediacy of the "here and now" and reflects the emotional state of the organism (i.e., hunger, fear, alarm, anger, loneliness, sexual arousal).

In part, at least among humans, limbic speech is hierarchically re-represented, refined, and elaborated by right (as well as left) hemisphere language structures. Thus, right cerebral language is social, melodic, emotional, inferential, and highly communicative of meaning and intent due to its more extensive interconnections with the limbic system, including the cingulate and amygdala. It is via the right (as well as limbic) brain language system that certain sounds, tones, and melodies come to have specific meanings which are understood even by people from wholly different cultures, hence the phrase: "I don't know what they're saying, but I sure don't like the sound of it."

Some of the tones and melodies of one's voice might convey different meanings in different cultures, for melody, like certain gestures, can be culturally and environmentally modified. Nevertheless, for the most part, the natural melody of one's voice also comes closest to what might be considered natural speech, the condition of all languages before they came to be stamped by temporal sequences and organized grammatically and subject to cultural variation. The right hemisphere and limbic system in fact mediate a natural and universal language which in its most evolved form is recognized as music.

Melodic and Musical Speech

We are all familiar with the "blues" and perhaps at one time or another have felt like "singing for joy," or have told someone, "You make my heart sing!" Indeed, music is not only "pleasing to the ear" but is invested with emotional significance. For instance, when played in a major key, music sounds happy or joyful. When played in a minor key, music often is perceived as sad or melancholic.

This is why certain melodics in the absence of speech, gesture, or even visual stimuli or context are recognized across most cultures as conveying a particular meaning and why unlike real life, all movies and television shows are accompanied by "mood" music; that is, to enhance or convince us as to what emotion we should be feeling.

This is also why composers in creating their musical compositions may rely on certain melodies or sounds produced by certain instruments to create a particular mood, feeling, or even visual experience. For instance, one is able to express joy or even triumph via the ascending notes of the major triad (1.3.5). In contrast, the notes of the minor triad and minor keys are used to indicate depression, melancholy, and even pain. Certain melodic sounds have also been associated with the devil or the divine, and this is why many composers have employed the tritone so as to indicate Satan or demonic or sinister forces.[7] It is for this same reason that the Catholic church for many centuries forbade the use of the tritone or augmented fourth in any compositions.

The melodies produced by certain instruments also have the same effect. Violins and harps are used to indicate angels, and trumpets and trombones announce priests and kings. Trumpets and drums can also signify military force and the powers of nature and war as well as that of divine. Similarly, a choral tune in an organ prelude is often employed at the beginning of a piece so as to indicate the theme as being religious.

Via music, one can produce the sound of bees in flight, a storm brewing, a babbling brook, the thunder of galloping horses, the calls of birds, mammals, and even insects. In fact, although insects communicate via olfactory cues, like humans they also produce music and often for the same purposes; such as to convey one's emotional and motivational state including sexual availability or desire and as a prelude and accompaniment to war.[8]

Boogying with Bugs

Like his male human counterpart one way for a sex-starved male insect to win his lady love is by producing a love song. That is, he sings to or makes music for her. He must also compete with other insect musicians for her attention.[9]

The most common method for an insect to produce music is via stridulation, the rubbing together of body parts.[10] Crickets and katydids will rub a filelike ridge on one wing against a scraper on the other wing, like a fiddle. By rapidly rubbing their wings together they can produce a trill or a loud rhythmic chirp which may go on for hours. Grasshoppers perform similarly with the exception that they rub their back legs against their wings. In contrast, cicadas will drum up a buzzinglike sound by vibrating air-filled abdominal chambers called tymbals. Like a drum, they are covered with a thin membrane.

Many female insects, including grasshoppers and cicadas, in turn will listen to these songs via specialized sensillae which are distributed over the body and which are covered by thin membranes forming a very primitive device similar to an eardrum. Crickets and katydids, in fact, have two eardrums located on the front of each tibia.[11] Others, such as mosquitoes, have hearing organs located at the base of each antenna.

Among crickets, the males may sing to attract a female, and once they make contact, they will change their song so as to get her in the mood. Moreover, after they have finished copulating, the males will sing yet a different song whose purpose is to dissuade the female he has inseminated from leaving. In effect he sings, "baby please don't go. . . . I need you!" Interestingly, among singing insects, there are males who refuse to sing from the outset and instead lurk just outside the singing territory of yet other males so as to intercept females when they come a'calling.

Like their human counterparts, many male insects will collect together and sing in a chorus. Although there may be a large aggregate of singing males, each has his own singing territory which is located and defended right next to that of his rivals. The song in fact, often defines his territory.[12]

Not all songs are produced to attract females. In some species of insect, it serves as a form of aggression between males. Many singing insects, like male birds, use their songs to defend and to lay out the boundaries of their territories and personal space. Even so, other males may enter the territory and attack these musical insects, thereby preventing them from singing and attracting females. In this way the newcomer not only usurps the territory of his rival but garners a stage upon which he can sing instead. This fighting over singing rights in turn results in the establishment of a dominance hierarchy such that the most dominant males do most of the singing. In fact, they produce certain songs whose function is to maintain dominance relations within a group of like-minded males.[13] The fighting songs, however, are completely different from songs of love.

Similarly, human beings sing not only for love but for aggressive purposes. Songs can be used to mock and express hate, as is demonstrated by the fast-clipped rhyming that is called "rap music," or the almost completely amelodic "punk rock."

Until recently, in the Bavarian, German–Austrian region of Europe, "song duels" were common, which contained obvious aggressive meanings.[14] Typically, the opponents would alternate in their singing of stanzas and via song hurl insults at their competitor or sing of their own praises. Similarly, Eskimos engage in song dueling in which two disputants sing lampoons, insults, and obscenities at one another while surrounded by a delighted audience which shouts encouragement and eventually takes sides. In opera, song duels are quite common and are used to convey any number of ideas and feelings.

Music and Nonverbal Environmental Sounds

As noted, individuals with extensive left hemisphere damage and severe forms of expressive aphasia, although unable to discourse fluently, may be capable of singing, swearing, or praying. Even when the entire left hemisphere has been removed completely, the ability to sing familiar songs or even learn new ones may be preserved—although in the absence of music a patient would be unable to say the very words that he or she had just sung. The preservation of the ability to sing has, in fact, been utilized to promote linguistic recovery in aphasic patients and acquisition of speech by the undamaged hemisphere, that is, melodic–intonation therapy.[15]

Similarly, there have been reports that some musicians and composers who were suffering from aphasia or significant left hemisphere impairment nevertheless were able to continue their musical career.[16] In some cases, despite severe receptive aphasia and although the ability to read written language was disrupted, the ability to read music or to continue composing was preserved.

One famous example is that of Maurice Ravel, who suffered an injury to the left half of his brain in an auto accident. This resulted in apraxia, agraphia, and moderate disturbances in comprehending speech, that is, Wernicke's aphasia. Nevertheless, he had little difficulty recognizing various musical compositions, was able to detect even minor errors when compositions were played, and was able to correct those errors by playing them correctly on the piano.[17]

Conversely, it has been reported that musicians who are suffering from right hemisphere damage (e.g., right temporal–parietal stroke) have major difficulties recognizing familiar melodies and suffer from expressive instrumental *amusia*. Even among nonmusicians, right hemisphere damage (e.g., right temporal lobectomy) disrupts the ability to perceive, recognize, or recall tones, loudness, timbre, and melody.[18] Right hemisphere damage also can disrupt the ability to sing or carry a tune and can cause toneless, monotonous speech, as well as abolish the capacity to obtain pleasure while listening to music, a condition also referred to as *amusia*. Similarly, when the right hemisphere is anesthetized, the melodic aspects of speech and singing become significantly impaired, and the volume and amplitude of speech is reduced as well.

Penfield and Perot have reported that musical hallucinations most frequently result from electrical stimulation of the superior and lateral surface of the temporal lobes, particularly the right temporal region.[19] Verbal hallucinations are most likely from left temporal lobe stimulation. In addition, the right hemisphere has been found to be superior to the left in identifying the emotional tone of musical passages and, in fact, judges music to be more emotional as compared to the left brain.[20]

I have postulated elsewhere that musical production is an outgrowth of the limbic system and related to the ability to mimic environmental sounds.[21] For instance, in addition to music, the right hemisphere has been shown to be superior to the left in discerning and recognizing nonverbal and environmental sounds.[22] Similarly, damage within the right hemisphere not only may disturb the capacity to discern musical and social–emotional vocal nuances but may disrupt the ability to perceive or recognize a diverse number of sounds which occur naturally within the environment such as water splashing, a door banging, applause, or even a typewriter in addition to emotional sounds and music in general.[23]

Findings such as these have added greatly to the conviction that the right cerebral hemisphere is dominant for most aspects of environmental and musical perception and expression. It is this relationship which in turn provides the foundation for the development of right hemisphere speech which is completely different from that employed by the left.

The right half of the brain is also dominant in regard to other aspects of communication, such as in discerning certain smells, the reading of facial emotion, and the comprehension of body language, including the manner in which one has been touched.[24] Hence, any variations in these gestures and paralinguistic variables generally reflect the mood, state, and attitude of the right hemisphere and limbic system. Sometimes these nuances in fact betray feeling and emotions the party would rather keep hidden, or which he or she does not even know he or she possesses.

The Rhythm of Life

Music and emotion are directly related to the body and can affect heart rate, and breathing, and can cause us to dance and sway, snap our fingers, or tap our feet. Our bodies are also bathed in music long before birth since we hear the beating of our mother's heart, the rhythmic pulses of her lungs filling and discharging air. Once we are born, musical sounds fill our ears, from the lullabies which were used to lull us to sleep to the musical sounds of our mother's voice, the words of which meant nothing. Eventually, however, we begin to recognize that certain sounds have specific meanings, and soon the left half of the brain, as it matures, begins to impose its own structure on the sounds that are heard as well as the sounds it is producing—grammar.

Music and human speech consist of a panoply of elements, yet perhaps their most salient and most easily recognized are rhythm and melody. In

Western cultures, periodicity and rhythm are common musical features and are emphasized in metrically performed music, an approach which has been refined and developed over the last several centuries. Metrically performed music was viewed as the "official" music of 19th-century Europe.

Of course, much of life is characterized by rhythmic and repetitive sequential pulses, be it the rising and setting of the sun, the flow of the tides, the action of the heart, or the breath of life. Although we may slither like a snake or roll like a stone, much of human movement is also rhythmical which in turn is in part a function of the symmetry of the body. When we walk it is generally right–left, right–left (or vice versa), when we hop it is up–down, up–down, and if we crawl it is generally forward–stop, forward–stop. However, if we could soar like an eagle or glide through the depths of the ocean like a fish, movement would be more like a nonrhythmical kinetic continuum. Movement is not always rhythmical or temporal–sequential. However, no matter how it is defined, be it sequential or continuous, movement occurs in space and is always geometrical.

Even when rhythmic movement is present, lock-step temporal sequences may not be. That is, not all rhythmical actions are repetitive, and recurrent elements may not occur or appear only randomly or rarely if at all. It could be said this is what makes us human, and not machines, and allows for the expression of individuality.

Similarly, music need not be rhythmical. Moreover, even when it appears so, it may be irregular and seemingly nonrational and almost out of control, if not chaotic, at least, insofar as the temporal–sequential regions of our brain are concerned.

Because of the pervasiveness of rhythm in life, rhythmic sounds and actions, like melody, can be easily mimicked and adapted for communication. Human beings are affected by rhythm and not only attend to but quite easily produce sequences which are strung together rhythmically. Indeed, if we examine the speech of former President Bush or even that of an individual who is fluently aphasic, we find that not only is the melody appropriate, but the rhythmic nature of the speech is maintained. If we were not listening to the order and arrangement of the words, we would probably not recognize the presence of a problem.

Just as speech contains melodic elements, conversely, music is generally comprised of rhythm. Just as it takes both halves of the brain to produce speech, we need both halves of the brain in order to appreciate musical production. There is always an interaction to some degree. It is this rhythmical aspect which makes music amenable to linguistic and mathematical analysis. In fact, when the sequential and rhythmical aspects of music are emphasized, the left hemisphere becomes increasingly involved, and music becomes more linguistic and languagelike in its production and perception. Hence, many professional musicians demonstrate almost an equal right and left cerebral contribution to musical perception and production, and some in fact, such as those still actively in training and thus still acquiring the "rules" governing its performance, demonstrate a left cerebral dominance.[24] It seems that when music is treated as a type of *language* to be acquired or when its mathematical and temporal–sequential features are emphasized, the left cerebral hemisphere assumes a much more vital role in its production and perception.

It is perhaps via the action of the parietal lobes that sequences and rhythm are imposed. The greater the parietal involvement, the more musical production seems more like a formal language and mathematical. In fact, some accomplished musicians who have willed their brains to science show an enlargement in the left (but not right) inferior parietal region which exceeds that of the norm.

It thus takes both halves of the brain to make music, the left providing rhythm and mathematical elements, and the right half of the brain providing the melody, the emotion, as well as the geometric visual spatial framework in which it is expressed. Although containing temporal–sequential rhythmic features, music in general is an outgrowth and specifically related not only to emotion and melodic speech, but the analysis of spatial relations and the location and source of sounds arising within one's spatial environment.[25] Like math, music, too, is geometric and spatial.

Melody, Math, and Geometric Space

Pythagoras, the great Greek mathematician, argued almost 2,000 years ago that music was numerical, the expression of number in sound. In fact, long before the advent of digital recordings, the Sumer–Babylonians, ancient Hindus, and then Pythagoras and his followers translated music into number and geometric proportions. By dividing a vibrating string into various ratios, they discovered that several very pleasing musical intervals could be produced. Hence, the ratio 1:2 was found to yield an octave, 2:3 a fifth, and 3:4 a fourth, 4:5 a major third, and 5:6 a minor third. The harmonic system which many composers employed in the nineteenth and twentieth centuries is based on these same ratios. Bartok frequently utilized these ratios in his musical compositions.

These same musical ratios that the Pythagoreans discovered also were found to have the capability of reproducing themselves. That is, the ratio can reproduce itself within itself and form a unique geometrical configuration which Pythagoras and the ancient Greeks referred to as the "golden ratio" or "golden rectangle." It was postulated to have divine inspirational origins. Music itself was thought by early man to be magical, whereas musicians were believed by the ancient Greeks to be "prophets favored by the gods."

This same golden rectangle is found in nature, that is, the chambered nautilus shell, the shell of a snail, and in the ear—the cochlea. The geometric proportions of the golden rectangle also were employed in designing the Parthenon in Athens, and by Ptolemy in developing the "tonal calendar" and the "tonal Zodiac"—the scale of ratios "bent round in a circle" and partitioned by the divine 12 solar discs. In fact, the first cosmologies, such as those developed by the ancient Egyptians, Babylonians, Hindus, and Greeks, were based on musical ratios.

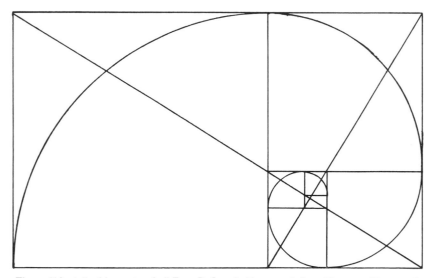

Figure 114. A "golden rectangle." From R. Joseph. *Neuropsychology, Neuropsychiatry, and Behavioral Neurology.* New York: Plenum Press, 1990.

Figure 115. A nautilus shell with a section cut away showing its spiraling chambers. From R. Fortey. *Fossils.* New York: Van Nostrand, 1982.

Pythagoras and then later Plato applied these same "musical proportions" to their theory of numbers, planetary motion, and to the science of stereometry—the gauging of solids. In fact, Pythagoras attempted to deduce the size, speed, distance, and orbit of the planets based on musical ratios as well as on estimates of the sounds generated (e.g., pitch and harmony) by their movement through space, that is, "the music of the spheres." Interestingly, the famous mathematician and physicist Johannes Kepler in describing his laws of planetary motion also referred to them as based on the "music of the spheres."

Thus music seems to have certain geometric properties, such as are expressed via the ratio. Pythagoras, the "father" of arithmetic, geometry, and trigonometry, heavily relied on the music of geometry. For Pythagoras, music was geometric.

As we know, geometry is employed in the measurement of land, the demarcation of boundaries, and, thus, in the analysis of space, shape, angles, surfaces, and configuration. The ability to perform all these forms of geometric analysis, however, is in turn dependent on the functional integrity of the right half of the brain and the superior and inferior parietal lobes.[26]

In nature, one form of musical expression, that is, the songs of most birds and many insects, are also produced for geometric purposes. That is, a bird does not "sing for joy," but to signal others of impending threat, to attract mates, to indicate direction and location, to stake out territory (e.g., shape, angles, surface, configuration, etc.), and to warn others to stay away who may attempt to intrude on his space.[27] Singing insects, as noted, use music for much the same purpose.

Hence, it is certainly possible that right hemisphere dominance for music and melodic speech may be an outgrowth or strongly related to its capacity to discern and recognize environmental spatial relations as well as its ability to mimic these and other nonverbal and emotional nuances. That is, it is somewhat probable that ancient man and woman's first exposure to the sounds of music was embedded in the environment for obviously musical sounds are heard frequently throughout nature (e.g., birds singing, the whistling of the wind, the humming of bees or insects). Bird songs, for example, can encompass sounds that are "flute-like, truly chime- or bell-like, violin- or guitar-like" and "some are almost as tender as a boy soprano."[28]

Perhaps in part, our musical nature is related not only to the limbic system but to our original relationship with nature and resulted from the tendency of humans to mimic sounds that arise from the environment—such as those which conveyed certain feeling states and emotions. Perhaps this is also why certain acoustical nuances, such as those employed in classical music, can affect us emotionally and make us visualize scenes from nature.

Moreover, if we may assume that long before man or woman sang his or her first song, the first songs and musical compositions were created by insects and especially our fine-feathered friends (sounds that inspired mimicry by humans), then it appears that musical production regardless of its source of origin, was first and foremost emotional, motivational, sexual, social, and directly related to the geometry of space—the demarcation of one's territory. Emotion and geometry are characteristics that music still retains today.

The Body Sings Electric and Soothes the Savage Breast

In many ways the perception and expression of right cerebral emotional and melodic speech is organized and parallels the neuroanatomical organization of left hemisphere language perception and speech output. In this regard, the parietal lobule plays a very significant role not just in sound and left brain speech perception, but in right brain musical and emotional perception as well. This explains the close relationship between music and geometric space, on one hand, and language, math, and temporal sequencing, on the other.

It is possible that this same right and left parietal (as well as limbic) melodic perceptual relationship also ties musical perception and production to the body. It is partly due to this somatic relationship that music comes to affect the body. Music can inspire us to clap our hands, snap our fingers, march or tap our feet, and can significantly alter body temperature and autonomic functioning, accelerate pulse rate, raise or lower blood pressure, and, thus, alter the rhythm of the heart's beat. Rhythm, of course, is a major component of music.

Given its intimate relationship to the body, it is not completely surprising that music has therapeutic influences and can influence our mood and emotional state. In fact, the ancient Greeks thought that music could be used to cure gout, sprains, and even insanity, and as recently as the 17th century, music would be employed in an attempt to banish depression and to bring relief from pain, a common prescription being that music was to be played above the injured or diseased body part. From the 12th century until the 17th, music coupled with dance was often prescribed or believed to banish diseases. However, the afflicted would often have to dance for days, resting only at night.

Prayer: Singing the Praises of God

It is via the control exerted by the limbic system and the right half of the brain over these aspects of language which allows an individual to continue to perceive and express emotional and melodic nuances even when severely aphasic. However, many such patients also retain the ability to pray and may do so spontaneously, or in accompaniment with a priest, or in response to religious programming over the radio. It is hard to see the connection between praying and geometry, or between prayer and the mimicry of environmental sounds. What praying is, however, is emotional, and it is most likely to be spontaneously stimulated when an individual feels they are suddenly facing death or catastrophe.

Because these abilities are retained even after a severe left hemisphere injury indicates that there is a neurological substrate which supports not only the ability to sing and pray but to experience and express religious feelings in general. Specifically, this tissue is located in the vicinity of the temporal lobe, including the amygdala and hippocampus within its depths. Like the rudiments of emotion and music, it appears that praying and religious thought may, too, be an outgrowth of that most ancient region of the brain—the limbic system.

In any case, emotional, melodic, musical, mystical, and religious feelings are a consequence of the great evolutionary refinement which occurred within the limbic and temporal regions of the right half of the brain in particular. Hence, perhaps the limbic system in conjunction with the neocortex of the temporal lobe, or maybe the right hemisphere in general, has made possible the ability to sing, curse, pray, to have spiritual and religious experiences, and to dream about and remember them as well.

III

THE NEUROANATOMY
OF A MEMORY

11 ═══════════════════════════════════

Remembrance of Things Past
The Limbic System and Long-Lost Childhood Memories

Well over a billion years ago, when the simplest of organisms swirled about or remained moored to the ocean floor, learning and memory remained restricted to almost reflexive reactions to the ultraviolet rays of the sun or the chemical composition of its environment. The organism either hid from or sought out nourishment from the sun or creatures secreting certain chemicals. Over the course of evolution, with the refinement of electrical and chemical means of communication, these tiny creatures became capable of storing more than just the energy from our great solar disc, but to convert it into information which could be recalled or transmitted to like-minded cells.

Worms, sea snails, and other such primitive creatures have long been capable of learning and remembering, and even passing such information to creatures similar to themselves.[1] However, rather than storing or responding only to sunlight, chemicals, or electric shocks, such as by learning to avoid or seek out these forms of stimulation, these and other animals are able to receive and transmit complex forms of information via chemical and pheromonal communication.[2]

Over the course of evolution, these chemical modes of information transfer have become more complex. In fact, chemical, pheromonal, and olfactory systems may well have provided the first truly refined means via which an organism could learn from experience and then recall this experience at its leisure. As noted, it was via expansions which occurred in the olfactory system that the forebrain and the limbic system, then later the first cortical motor systems (the basal ganglia), and then the telencephalon and neocortex of the brain would be formed.

However, as pertaining to learning and memory, two important expansions occurred in the olfactory–limbic system almost 500 million years ago. These developments took place within the amygdala and hippocampus,[3] nuclei which continue to serve attention, and information retention and recall even within the modern human brain.[4]

Darren had been driving with his friends early in the morning when upon turning onto a rain-soaked street, their car slid over an embankment. His friends were killed, but Darren remembered nothing of it. In fact, he remembers nothing after coming home from school the previous day and nothing since. If you introduce yourself to him and then leave the room only to return a few minutes later, Darren behaves as if he met you for the first time. Every time he is told that his father has passed away, he shows the same grief and the same shock. Many friends and family had told him that he was "lucky," but he quickly forgets that, too, for although he escaped with his life, he lost his memories. When he was thrown out of that car he suffered massive head injuries and skull fractures which all but destroyed much of both anterior and inferior temporal lobes and the memory centers within their depths—the hippocampus and amygdala.

Hippocampus

The hippocampus is an elongated structure located within the depths of the temporal lobe (behind the amygdala). In humans it consists of an anterior and posterior region and is shaped somewhat like a telephone receiver.

One of the three major neural pathways leading to and from the hippocampus is the **entorhinal** (or nose) area which is sometimes referred to as the gateway to the hippocampus. As we've noted, the hippocampus evolved from tissue concerned almost solely with olfactory input. It is via the entorhinal area that the hippocampus continues to receive olfactory and amygdaloid projections. Indeed, the hippocampus is greatly influenced by the amygdala which in turn monitors and respond to hippocampal activity. Together the hippocampus and amygdala complement and interact in the generation of emotional and other types of imagery, as well as attention, learning, and memory.[5]

Hippocampal Arousal, Attention, and Inhibitory Influences

Like the amygdala, with the evolution of the neocortex, the hippocampus evolved as well and maintains extensive interconnections with the new brain over which it appears to exert considerable influence. Indeed, the neocortex essentially consists of cells whose evolution and formation was provoked by the limbic system. In part, the neocortex served to perform analyses that the limbic system was not capable of, and in this regard it evolved out of and continues to serve the limbic system.

The hippocampus acts to influence the reception and probably the filtering of information reception throughout the brain and within the neocortex, with which it maintains extensive interconnections.[6] It also appears to act so that the neocortex is not over- or underwhelmed when engaged in the processing of information. This is because high or very low states of excitation are incompatible with alertness and selective attention as well as the ability to learn and retain information. The hippocampus therefore acts to reduce or increase arousal levels.

Attention and Inhibition

When exposed to novel or noxious stimuli, the hippocampus becomes highly aroused, which suggests that it is processing or learning the information it is being exposed to. With repeated presentations of a novel or unpleasant stimulus, the hippocampus habituates, which suggests that once learning has occurred and the information is stored, hippocampal participation ceases.[7] Thus, as information is attended to, recognized, and presumably learned and stored in memory, hippocampal participation first increases and then steadily diminishes. However, it is also very responsive to events which are emotionally significant or unpleasant.

How Axons and Dendrites Learn to Remember[8]

As noted, different regions of our brain communicate via nerve cells called neurons. The neuron (or nerve cell) transmits messages to other nerve cells via a nerve fiber called an axon. Neurons receive messages via their dendrites.

Each neuron contains an axon and several dozen or more dendrites. Axons can be quite long and sometimes travel considerable distances to other cells. The corpus callosum, for example, is made up of millions of axons. Dendrites, although very short in length, are quite bushy and form what have been referred to as dendritic trees with many spines and branches.

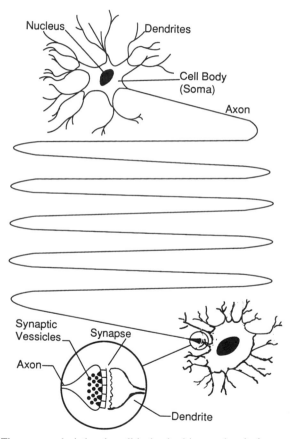

Figure 116. The neuron, depicting the cell body, dendrites, and a single transmitting axon which travels sometimes considerable distances to make contact with the dendrites of other neurons at a gap junction, the synapse. When sufficiently activated, the axon will release neurotransmitter substances which are stored in synaptic vesicles.

Axons make contact, at the synaptic junction, with the dendrites of other cells, and sometimes the soma of the cell itself, or more rarely, with another axon. Dendrites, because of their extensive branches, can receive messages from many axons as they have very extensive receptor surfaces. Axons, however, transmit impulses only to a single dendrite or cell.

Axons fire in response to an "all-or-none principle." That is, it either becomes sufficiently activated and fires, or it does not depending on if its firing threshold has been surpassed. For this to occur, the "resting potential" (e.g., its base level of activity) must be increased in order to surpass this threshold.

When the axon is triggered and fires, it releases a chemical neurotransmitter. The dendrite has a receptor surface which acts to receive neurotransmitters released from the axon of a different cell. If sufficient transmitter is released, the dendrite becomes activated and will allow information to be transmitted to its cell body which via its own axon may relay the message to the dendrite of the next cell. However, although a dendrite and cell may become stimulated, if the threshold for axonal activation is not surpassed, the axon leading from that cell to the next dendrite will not fire. If the axon is sufficiently activated, it will release its neurotransmitters.

The chemical transmitter acts on the dendrite, the cell body, and its axon, by increasing activity until the firing threshold is reached, at which point it ignites and discharges. If the receiving dendrite, cell body, and its axon are insufficiently activated, the impulse received fades such that information transmission stops with that cell. However, for a brief time period, the cell remains slightly activated and remains close to firing.

Due to this principle, although a single impulse may not cause the cell to fire, since the dendrite and neuron have been partially activated (such that it is well above its resting potential), if a second or third low-level impulse is also received (such as from the same axon or a different axon making synaptic contact at a different branch of the dendrite), there is an additive effect, and the threshold will be easily surpassed, and it will fire.

In some ways, this parallels the attention-getting response of the telephone, or the flushing responses of a toilet. That is, a phone will draw a person's attention and results in a response not just by how loud it rings, but by the number of times it rings. Conversely, if a toilet is only partially flushed (such that the lever is not pushed all the way down), then it fails to discharge its cargo. However, if small portions of water are quickly and repetitively added to the contents of the toilet bowl, via this additive effect, the toilet, too, will flush.

Learning, Memory, and Neuronal Growth

When brain cells are insufficiently utilized, they tend to die and drop out, or their dendrites lose their branches and receptor surfaces. We lose thousands of brain cells every day. However, those which are repeatedly exercised often grow larger and the dendrites they possess become more extensive. Moreover, new axons may grow connections to the new dendritic branches of these very active cells. This is, in part, how learning and memory are made possible.

As more and more information is assimilated and learned, more and more cells become associated and interlinked thus enabling mental processing to become more complex and elaborate. When one considers that an adult brain contains over 100 billion neurons (each about half the size of the period at the end of this sentence) and that a dime-sized area of neural tissue contains over 10 billion interconnections, it is much easier to comprehend how a considerable amount of learning can occur via these interactions.

Potentiation

An axon which is repeatedly utilized for information transmission will increase its supply of neurotransmitters. Conversely, a dendrite which is repeatedly stimulated not only becomes more complex, but each individual receptor surface (at the synaptic junction) may become more extensive so as to take advantage of the increased amount of neurotransmitter available.

A dendrite and neuron that is repeatedly stimulated will also become **potentiated**. That is, it remains activated (albeit below firing threshold) for long time periods. When the next stimulus is received, the neuron is more likely to fire.

Because of these many changes, although the firing thresholds remain the same, when the receiving cell is already partially active and now much more sensitive, its threshold can be surpassed more easily. Moreover, due to **potentiation**, and the increased levels of available neurotransmitter, a highly active, potentiated neuron can fire with minimal stimulation. The cells involved become activated with ease due to the physical and chemical changes which have occurred between them. In a learning situation, it could be said that the person can now perform the task without thinking about it.

These neural alterations can be produced artificially as well. For example, it is well known that a drug addict eventually requires a stronger and then stronger dose of narcotic in order to "get off." This is because after each dose and following repeated dendritic and cellular activation, the dendritic receptor surface will increase in size so as to take up the repeatedly available chemicals. In fact, within the limbic system there are receptors which are specialized to respond to narcotic substances, and there are cells which produce them. These opiatelike neurotransmitters are called enkephalins and endorphins.

However, in regard to an addict, since an axon is not producing these chemicals which are being supplied externally, this requires that the drug be administered more often and at higher doses so that the now-larger dendritic surface will become activated, as there are now more dendrites which have grown in response to this activity. It is in this manner that some people become addicted to drugs such that they soon need more and more. This is the result of increase in the nerve-cell receptors which respond to that drug which now have to be periodically replenished in greater amounts.

Nevertheless, the very process of potentiation, and the increase in receptor surface and the number of dendrites is an important part of learning and memory. For example, some scientists view memory as consisting of at least three separate stages: immediate memory (which may last a few seconds), short-term memory (which may last from seconds to a few minutes, hours, or even days), and long-term memory (which may last up to a lifetime).

It has been found that axonal–dendritic interconnections are likely to become more extensive in correspondence with the learning of events which are stored in long-term memory. However, the same can occur within a few seconds of learning something new, the memory of which may last for months. Many of these synaptic and activational changes, in turn, are most apparent within the anterior regions of the hippocampus (which maintains rich interconnections with the amygdala). Moreover, this same region of the hippocampus will become electrophysiologically potentiated during learning tasks. If the hippocampus is injured or functionally suppressed, no new learning occurs, and potentiation does not appear.

This suggests that the longer the potentiation, either at the cellular or hippocampal level, the stronger might be the memory and the more likely it will persist over time. In fact, long-term potentiation lasting up to several days has been noted in the hippocampus following successful learning trials, which in turn may reflect the transition of information from short-term to long-term memory.

Dendritic Learning and the Association of Ideas

As we've noted, in learning situations there often results increased intracellular complexity. Other cells may grow nerve fibers into this area of learning activity, and all cells involved may become more complex as they undergo physical changes to accommodate the increases occurring. That is, a dendrite will attract the axons of other cells to it which then makes contact so that even more information can be exchanged and processed.

With repeated use and with the alterations in dendritic growth, changes in the receptors surface and the amount of neurotransmitter available, and the number of axons contacting a particular dendrite, it now takes very little to cause one cell to activate a second cell and so on, so that the same message becomes more complex as more details or associations are added. There is an increase in the number of synapses, the number of different cells making contact, the amount of transmitter available, as well as an increase in the resting potential of the nerve and its level of potentiation, all of which requires that fewer or less-powerful signals be transmitted in order to cause activation. By repetition and practice it becomes easier to perform a certain action until finally it becomes like a reflex as it takes very little to trigger a response. Indeed, this is how some habits are also formed. Practice makes perfect.

However, eventually, with rapid repeated firing, the amount of available neurotransmitter begins to deplete and cellular activity ceases or becomes haphazard. At a behavioral level, we could say that fatigue has set in.

The same thing could be said for certain thoughts and even complex actions. The more often we produce the same thoughts and entertain certain feelings, the more often they will be experienced in the future as it now takes very little to set off the whole chain of events. Eventually, however, we reach a point where we can think about the subject no more, at least for a while, only to discover later that we are again dwelling on the same subject matter.

On the other hand, with an increase in the number of cells which are now linked together, the more complex is the resulting thought or the level of analysis as yet more and more associations, sensations, feelings, and past experiences are brought to bear on whatever is being processed or learned. Those with the greatest number of neuronal cells that are interconnected, are at a tremendous intellectual advantage as compared to those that are not similarly endowed.

Hence, when certain cells repeatedly communicate, a circuit of experience is created. An assembly of cells becomes associated via their interconnections (which become stronger with use). Due to the physiological and chemical changes which have occurred, complex actions can be initiated in an effortless and routine fashion. For instance, in getting dressed, a whole sequence of associated actions takes place which are so well learned that one need not even think about the different steps involved; for example, fastening clothes and so on. The entire circuit of experience is activated and occurs almost in reflex fashion.

However, because learning is a continual process, not only can these circuits change in configuration and number, but memories can be altered accordingly. This is because a new memory can become associated with an old memory, if they are both thought about together. Something like this occurs when after an argument someone thinks about what should have been said, what might have been said and by whom, what this could have meant, and so on. When this occurs, old memories can become contaminated and reshaped accordingly. As such, one spouse is likely to claim that he or she remembers things that didn't even happen. There are a number of ways that memory can be altered, as well as completely deleted.

Memory Confabulations

All memories and perceptions are influenced to varying degrees by our own personality and emotional status, including what we find objectionable, pleasurable, and so on. Similarly our perception or understanding of another person's status or intentions not only effects memory but can change it. If we know someone is a banker and not a construction worker, our stereotypes as to the appearance of a person in these professions can cause alterations in memory and subsequent recall of what they look like, even immediately after we have met such an individual such that details may be added or subtracted.[9] In one experiment, students who were subsequently given personal details and were then asked to describe a person they had observed earlier were astonished to discover that their actual appearance was completely different from what they remembered.

It sometimes happens in criminal trials that witnesses will recall seeing someone and may identify him or her as the culprit when in fact he or she is not, although they did in fact see him or her, but in a different place or time. This is particularly important when people are first shown mug shots and then exposed to a lineup including an individual whose mug they viewed.

That is, when they seem him in the lineup, sure enough he will look familiar and they are likely to recall seeing him. Unfortunately, they may then misidentify the time and place, putting him at the scene of the crime when in fact it was in the mug book where he was first viewed. Or a witness may identify as the culprit someone who came into a store just minutes before the crime occurred simply because they remember seeing him. In one case, a storeclerk who had been robbed and beaten later positively identified a man who had come into the store at the same time the previous day but who nevertheless had an airtight alibi as to his whereabouts when the crime occurred (he was in court contesting a ticket).[10]

In experiments conducted by E. Loftus and colleagues,[11] people were told the details of a crime and were shown pictures which included not only the criminal but bystanders. Later half the group were shown mug shots that included the criminal and the other half mug shots that did not include him but that of a bystander. Over 60% of the latter group misidentified the bystander as the criminal. Considering the widespread use of police mug shots, this certainly puts some people at a grave risk of being misidentified due to this contamination based on familiarity.

In criminal cases this contamination sometimes occurs during the questioning period. In another experiment by Loftus and colleagues, half the subjects who watched a film of a car accident were asked, "How fast was the white sports car going when it passed the barn while traveling along the country road?" However, there was no barn in the film. Nevertheless, those who heard the "barn" question were six times more likely to recall seeing a barn when there was none.

In a similar experiment, half the subjects saw a film where a car ran a yield sign. When those who saw the yield sign were asked about the car and the "stop sign," 80% of the subjects later remembered a stop sign when in fact it was a yield sign that they had observed.

Similar forms of contamination can occur just by thinking and remembering. As many husbands and wives have discovered following ar argument, often the spouse will later recall details and other items that the other spouse is positive did not occur. This is because by thinking and rehearsing, other variables sometimes come to be considered and incorporated. "I should have said . . .," "I bet he was going to say . . .," and so on. Unfortunately, this information often comes to be associated with the original memory such that memory becomes contaminated by erroneous material.

Nevertheless, it is because memories can become altered and contaminated that learning and creativity have been made possible. Essentially,

adding new information results in an alteration in the synaptic, axonal–
dendritic neural circuitry involved in that particular type of learning, and
thus an alteration in the configuration and complexity of the entire circuit
of all associated experiences. It is via activation of these nerve-cell circuits
that we learn and remember what we have learned, and this is made possible
by neuronal cell assemblies located in a number of specific brain regions,
most notably the hippocampus and the amygdala.

Learning and Memory: The Hippocampus

The hippocampus is most usually associated with learning and memory
encoding (e.g., long-term storage and retrieval of newly learned information), particularly the anterior regions. Bilateral destruction of the anterior
hippocampus results in striking and profound disturbances involving memory and new learning (i.e., **anterograde amnesia**). For example, one such
individual, "Henry," who underwent bilateral surgical destruction of this
nucleus, was subsequently found to have almost completely lost the ability
to recall anything experienced after surgery.[12] If you introduced yourself
to him, left the room, and then returned a few minutes later he would have no
recall of having met or spoken to you. Dr. Brenda Milner has worked with
H. M. for almost 20 years, and yet she is an utter stranger to him.

Presumably the hippocampus acts to protect memory and the encoding
of new information during the storage and **consolidation phase** via the
gating of afferent streams of information and the filtering/exclusion (or
dampening) of irrelevant and interfering stimuli.[13] When the hippocampus
is damaged or overstimulated, there results input overload, the brain is
overwhelmed by neural noise, and the consolidation phase of memory is
disrupted such that relevant information is not properly stored or even
attended to. Consequently, the ability to form associations (e.g., between
stimulus and response) or to alter preexisting schemata (such as occurs
during learning) is attenuated.

Hippocampal and Amygdaloid Interactions: Memory

The hippocampus plays an interdependent role with the amygdala in
regard to memory.[14] The role of the amygdala in memory and learning
seems to involve activities related to reward, orientation, and attention, as

well as emotional arousal. If some event is associated with positive or negative emotional states, it is more likely to be learned and remembered.

The amygdala seems to reinforce and maintain hippocampal activity via the identification of motivationally significant information and the generation of pleasurable or aversive feelings.[15] That is, a positive or negative feeling increases the probability of attention being paid to a particular event and of its being stored in memory.

Due to its intimate involvement in almost all aspects of emotion, the amygdala also acts to provide the hippocampus with complex associations involving the fusion of different sensory qualities. In this manner, we can recall, for example, that a certain person had a lovely, sweet, syrupy voice, or conversely, feel bitter about the manner in which he or she may have acted toward us at one time long ago.

Hence, the hippocampus acts to reduce or enhance extremes in arousal associated with information reception and storage in memory, whereas the amygdala acts to identify the social–emotional–motivational characteristics of the stimuli as well as to generate appropriate emotional rewards so that learning and memory will be reinforced. The amygdala also provides complex associational images and impressions for memory storage. Thus, we find that when both the amygdala and hippocampus are damaged, striking and profound disturbances in memory functioning result.

Conversely, when these limbic-system nuclei are activated, such as through smell, touch, and so on, learning and memory can be profoundly and permanently affected. It is in this manner that a single sniff of a long-forgotten fragrance can elicit the remembrance of things past, and all the attached emotions, be they love, heartache, passion, or sexual longing.

It is also via these interactions that certain auditory, visual, and tactual sensations not only trigger certain memories but can result in the neocortex being completely overwhelmed by these ancient limbic structures when the emotionally significant or traumatic is recalled. As such, we may become overwhelmed by love or anger, or the forgotten, the remembered, or the unknown, and then later not know what came over us.

Right and Left Hemisphere Learning and Memory

The right and left cerebral hemispheres are specialized in regard to the type of material they can receive, process, and respond to and in fact speak different languages. Due to these specializations, some types of information cannot be transferred or even recognized by one versus the other half of the brain. One half of the brain may perceive, process, and store in memory certain aspects of experience that the other half knows nothing about. This is particularly true regarding negative emotions and emotional trauma which may stay confined to the right half of the brain.

Moreover, due to these specialties, the two halves of the brain may reach different and conflicting conclusions about what ostensibly seems to be the same piece of information. This is a major source of creative thought, insight, and profound intuitive awareness. It also provides the foundations for intrapsychic conflict as the two halves of the brain may not only perceive things differently but draw different conclusions and have different memories triggered. Based on these different memories and experiences, they may then act in an oppositional manner. One half of the brain may respond in an emotional manner and attempt to do one thing, the other half, not sharing in these feelings or these memories, may attempt to accomplish something entirely different.

Lateralized and "Unconscious" Memories: Right- and Left-Brain Cerebral Memory

In 1990, a retired California fireman, George Franklin, Sr., received a life sentence for the 1969 murder of an 8-year-old girl, Susan Nason. Mr. Franklin was convicted in San Mateo Superior Court on the testimony of his daughter, Eileen Franklin-Lipsker, who reported to police in 1989 that she had suddenly recalled memories that had been repressed and forgotten for 20 years. Specifically, she recalled watching her father rape and then kill the victim, her girlfriend, by smashing her skull with a large rock. She claimed that after almost 20 years the memory had come to her when certain actions and facial expressions of her own daughter suddenly triggered their recollection.

Since the right and left cerebral hemispheres process information differently, the manner in which it is represented in memory will also be lateralized. This is because the code or form in which a stimulus is represented in the brain and memory is determined by the manner in which it is processed and the ensuing transformations which take place.

It is well known that the left half of the cerebrum is responsible for the encoding and recall of verbal memories whereas the right brain is dominant in remembering visual–spatial, nonverbal, and emotional memory functioning. When listening to someone who is angry, the left brain might encode the words he or she is saying. The right half of the brain may perceive and store in memory the look on his or her face, tone of voice, the angry gestures used, as well as the overall emotional gestalt of the situation including one's reactions to what is being said.

If the right half of the brain were badly damaged (e.g., destruction of the right temporal lobe which contains some of the memory centers), the patient (i.e., his left brain) would be unable to store in memory most aspects of emotional, visual–spatial, and related stimuli. This particular person (i.e., his left brain) would not be able to remember where he laid his wallet or car keys, how to get (drive through space) to the dentist's office, would fail to recognize him if he ran into him in the parking lot, and would forget the argument he had had with his receptionist about the overdue bill. However, this person (his left brain) would be able to remember the dentist's name, his phone number and address, the amount of the overdue bill, and the conversation he had had with the officer manager about the bill.

If the memory centers of the left hemisphere were damaged (which are located in the left temporal lobe), this particular person would have exactly the opposite problems with memory. He would forget the dentist's name, the amount of the bill, but recall the argument, his face, and so on.

Responding to Hidden Memories

Lateralization of memory in turns affects complex behaviors, for onehalf of the brain may experience and store certain information in memory, and at a later time in response to certain situations act on those memories, much to the surprise, perplexity, or chagrin of the other half of the brain.

When one half of the brain learns, has certain experiences, and stores information in memory, this information may not be always available to the opposing cerebral hemisphere; one hemisphere cannot always gain access to memories stored in the other half of the brain. However, partial access can be obtained if observable clues can be associated with internal feelings.

When Carol was a little girl, she was molested on several occasions by her stepfather. The first time she was nearly 4 years old and had been sitting next to him on the couch watching TV when he began to stroke and run his fingers through her hair. He continued this action while he cajoled and intimidated her into performing fellatio. He did this to her on 10 or more occasions over a 1-year time period until he moved away. Somehow she managed to forget all about this until many years later while in college.

She was in bed with her boyfriend, and they had just finished making love when he began to stroke and run his fingers through her hair as he exerted gentle downward pressure on her head, directing her to his genitalia. All at once she began to panic, became quite hysterical, and started crying and trying to strike her boyfriend. Then, grabbing her clothes and quickly getting dressed, she ran from his apartment.

For the next several weeks, she refused to talk to him, hung up when he called her, and began to feel an overwhelming aversion toward men. She sought counseling, but to no avail.

It was only a year later while watching television that the entire memory of what had happened to her, so many years before, unraveled. In the show a father walked into his crying daughter's bedroom and while trying to soothe her began to brush and run his fingers through her hair as he pulled her next to him. Immediately Carol began to feel angry and upset, and then she remembered, her stepfather, the television, the fellatio, his fingers in her hair. The forgotten images began to flood her mind.

Essentially, the memory of what her stepfather had done had been stored in the memory banks of the right half of her brain and had been forgotten by the left, which does not store negative but only positive memories. Years later, when her boyfriend began to stroke her hair after having sex, and while encouraging her to perform fellatio, the association between sex and having her hair fondled was reactivated. The right half of her brain then began recalling the visual and emotional images of what had happened and became terribly upset.

Her left hemisphere, however, having little linguistic memory of the tactual, sexual act and its associated negative emotions, had "no idea" as to why she was behaving in this manner. Nevertheless, her right brain prevailed, and she ran way in hysterics.

It was only a year later that the left brain was clued in to what had triggered her hysterical reaction. When both the right and left brain observed the father on TV brush and fondle the little girl's hair, Carol's limbic and right brain became upset and the images of what had happened to her spilled forth. Looking at the movie and feeling the revulsion the left cerebrum suddenly was given a tremendous amount of information as to what was bothering her and was then offered access to the horror that had so long been locked away seemingly forgotten. Transfer was made possible because they were both observing the same event and feeling the same feeling, upset.

Indeed, I have frequently had patients tell me that long-forgotten memories of incest, molestations, and other traumatic as well as quite happy events from childhood have been triggered by similar seemingly innocuous occurrences. They may be watching children play, putting their own kids to bed, becoming angry when they have broken something, or merely overhear a child saying something innocent in another room when suddenly a long-suppressed memory unfolds before their eyes. Essentially the memory is being regurgitated by the right, and the left, being cued by the emotional response coupled with the scene being observed, gains access to what had been a well-kept secret of the right half of the brain.

Some Memories Are Stored Only in the Right or Left Hemisphere

In the normal brain, memory traces appear to be stored unilaterally (in one half of the brain) rather than laid down in both hemispheres (i.e., bilaterally), depending on if they are visual–spatial, emotional, linguistic, and so on. For the opposing hemisphere to gain access to these memories, it has to activate the memory banks of the other brain half via the corpus callosum, the rope of nerve fibers which interconnects the left and right halves of the brain. As demonstrated by R. W. Doty and W. H. Overman,[16] if the corpus callosum is surgically severed, the memory remains confined to the half of the brain where it is stored and the opposing hemisphere now knows nothing about it.

In one study,[17] Drs. G. L. Risse and Michael Gazzaniga injected sodium amytal (an anesthetic) into the left carotid arteries of various neurosurgical patients in order to anesthetize the left half of the brain. After the left cerebrum was inactivated and essentially asleep, the awake right hemisphere, although unable to speak, was still able to follow and respond to commands, for example, holding and palpating an object with the left hand or looking at pictures. They then gave the right half of the brain (via the left hand) a number of objects to hold and showed it several pictures.

Once the left hemisphere recovered from the drug, as determined by the return of speech and right-handed motor functioning, such that both halves of the brain were now awake, none of the eight patients studied was able to recall *verbally* which pictures had been shown to the right brain or which objects had been held and palpated with the left hand, "even after considerable probing."

Although encouraged to guess, most patients (i.e., their left brains) refused to try and insisted they did not remember anything. When they were shown several objects or pictures and asked to guess by pointing with the *right hand* (which is controlled by the left brain), patients continued to state they could not recall any of the pictures or objects or pointed at the wrong items.

When they were asked to point with their *left hand*, most patients immediately raised the *left hand* and pointed to the correct object. This is because the right cerebrum, which controls the left hand, was fully able to recall what it had been shown, although this knowledge was not translated into words, and the right cerebrum was unable to talk about it (talking being controlled by the left brain). The left cerebrum was unable to gain access to these memories even though they were stored in the right hemisphere. The right half of the brain not only remembered but was able to act on its memories.

Surprisingly, once the person pointed or grasped the correct object with the left hand, such that both halves of the brain were able to see what only the right hemisphere had learned and remembered, the left brains of most of these patients immediately claimed to now recall having been shown the item, just like the girl whose hair had been stroked while performing fellatio as a child.

This indicates that when exchange and transfer is not possible, or is in some manner inhibited, or if, for any reason, the two halves of the brain become functionally disconnected and are unable to share information, the possibility of information transfer at a later time becomes extremely difficult. The information can become lost to the opposite half of the cerebrum. Nevertheless, although lost, these memories and attached feelings can continue to influence whole brain functioning in subtle as well as in profound ways.

Nevertheless, once the right half of the brain acts on its memories, the left hemisphere may then gain access to some of this information via observation and guesswork. If the left hemisphere guesses correctly, the limbic system and the right cerebrum may reward and reinforce it emotionally or give it clues by generating various emotions until it guesses correctly.

In very young children, however, sometimes these emotions and memories cannot be expressed or accessed by the left cerebrum because of limited language abilities and its considerable difficulty verbally describing negative but not positive emotions (as well as the immaturity of the corpus callosum, as discussed below). Consider, for example, a child who has been molested. Although she may never talk about it, and may deny it if asked, she may nevertheless go to school and pull up her dress, pull down her panties, play with the genitals of her friends, or allow them to play with hers. In other words, she acts out the memories and experiences selectively stored within the right half of her brain. Because her right brain cannot talk about what has happened, it acts it out instead.

Long-Lost Childhood Memories

For most individuals, events which occurred before age 4 are very difficult if not impossible to recall. There are several explanations which can account for what appears to be an age-related amnesia. For one, information processed and experienced during infancy versus adulthood is stored in memory via certain cognitive transformations and retrieval strategies which are quite different. As the brain matures and new means of information processing are learned and developed, the manner in which information is processed and stored becomes altered.

Although these early childhood memories are stored within the brain, the adult brain no longer has the means of retrieving them, that is, the key no longer fits the lock. That is, early experiences may be unrecallable because infants use a different system of codes to store memories whereas adults use symbols and associations (such as language) not yet fully available to the child.

Much of what was experienced and committed to memory during early childhood took place prior to the development of linguistic labeling ability and was based on a pre- or nonlinguistic code. Hence, the adult, relying on more sophisticated coding systems, cannot find the right set of neural programs to open the door to childhood memories. The key does not fit the lock because the key *and* the lock have changed.

Limitations in Right and Left Hemisphere Communication. Much of what is experienced and learned by the right hemisphere during these early years is not always shared or available for left hemisphere scrutiny (and vice versa). That is, like an adult, a child's two cerebral hemispheres are not only functionally lateralized, but limited in their ability to share and transfer information. More important, however, infants and young children have right and left hemispheres which are not fully interconnected. Their right and left hemispheres thus have considerable difficulty communicating; much more so than in an adult. This is due to the immaturity of the corpus callosum and the limited ability of the left half of a child's brain to process and store negative social–emotional experiences.

The Immature Corpus Callosum

The great nerve-fiber bundle, the corpus callosum, is the main "psychic corridor" via which information can flow between the right and left cerebral hemispheres. The corpus callosum, however, takes over 10 years to grow completely and mature.[18] The process of development and maturation is not complete until the end of the first decade. This greatly limits information transfer between the hemispheres, particularly in very young children.

Like adults who have undergone surgical splitting of the corpus callosum (called "split-brain" surgery), it has been shown (by David Galin, R. Kraft, Dan O'Leary, A. Salamy, myself, and others[19]) that communication between the two halves of the brain is so poor that children as old as age 4 have difficulty transferring emotional, tactile, cognitive, or complex visual information between the hemispheres, for example, describing complex pictures shown to the right brain, or indicating with the right hand complex objects and shapes which have been felt by the left hand (and vice versa).

Indeed, when a child is questioned about a picture shown selectively to the right brain, the left (talking half of the) brain may respond with information gaps which are erroneously filled with confabulatory explanations. The left will make up explanations as to what was seen by the right since it does not know.[20]

Thus the left brain of a very young child has incomplete knowledge of the contents and activity occurring within the right. This sets the stage for differential memory storage and a later inability to transfer information between the cerebral hemispheres. That is, just like the experiments of Drs. Doty, Overman, Risse, and Gazzaniga, when the two hemispheres are unable to communicate and one brain half learns and stores certain experiences in memory, the other half of the brain cannot always gain access to those memories even when communication between the two brains is restored or established.

In young children, their two hemispheres cannot fully communicate and cannot share information due to corpus callosum immaturity and because each half the brain is specialized to process different types of information. Later in life, although the corpus callosum has grown and now allows for more efficient communication, these early memories still cannot be shared.

As discussed earlier, if one hemisphere learns, and at that time memory and learning transfer was not possible, then transfer later in life becomes very difficult. The information and memory become stored only in one half of the brain. They remain well-kept secrets.

When Well-Kept Memories Become Activated Secrets

Because of lateralization and limited exchange, the effects of early "socializing" experience can have potentially profound effects. As a good deal of this early experience is likely to have its unpleasant if not traumatic moments, it is fascinating to consider the later ramifications of early emotional learning occurring in the right hemisphere unbeknownst to the left; learning and associated emotional responding which *later* may be completely inaccessible to the language axis of the left cerebral hemisphere.

Although (like the adult brain) limited transfer in children confers advantages, it also provides for the eventual development of a number of very significant psychic conflicts—many of which do not become apparent until much later in life. This is because the two brains not only have different memories which cannot be shared, the hemispheres may independently recall certain experiences and all associated feelings, and then act on them. As such, people may respond nervously, anxiously, angrily, fearfully, and not know what is really bothering or even "what came over" them.

Moreover, due to the immaturity of the callosum, children can frequently encounter situations where the right and left halves of the brain not only differentially perceive what is going on but are unable to link these experiences so as to fully understand what is occurring in order to correct misperceptions.

This is what sometimes happens with young children who are molested. Not only are they confused as to what is happening to them (or as explained by some, they pretend it is happening to someone else), but sometimes they seemingly forget what had happened or even that it had happened at all, although they may be plagued by the associated trauma for years. The molestation may stay a secret of the right half of the brain until the memory is accidentally triggered by some action or event witnessed later in life by both halves of the brain.

Limbic Memories and Unconscious Origins

Given the fact that the right and left hemisphere and the limbic system each literally has a mind of its own, this curious asymmetrical arrangement of brain function and differential manner of forming memories may well predispose the developing individual later to come upon situations in which he finds himself responding emotionally, nervously, anxiously, neurotically, sexually, angrily, hungrily, but without linguistic knowledge, or without even the possibility of linguistic comprehension as to the cause, purpose, eliciting stimulus, or origin of his behavior. Fortunately, this neurological organization involving the three brains and the three minds also provides the source of creativity, the arts, language, music, and thought, all of which, of course, are also dependent on the ability to learn, remember, and communicate.

IV

ORIGINS

12

Unconscious Origins
The Evolution of Mind and Brain

Unconscious Origins

About 1 billion years ago, the first oxygen-breathing multicellular species began to proliferate beneath the nutrient-rich, primeval seas. Although without brain, nose, mouth, ears, or eyes, all were able to sense and chemically communicate with their environment. Soon a variety of invertebrate and boneless creatures began to flourish in great numbers, and the first neurons appeared. From these creatures a plethora of aquatic forms diverged, and their brains continued to grow and develop, particularly those cerebral structures concerned with the sense of smell and the detection of pheromones, the olfactory bulb.[1] The brain and the ability to communicate had by then become quite complex and enabled creatures to perform a detailed pheromonal and chemical analysis of the environment as well as determine the edibility, sexuality, intent, and status of one's fellow creatures.

Boneless fish were followed by the first vertebrate armored fish about half a billion years ago, followed by sharks 50 million years later. With the appearance of these animals, the brain and limbic system had undergone considerable differentiation and growth. Now not just a network of nerve cells but the entire brain was able to share, store, remember, and communicate information.

The first bony fish who splashed about in the ancient seas half a billion years ago appear to have had no jaw and were not very good swimmers, lacking the paired fins characteristic of later and modern-appearing jawed fish who presumably evolved from this line. A variety of jawed species soon evolved, however, one of which developed "lobe fins" that were reinforced by bony spines protruding from inside the body. Those fish blessed with lobe fins were able to use these appendages to probe and root among the fertile ocean floor and eat of its rich organic life. It is presumed that over the course of time that lungfish equipped with these same lobe fins were enabled to venture along riverbanks, oceanfronts, and onto dry soil. That is, it is believed that these lobes were the evolutionary precursors to what would eventually become limbs. Soon creatures would be able to employ complex chemical signals as well as movements extending beyond the head and whole body to communicate their fears and desires in the open air and on dry land.

It was during this period, about 350 billion years ago that there was an explosion of life with larger and increasing complex forms appearing rapidly. Many not only cast their hungry gaze on the dry shores where plants grew in wild abandon, but were now able to venture forth on their own.

These primitive insects and amphibious creatures also began to diverge into multiple species. This evolutionary metamorphosis was accompanied by important alterations in their brain and their ability to produce young. The ability to communicate by postural gesture, movement, and via phero-mones was now accompanied by auditory signaling.

It is from these shore-hugging amphibians that reptiles emerged. However, reptiles soon diverged into a staggering number of species includ-ing reptomammals and what would become dinosaurs, birds, and true mammals. In fact, these first reptomammals, the therapsids, whose develop-ment preceded the dinosaurs, began to venture forth and grow to huge sizes well over 250 million years ago.[2]

Therapsids: The Reptomammals

With the appearance of the therapsids, auditory and visual sensitivity increased dramatically, and creatures began to care for their young. More-over important physical alterations occurred not just in their brains but in their body structure, stance, gait, limb development, and in thermoregula-tion. Moreover, the ability of these and other multicellular organisms to communicate underwent a tremendous expansion for just as the insects began to form their first nations, therapsids and the dinosaurs were nursing and caring for their young, and those who were predators began to run in packs, whereas those who were herbivores grew to gigantic sizes and traveled in herds.[3]

In contrast to lizards and amphibians, the legs of the therapsids and the dinosaurs were now located beneath rather than alongside the body. This improved their running ability and enabled them to run for long time periods without affecting their ability to breathe. Reptiles must stop in order to breathe since their legs, situated alongside their body, constrict the expansion of the lungs as they run. Moreover, the ear underwent important modifications, and vocalized communication assumed a new importance with the appearance of the therapsids.[4]

The brains of the therapsids, although larger, remained somewhat lizardlike as there was absolutely no neocortex. However, it is likely that in addition to the old limbic cortex, that a transitional form of limbic cortex had begun to develop, the cingulate gyrus. Coinciding with these developments, their ability to communicate expanded beyond simple gestures, posturing,

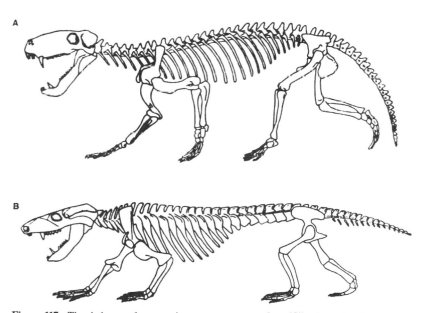

Figure 117. The skeletons of two carnivorous reptomammal, wolflike therapsids. Reproduced with permission from P. Maclean. *The Evolution of the Triune Brain.* New York: Plenum Press, 1990.

Figure 118. Two saber-toothed, wolflike therapsids are feeding on a herbivorous therapsid, Styracocephalus. From G. S. Paul. *Predatory Dinosaurs of the World*. New York: Touchstone, 1988.

as well as olfactory signaling, and now included the capacity to produce a variety of complex meaningful sounds, such as perhaps the separation cry of their infants.

Two hundred million years ago, the dinosaurs, too, however, enjoyed similar advances in their ability to compete, communicate, and survive. Nevertheless, the dinosaurs remained subordinate to the therapsids and quite small until a cataclysmic event brought forth a new world order which provided them with a temporary competitive advantage and which enabled them to flourish at the expense of the larger therapsids who had been quite suddenly killed off. About 225 million years ago, the Earth was struck by a giant meteor,[5] which may well have broken up the giant southern continent, Gondwanaland, splitting it into what is today South America, Africa, Antarctica, India, and Australia. In consequence, the largest of the therapsids and almost 90% of the life on this planet became extinct.[6]

Not all of the therapsids became extinct, however, for many of the small, rabbit-sized members of their species were spared as were many other tiny creatures, including the dinosaurs who quickly grew in size and completely took over the planet. In order to withstand and survive this

Figure 119. An Allosaurus considers snacking on several tree-living, primitive, multituberculate mammals who voice their objections. From G. S. Paul. *Predatory Dinosaurs of the World.* New York: Touchstone, 1988.

Figure 120. Gondwanaland. South America, Africa, Antarctica, India, and Australia made up one giant continent which began to break apart 225 million years ago, a time at which a giant meteor is believed to have struck, killing 90% of all life and possibly fracturing the continent, which then began to drift apart. From P. Maclean. *The Evolution of the Triune Brain.* New York: Plenum Press, 1990.

oppression, these small, surviving reptomammals over the next 150 million years became smarter, and their brains became larger and more complex, until finally they evolved into the first true neocortically equipped mammals.[7]

Many of the these ancient mammals were probably neurologically similar to possums and hedgehogs, whereas others were as complex and dangerous and savage as a small mountain lion or wolf; however, all possessed neocortex. By 65 million years ago, these catlike lions and wolf-dogs probably hunted and feasted occasionally on protoprimates, our ancient ancestors, the prosimians, creatures in some ways neurologically similar to modern-day tree shrews.

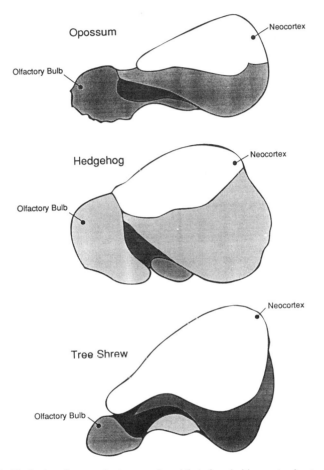

Figure 121. The brains of two ancient mammals and that of a primitive protoprimate, the tree shrew. The shroud of neocortex progressively expands to cover the forebrain and the olfactory bulbs.

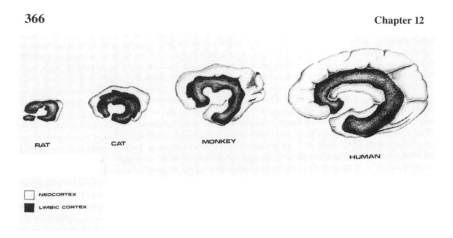

Figure 122. The limbic cortex (including the cingulate gyrus) in four species. Note similarities in size, location, and extent and progressive expansion of the neocortical mantle. From R. Joseph. *Neuropsychology, Neuropsychiatry, and Behavioral Neurology.* New York: Plenum Press, 1990.

The Demise of the Dinosaurs and the Reign of Mammals

Although dinosaurs continued to roam the planet, they were in decline by the time the true mammals and protoprimates appeared on the scene, about 75 to 100 million years ago, and long before the last giant meteor struck. Corresponding with this decline of the dinosaurs was the continued expansion and development of neocortex in both the prosimian primate and mammalian lines which in turn conferred upon these creatures extraordinary powers of intelligence, foresight, planning, and communicating.

Certainly, as is evident from history, amphibians, lizards, dinosaurs, and many birds were no match for this neocortical development, and many species were wiped out once mammals (like the dinosaurs before them) were given a temporary competitive advantage.

The demise of the dinosaurs and the rise of the mammals was also hastened by a giant asteroid or meteor which struck the planet near the Gulf of Mexico about 65 million years ago,[8] leaving a crater 110 miles wide. The result was a global catastrophe coupled with immediate and drastic global changes, cooling, and ultraviolet irradiation; consequences that had their most profound effects, however, on the largest creatures as they are the most vulnerable to the neutralizing effects of ultraviolet radiation and the most severely affected by the loss of most of their food sources. Rabbit-sized mammals, lizards, birds, small dinosaurs, and other tiny creatures would have been more or less spared, just as were other small creatures following the previous calamity, 225 million years ago. These smaller creatures would have had little difficulty finding shelter and surviving on the rotting remains of their giant cousins until the climate returned to normal.

The first prosimian primates to scurry about this planet apparently diverged from the mammalian line some 70 to 100 million years ago. In fact there may have been multiple mammalian metamorphoses and multiple protoprimate lines as well, including those which would gave rise to the quite distinct and less evolutionarily advanced New World monkeys of the Americas, who kept their tail. In contrast, Old World monkeys and apes lost this prehensile appendage.

Primate Evolution

These first protoprimates were possibly little creatures with long snouts and whiskers who devoured insects and plants and raided the nests of the dinosaurs who although in decline still ruled the planet when they first appeared. It was presumably from one of these first primates stocks that Old World monkeys, apes, and humans branched off and allegedly descended.[9]

For many millions of years, these primitive primates most assuredly scurried about in the shadows or in the cool of the day or evening, attempting to compete with ground-hugging mammalian carnivores, for living space. Others took to the trees where living was less dangerous but much more complicated.

Those who remained on the ground did well for almost 10 million years, but eventually they died out as they were no match for the flourishing multitude of mammalian predators that had wiped out the last of the remaining tiny dinosaurs. Only the raptors survived as they took to the sky and became modern-day birds.

Those primates who took to the trees adapted, flourished, and rapidly evolved to living among the branches of the forest. In consequence, tremendous alterations occurred in hand–eye coordination. Within the brain there were tremendous expansions of the visual, auditory, tactual–gestural, and fine motor neocortex. It was from this tree-loving stock that monkeys evolved in Africa about 40 million years ago.[10]

From 70 to 40 million years ago a large variety of primates continued to evolve, adapt, and prosper, and soon they gained dominion over the trees. However, about 35 million years ago their numbers suddenly began to drop precipitously. It was not a plague or some cosmic calamity which wrought these changes, but their own success at reproduction and evolution. Primates began to compete among themselves for living space, and some species died out or were killed off.

The monkeys, like the primates who came before, predominantly ate plants with insects as a secondary diet. It was the monkeys who gained dominion over the trees and whose brains became increasingly refined for controlling movement, and even flight through space. From this stock arose the apes about 30 million years ago and from which the ancestral line leading to human beings diverged about 10 to 15 million years ago. Primates now not only had established dominion over the trees but were proceeding to accomplish the same upon the forest floor.

Thirty million years ago, when apes appeared, a tremendous alteration in the ability to grasp and manipulate objects with the hand resulted. Eye–hand and fine motor coordination became exceedingly advanced as did the ability to communicate one's feelings and expectations through the hand and touch as well as facial expression and posture. They were also increasingly able to lurch about in a semierect posture. It was with this development which eventually led to upright posture in humans.[11]

Evolutionary metamorphosis is most likely to occur when an organism is exposed to a multiplicity of environments, or where two divergent worlds meet. For the primates, the netherworld of change was found where the forest ended and the savanna and grasslands began. It was exactly in such an environment that a new type of primate evolved and the line that would eventually give rise to *Homo sapiens sapiens* (the wise man who knows he is wise) appeared. A new breed of animal was on its way to developing spoken language and to threatening a good part of the planet's multiple life forms with death and extinction.

Probably more than a dozen types of apes evolved in a brief time span extending from 30 to 15 million years ago. However, one of them was quite unique and may have been one of our first real and most earliest ancestors. This man-ape is referred to as Ramapithecus, and male and female alike, they dwelled and ranged throughout the African and Asian savannas.[12] Ramapithecus was the most successful ape of its time. They were able to walk for short distances on two feet and were probably able to make small and primitive tools such as digging sticks.

Ramapithecus stood about 3 feet high, had a low forehead, a flat, wide nose, and had a face, nose, and jaw shaped more like a muzzle. However, unlike the mouth of a predator he had very small canines and a jaw that moved from side to side, like a cow. Hence, like other herbivores, this little preman was best adapted to eating vegetable matter and other tough fibrous foods, such as roots, stems, nuts, bulbs, and seeds. Like his ape cousins, the males probably occasionally hunted, captured, and killed by hand small game and possibly other primates, whereas females and their young obtained the bulk of their food by gathering. Nevertheless, to fulfill their dietary requirements they had to leave the trees and spend a greater and greater amount of time on the ground so that they could eat. Like other herbivores, it is likely that these creatures spent an inordinate amount of time in the pursuit of food.

Over the next few million years, the primate line continued to evolve as a variety of selective and different environmental pressures were being exerted on them in different parts of the world. It was while on the ground that Ramapithecus or some other closely related primate underwent further evolutionary metamorphosis around 5 to 10 million years ago. It was during this time period that several subtypes of Australopithecines appeared in various parts of the world, all of whom coexisted for some time.[13] It was also about 4 million years ago that the first of the line of *Homo*, that is, *Homo habilis* (the handy man), appeared.

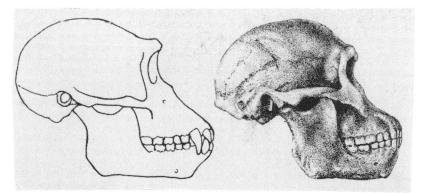

Figure 123. The skull of an Australopithecus transvallensis (right) compared to the skull of a chimpanzee (left). From G. Clark. *The Stone Age Hunters*. London: Thames & Hudson, 1967.

Australopithecus and Homo Habilis

Australopithecus and *Homo habilis* were extremely advanced for they had learned to stand on their hind legs and to walk in an upright manner.[14] Similarly, their feet evolved so as to better accommodate standing, walking, and running on two legs rather than on all four as do monkeys. In consequence, the arms and hands were freed of the necessity of holding or hanging on to a tree branch in order to move about. This conferred enormous evolutionary and neurological advantages with increased parietal representation of the hand. Moreover, those who did not stand upright probably had many accidents where they slouched through the tall grasses right into the jaws of a waiting predator. Those who stood upright had the advantage of being able to see farther as well as over the tall grasses and could thus spy the presence of predators at a distance to avoid them.

Moreover, now they could now carry things such as infants and engage in new activities that involved gripping with the hand in a wholly new manner. In consequence, they gained the ability to not only hold, grasp, explore, and manipulate objects, but also perhaps to make simple stone tools. In a few million years, their descendants would be gripping pens and pencils and typing their thoughts.

It is likely that Australopithecus was sufficiently resourceful so as to use sticks for digging and perhaps even as a weapon that it could brandish in the face of an enemy or competitor. In fact, according to Raymond Dart,[15] they were actively killing and cannibalizing one another, though this has also been disputed.[16]

However, in this regard, these protohumans were not unique for we know the apes will hurl rocks at monkeys or baboons, will wave sticks and threaten their enemies and competitors, will use rocks as primitive hammers to smash open nuts, and will attack and kill their fellows.[17] Apes also use sticks to dig out insects and steal honey, and they make sponges of leaves by which they can soak up water. Even nonprimates such as birds will use sticks to root out insects, whereas sea otters will use a rock to smash open a shellfish. However, although apes, mammals, and birds can make use of tools, only humans use tools to make other tools or to make clothes.

These alterations coincided with tremendous developments in the brain and also promoted social awareness and eventually the formation of close bonds with mates. These changes also enhanced the art of communication. For example, in addition to holding hands, the hands would also be used for signaling among other primates so as to convey feelings of anger (such as angrily swinging or throwing a tree branch), a desire to share (such as begging for a small piece of meat), to indicate friendliness or the desire to form a social bond, or even to soothe an angry neighbor by grooming his coat.

Like his ape cousins, Australopithecus, as well as *Homo habilis*, was also fairly social and probably lived in a small social or family group.[18] However, his language capability probably differed only marginally from that of modern-day apes. In this regard, it should be recognized that apes, like humans, have also continued to evolve, albeit quite slowly, and that a modern ape (such as Washoe or Koko) cannot be compared to an ape of 5 to 10 million years ago. Rather, modern-day, evolutionarily advanced apes are probably more representative of what constituted the human mind and behavior just a few million years ago.

Ancient chimpanzees and gorillas continued to lead an arboreal life and thus continued to evolve in the forests where evolutionary pressures remained slight and were thus slow to take place. Ramapithecus, Australopithecus, *Homo habilis*, and then humans instead took their next evolutionary step in the savannas and managed to spread throughout Africa and into Southern Europe, the Middle and Far East including India and China where tropical climates continued to dominate the land.[19]

Some have argued that it was from the Australopithecus stock that humans eventually arose[20]; others have suggested that it was *Homo habilis* who lived at the same time and in the same places.[21] It is also possible that the lineage went as follows: Ramapithecus to Australopithecus and *Homo habilis*, and both to distinct branches of *Homo erectus* and finally the first primitive *Homo sapiens* about 500,000 years ago in Northeast Africa and what is Israel and Palestine.[22]

That is, it is possible that Australopithecus and *Homo habilis* gave rise to different and distinct lines of human being.[23] Perhaps one branch of the Australopithecines gave rise to a particular stock of *Homo erectus* which in turn gave rise to *Homo sapiens* and then the Cro-Magnon in Southern and Northern Africa. In contrast, one branch of *Homo habilis* who lived at the same time as the Australopithecines may instead have given rise to a different stock of *Homo erectus*, such as those who lived in Eurasia and the more northern regions of the planet and who in turn gave rise to the Neanderthal dead end.

On the other hand, it is quite possible that a single line of *Homo erectus*

Chapter 12

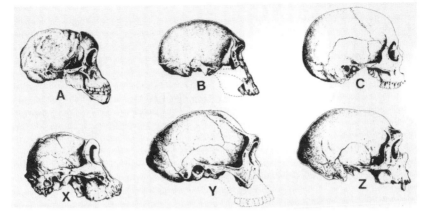

Figure 124. Two possible separate lines of evolution, one of which led from (A) Australopithecus africanus to (B) *Homo habilis* to (C) Cro-Magnon. Note enlargement of frontal region. The other led from (X) Australopithecus robustus to (Y) *Homo erectus* to (Z) the Neanderthal dead end. Reproduced with permission from P. Maclean. *The Evolution of the Triune Brain*. New York: Plenum Press, 1990.

may have undergone this metamorphosis in Northeast Africa and Palestine about 500,000 years ago,[24] and then diverged into what would become the Neanderthals who migrated north and east, and the Cro-Magnon people who initially migrated south.

Homo Erectus

In any case, following on the heels of *Homo habilis* and Australopithecus was a wide range of quite different individuals collectively referred to as *Homo erectus*, who lived throughout Africa and Eurasia from approximately 1.7 million until about 300,000 years ago. As based on physical differences and estimates of brain size, as well as geographic location in which they lived, it is also possible that there were at least three branches of *Homo erectus*, each of which may have given rise to a distinct branch of *Homo sapiens*.

The *Homo erectus* were the first individuals to have harnessed fire and the first who developed crude shelters and home bases.[25] They also utilized various earth pigments (ocher) for perhaps cosmetic or artistic purposes. In this regard, such individuals were beginning to experiment with individual

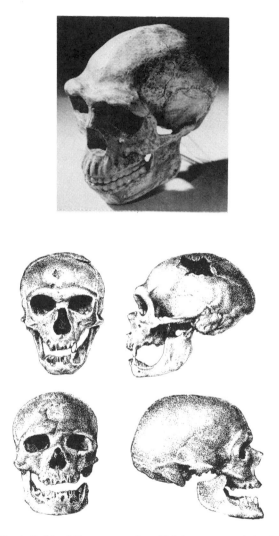

Figure 125. The skull of (top) *Homo erectus* (note high brow over eyes), (center) Neanderthal (note high brow over eyes), and (bottom) Cro-Magnon (note expansion of frontal region of skull). From G. Clark. *The Stone Age Hunters*. London: Thames & Hudson, 1967.

creative and artistic expression. They may well have also been entertaining the first true thoughts of the supernatural as they gathered together in the still of the night before the hallucinogenic flames of their cave fires.

It was during the later stages of *Homo erectus* development, about half a million years ago, that the human brain appears to have become significantly enlarged.[26] It was also during their time that big-game hunting had its onset, and sexual, physical differentiation in the size of the hips and pelvis occurred which in turn limited the ability of females to run and maneuver about in space, at least as compared to males.[27] As such, an additional divergence in the mind and brain of man and woman may have resulted as each became increasingly specialized to perform certain tasks.[28] Males increasingly engaged in big-game hunting, whereas females continued to spend a considerable amount of time gathering, engaging in child care, and maintaining the home base.

Hence, a tremendous expansion had occurred in the brain and tool-making skills of *Homo habilis*, and the sexual revolution and the origin of the family probably began with these people. However, it is not likely that they were capable of producing anything resembling a complex spoken language. On the other hand, they were no doubt quite vocal and capable of conveying a wide range of nuances and even direct commands and requests via verbalizations. Hence, the first rudiments of vocabulary and a denotative spoken language may well have appeared by at least the last few hundred thousands years of their reign until they died or were killed off around 300,000 years ago.

Neanderthal and Cro-Magnon

About 500,000 years ago, the first primitive and archaic *Homo sapiens* began to appear in increasing numbers of North Africa extending from Palestine, Israel, and into northern Egypt along the tributaries of the Nile. Like those before them, the brain of these people continued to expand, and they soon began to infiltrate Eurasia and southern Africa. Wherever they wandered they proved to be far more intelligent and resourceful as compared to the last remnants of *Homo erectus*, who over the next several thousand years were eradicated or simply died out as a species.

As noted, over the ensuing eons, these first *Homo sapiens* appear to have diverged into two species of human, the Neanderthal and the Cro-Magnon, both of whom appeared as distinct racial and cultural groups about

130,000 years ago but on different continents, Eurasia and Africa, respectively.[29] The Neanderthal flourished and continued to proliferate in Europe, Palestine, and Iraq from almost 130,000 to about 40,000 years ago only to be completely wiped off the face of the Earth, most likely by the Cro-Magnon, who in successive waves of humanity increasingly encroached, invaded, and probably raped, enslaved, and killed them off. The descendants of Cain had returned to slaughter the children of Abel.

The Neanderthals were no match for the superiorly endowed Cro-Magnon people and their massively developed brain, the frontal lobe in particular, the senior executive of the brain and personality. The Cro-Magnon also had much longer arms and limbs which along with their larger size gave them a tremendous physical advantage over the not especially well-endowed Neanderthals.

Neanderthals were heavily muscled, especially around the neck and

Figure 126. Forty thousand years after he was buried and covered with flowers, a Neanderthal seemingly glowers from the grave at those who have disturbed his rest. Reproduced with permission from R. Solecki. *Shanidar: The First Flower People*. New York: Alfred A. Knopf, 1971.

shoulders so that they must have looked in some respects, at least from the back, like modern bodybuilders.[30] From the front they may have more closely resembled someone with a big head and suffering from mental retardation due to the simian bulge over their eyes and the flattened frontal region of their skull. Nor were the mental capabilities much more advanced. Their bones were also thicker and their legs and arms were stubby, which in turn limited their ability to throw rocks, spears, or other objects long distances. As such, although they liked to hunt, they were not terribly formidable. The average height of the male was about 5'4" but they would have weighed more than 30 pounds heavier than a modern man of the same height.

Like *Homo erectus*, it appears that the Neanderthals loved the ocher pigment, and in fact liked to paint the insides of their caves completely red.[31] Near the end of their reign, they also appear to have developed an extremely primitive culture with ritual beliefs. For example, about 60,000 years ago they began to bury their dead.[32] However, like perhaps the Australopithecines some 3 million years before, the Neanderthals may have also cannibalized their fellows by extracting and eating the brain from the base of the skull.

The Neanderthals, like *Homo erectus*, were probably extremely limited in their ability to produce and articulate many sounds common among most languages today simply because the voice box they possessed was quite rudimentary.[33] As a result, they were probably unable to create many vowel sounds and could not form certain consonants such as *g* and *k*. Indeed, their ability to produce speech sounds was probably only about 10% of that of modern humans. Moreover, their mouth was quite big and was used as a tool, reducing further their ability to articulate.

Rather, the language of Neanderthal like the primitive *Homo sapiens* who came before may possibly have consisted of clicks and the rising and falling of tones, as well as occasional screams and yells. Although they may well have employed a limited vocabulary of words, it is likely that predominantly they communicated via gesture, facial expression, body posture, mimicry of animal and environmental sounds, as well as through the melodic, musical, and tonal qualities of their voice.[34] Olfaction probably also continued to play a significant role in signaling and communication.

Although probably quite fluent in natural gestural and emotional vocal communication, it is not likely that they developed complex sign language since their ability to engage in complex temporal–sequential motoric abilities (such as is involved in tool making) was very primitive and crude, their tool kit ranging only from between 30 to 60 very crude items made of stone.[35] Of course, whatever they constructed of wood has long turned to dust. Nevertheless, from the time they first appeared around 130,000 years ago until contact with the Cro-Magnon, there was little significant change in their tools, or culture except near the very end, nor is there evidence of complex language or artistic capability, or even any suggestion that they wore clothes. They displayed little innovation except near the very end when they came into increasing contact with the invading hordes of Cro-Magnon peoples. Hence, even those aspects of the left brain concerned with complex temporal sequencing and spoken language may have been only poorly developed until the Cro-Magnon people appeared on the scene.

In part, not only their shape of their throat and mouth conspired against them in developing language and advancing of their culture, but their life span. The Neanderthals generally lived to be about 40. Hence, there were probably no grandparents around to take care of and help in the training of their grandchildren. Moreover, as children reached adulthood, the adults who had populated their early lives were all dead. This would put quite a drain on the development of history, sagas, personal identity, and the learning from the mistakes of one's ancestors. They were not around long enough to pass this information on, for soon after it was acquired, they had died. In contrast, the Cro-Magnon lived into their 60s.

The Cro-Magnon and Spoken Language

The ability to engage in complex conversational speech probably remained severely limited until the appearance of the Cro-Magnon over 130,000 years ago. The Cro-Magnons stood as tall, had fully developed vocal capabilities, and had a larger brain than present-day *Homo sapiens sapiens*.[36] They also brought to the world complex spiritual beliefs, vivid, colorful imagery, finely developed artistic expression, and pictorial language. Indeed, a wild acceleration in the evolution of the mind and brain possibly began to escalate, and mental functioning become quite complex and profound. Humans appear to have become self-conscious, and the two halves of the brain seem to have increasingly become adept at performing new functions, including the art of creativity.

Many built large houses with stone foundations and lived in villages containing as many as 500 individuals. They wore makeup, jewelry, and personal decorations, and sewed their own clothes. They also believed in an afterlife, had complex supernatural beliefs and rituals which they regularly performed, and buried their dead with flowers, clothing, beads, headbands, necklaces, weapons, and offerings of food. They became extremely proficient at hunting, gathering, and harvesting, tool making, and temporal–sequential processing, and no doubt probably had names for each other and their tools. The Cro-Magnon were capable of engaging in complex conversations and due perhaps to the increased leisure time they now enjoyed, a consequence of their superior hunting and food-gathering abilities, were painting and sculpting in profusion. Indeed, they were accomplished artists

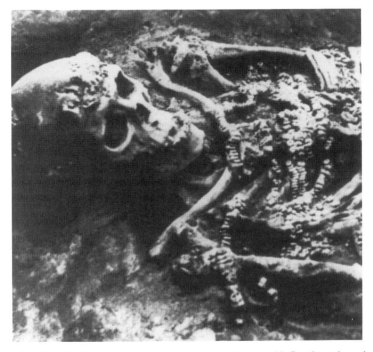

Figure 127. The skeletal remains of a Cro-Magnon man discovered in Russia, estimated to be over 25,000 years old. Note the beads, bracelet, and headband. The grave and body were sprinkled with ocher (a pigment). Courtesy of the Novosti Press Agency.

and left thousands of paintings, drawings, etchings, and fine sculptures of bisons, deer, wild horses, and bears in caves and rocky cliffs everywhere they dwelled.[37]

Before the Cro-Magnon and their frontal lobe arrived on the scene, previous species of *Homo*, like most other animals had lived pretty much in the "here-and-now." Their concern for the future probably did not extend much more than a few days. Because their lives were rooted in the present, rather than the future, they had to constantly refer to memory and to the past in order to survive. This is also why we see little or no ingenuity or change in the design or number of their tools among the Neanderthals. Cro-Magnon, too, relied on memory but was able to combine this with language and a sense of the future so as to use memory not just to survive from day to day, but to aid in planning for tomorrow.

With the appearance of the Cro-Magnons, not only did language appear and the frontal lobes dramatically increase in size, but the two halves of the brain probably became highly dissimilar in function. As I have argued in detail elsewhere,[38] with the appearance of language and then reading and writing and math, older, nonlanguage functions formerly associated with the left half of the brain were displaced. This is because there is only so much neocortical space available, and these new functions drove out the old. However, functions associated with the right half of the brain not only were retained, they, too, developed and evolved. The right and left halves of the brain had become specialized and in so doing were able to accomplish twice as much, which in turn gave birth to the art of creativity.

Where as before there had been a more or less unified, original mind, now there was a new and additional form of mental processing which in conjunction with the old gave birth to a profound creative spirit as well as the capacity to reason and form complex thoughts. Creativity, this fusion of image, sound, emotion, and thought, is also, in part, a product of this split in psychic functioning and the miscommunication that results as one half of the brain attempts to interpret, extrapolate, and make sense of and to fill the gaps in the limited information received from the other.

However, with the appearance of language, profound artistic expression, self-consciousness, and right- and left-brain functional specialization, a schism had formed in the psyche of woman and man. The mind had now become subject to fragmentation. Indeed, another product of this great change in cerebral organization and the invention of complex spoken language was the creation of intrapsychic conflict and confusion. Some thoughts, memories, and feelings possessed by one half of the brain are sometimes not understood or recognized by the other. When they are expressed by or when they serve only to arouse one half of the brain, the other half may be completely mystified or embarrassed by what is occurring.[39]

It is because of this unique relationship that one might say one thing, but through gestures, body language, facial expression, melody and tone of voice, in fact say something entirely different. It is also for this reason that sometimes the limbic system and the right half of the brain engage in certain behaviors which the left half of the cerebrum and the language-dependent conscious mind finds puzzling, disturbing, and which causes it to sometimes proclaim, "I don't know what came over me . . . why I said that . . . why I did that . . ." or even that anything amiss occurred at all.

Handedness, Language, and Temporal Sequencing

It has been repeatedly argued by a number of investigators that language, in particular its grammatical and syntactical components, is directly related to handedness and motor control. Among the majority of the population it is the right hand which is dominant for grasping, manipulating, writing, creating, destroying, and communicating. Although the left hand assists, it is usually the right which is more frequently employed for orienting, pointing, gesturing, expressing, and gathering information on the environment. The right hand appears to serve as a kind of motoric extension of language and thought insofar as it often acts at the behest of linguistic impulses. They are intrinsically linked.

While talking, most individuals display right-hand/arm gestural activity which appears to accompany and even emphasize certain aspects of speech. When speaking, the areas of the brain controlling right-hand use become activated. In part, this is due to the spread of neural excitation from the speech area to the immediately adjacent cortical regions which control hand movement. Indeed, both occupy to some degree the closely aligned neocortical space in the frontal lobe and rely upon similar neural centers in regard to motor programming. Hence, because they are so intimately linked, speaking triggers hand movement due to the spread of neural excitation.

Naming, Knowing, Counting, Finger Recognition, and Hand Control

A variety of theories have been proposed to explain the evolution of handedness and language. Nevertheless, from an evolutionary, phylogenetic, and ontogenetic perspective, handedness and temporal–sequential motor control probably preceded the development of language-specialized nerve cells.

It is first via the hand that one comes to *know* the world so that it may be named and identified, as we've noted. The infant first uses the hand to grasp various objects so they may be placed in the mouth and orally explored. As the child develops, rather than mouthing, more reliance is placed solely on the hand (as well as the visual system) so that information may be gathered through touch and manipulation. It is this same developmental pattern which characterizes evolutionary metamorphosis and the capacity to gather and communicate knowledge.

As the child and its brain matures, instead of predominantly touching, grasping, and holding, the fingers of the hand are used for pointing at and then naming the object indicated. It is these same fingers which are later used for counting and the development of temporal–sequential reasoning; that is, the child learns to count on her or her fingers, then to count (or name) by pointing at objects in space.

In this regard, counting, naming, object identification, finger utilization, and hand control are ontogenetically linked. In fact, these capacities seem to rely on the same neural substrates for their expression, the left inferior parietal lobe and the angular gyrus within the parietal lobe, and the motor centers corresponding to the hand in the frontal area of the brain.

It is relationships such as these which lend considerable credence to the argument that over the course of evolution the predominant usage of the right hand enabled the left brain to develop nerve cells specialized for counting, naming, temporal–sequential processing and thus for the mediation of grammatical–syntactical speech and language.

Speculations on Tool Making and Language

There is some evidence that nonhuman primates and certain mammals, including some birds, may have functions which tend to be more greatly concentrated in one versus the other half of their brains. However, among almost all other primates and animals, within each species about 50% tend to use their left and about 50% prefer their right limbs. Animals, as a group, are neither right nor left hand dominant as opposed to humans. Our own human ancestors were probably without preference until at least 3 million years ago.[40]

Once this preference became established, the left half of the brain began to evolve and adapt as it increased its proficiency in right hand motor control which in turn affected brain development. Tool making also evolved as did the ability to make more efficient weapons and hunting implements. Indeed, over around 2.4 to 2.6 million years ago simple stone tools were being struck by *Homo habilis* or Australopithecus or both, and by 100,000 years ago, with the appearance of our remote *Homo sapien* cousin, Neanderthal, humans were making a variety of very simple stone tools in abundance. However, it was not until the Cro-Magnons appeared that tool making became literally an art and evolved beyond the use of rock and stone.

Tool making and tool factories be they primitive or modern, require repetition and order as the implements are fashioned. First you take your bone and scrape, then you heat, then you sharpen, and so on, such that a temporal sequence is established and the same type of tool can be made over and over again. Hence, by 100,000 years ago the left hemisphere was not only dominant in regard to hand and motor control but was beginning to become increasing proficient at temporal and sequential processing. In part this increasing specialization was also related to gathering activities, an intensive full-time affair which involves repeated temporal and sequential hand movement.

Sex Differences in Languages of the Brain

Foraging, scavenging, and the chasing and hunting of small game have probably been a dominant activity of human beings and their ancestors for several million years. Naturally, our mind and brain have been tremendously influenced by these activities and have evolved accordingly.

Initially, both the male and the female of the species probably engaged almost equally in scavenging and gathering. However, the hunting of small game was an activity more or less dominated by males just as it is among apes.[41] This was the case until just a few hundred thousand years ago with the onset of big-game hunting. It was possibly this event, coupled with the rapid and progressive development of men and women's big brain several thousand years earlier, that a divergence in the mind of man and woman appears to have increasingly occurred.[42] The hunting of large game animals became an activity at which *Homo erectus* became increasingly proficient. However, it appears to have become the dominant domain of the male, although they probably continued to assist in gathering.

Presumably females continued to engage predominantly in gathering, and to a lesser extent fishing and the hunting of small game, and only occasionally assisted males during the hunting and stalking of large animals. In part this was a consequence of the presence of children and the changes in the female hip structure which reduced the ability to run and maneuver in space as effectively as males. However, among apes, it is also the males who engage in hunting and killing, whereas females in fact spend more time than males in the gathering of edibles, which in turn is probably a function of women being accompanied by their infants with whom they share their food.

Similarly, with the exception of humans who lived in the very coldest of climates where a gathering way of life would be quite difficult due to the scarcity of edible vegetable matter, for at least the last 100,000 years until perhaps about 10,000 years ago, females, and not males, appear to have been the main providers of food.[43] Big- and even small-game hunting has always been (except in the much colder far northern climates) a supplementary means of acquiring an adequate food supply.

Even among the great majority of the very few modern hunting and gathering societies in existence today, spoils from the hunt account for only about 35% of the diet. Gathering, which, we are assuming, has been the predominant domain of the female for the last 300 or so thousand years, accounts for the remainder. Even among the Cro-Magnons, for whom hunting was the center of religious and artistic life, 60% to 80% of their diet consisted of fruits, nuts, grains, honey, roots, and vegetables, which were probably gathered by the females. Hence, it probably was the female and not the male who wielded economic dominance for a significant period of our history.

Nevertheless, with this general division of labor, one might suppose that over the course of several hundred thousand years, tremendous differential influences on mental and brain functioning resulted due to selective evolutionary adaptions and pressures on survival. That is, those who were best able to adapt were most likely to pass on their genetic traits to the next generation. In consequence, the brains and minds of men and women probably became adapted and molded accordingly, each in accordance with the activities they were best at.

In this regard, as gathering and the harvesting of wild foods involves perseverative, temporal–sequential hand movements, those who were most successful at these activities probably had brains that were best adapted to these tasks. However, if the right hand is being used predominantly for picking and gathering, whereas the left hand is more likely to hold a receptacle, one might suspect that the left half of the brain would not only be more adept, but through what is known as learning, would become even more adept at temporal–sequential processing.

In contrast, as the searching and stalking and killing of prey would require good visual–spatial skills, sensitivity to environmental and nonverbal nuances, and the ability to hold one's tongue or to master one's fears, those who were most successful at these activities would in turn continue to develop and pass on to the next generation these same capabilities.

Tools, Language, and Gathering

The basic skills necessary in the gathering of vegetables, fruits, seed, berries, and the digging of roots include the ability to engage in fine and rapid, temporal–sequential physical maneuvers with the arms, hands, and particularly the fingers. As gathering has been a dominant activity for such a long time period in our prehistory, it is thus not surprising that the brain has possessed rudimentary temporal–sequential capabilities for several million years.

In addition to gathering, our ancient women ancestors made tremendous use of tools and *may* have been the first tool makers. Tool making, like gathering, was possibly a major aspect of their lives, and one they may have spent considerable time engaged in. In fact, the first tools were made not for hunting but for gathering and rooting plants; that is, if we suppose that gathering predated hunting and scavenging. For example, in grubbing for roots and bulbs, gatherers would need a digging stick which they periodically sharpened by using stone flakes. They would also carry a hammerstone for cracking nuts and for grinding the various produce collected during the day.

These ancient women did not spend their time solely gathering, for foods had to be prepared, and clothes had to be fashioned out of hides. Their duties may have included cleaning the hides via the use of a scraper, drying, and curing the skin over the smoke of a fire, and then using a knife or cutter to make the general desired shape, and then a punch to make holes through which leather straps or vine could be passed so as to create a garment that could keep out the cold. By Cro-Magnon times they were weaving and using a needle to sew garments together. Although there is no way of knowing if a woman was the first tool maker, women certainly regularly engaged in these endeavors as more or less ongoing activities.

In tool making, technique comes to have preeminence. With the invention of technique, a linear, temporal, and sequential approach long developed over eons of gathering, probably came to characterize the process. Certain tools are made a certain way with special instruments with particular movements, and with a certain degree of muscular power and considerable precision. To make and utilize tools requiring a precision grip required that the manufacturer not only have hand and fingers capable of such feats, but a brain that could control this hand and which could use foresight and planning in order to carry out the manufacture.

Due to selective pressures and the survival and breeding of those who were successful at these activities, the left half of the brain, which controls the right hand, became increasingly adapted for the control of temporal sequencing be it for the purposes of tool making, the manufacture of clothing, or for gathering. Hence, it was not only tools, but the neural substrate for the temporal–sequential and grammatical aspects of what would become spoken language that was being forged and passed on to succeeding generations, a process that had its onset several million years ago.

As females have engaged in gathering as a dominant mode of activity for time periods much longer than males, coupled with their possible role in tool manufacturing and tool use (e.g., skinning, clothes making, etc.), it might be assumed that these changes were maximal in the brains of women, particularly in the motor areas controlling speech (Broca's area) and the hand.

In fact, females tend to display many superiorities in regard to left-hemisphere language skills. Women excel over males on word-fluency tests, for example, naming as many words containing a certain letter, or words belonging to a certain category.[44] Females also vocalize more as infants, speak their first words and develop larger vocabularies at an earlier age, and improve their articulation skills at a faster rate. Similarly, among children, the speech of females is easier to understand. Fine motor skills, such as those involving rapid temporal sequencing, are also abilities at which females tend to excel as compared to males.[45]

Men are far more likely to suffer language-related disorders and to lose language abilities at a faster rate than women as both groups age. It is also well known in clinical practice that when a patient becomes aphasic following a stroke, women tend to recover their language more quickly and more fully.

These and the other findings mentioned above thus suggest possible neuroanatomical differences in the brains of males and females. Indeed, not only is the limbic system sexually differentiated, but possibly the language axis of the left hemisphere as well. That is, when we study the anatomy of the brain, it is apparent that those areas which subserve language are larger on the left versus the right side of the cerebrum; that is, the left superior temporal lobe.[46] However, there are some studies which indicate that when these differences are statistically analyzed, that they are maximal in females. They appear to have more brain space devoted to language functioning. However, it is important to note that some well-respected neuroscientists have also claimed that there are no gender differences in the size of the left superior temporal lobe. Nevertheless, as both sexes age, males tend to lose more cortical tissue in the brain areas subserving language, indicating that this modality is more fragile in males.

There is some suggestion that functionally, Broca's area is more well developed in females than males, whereas the left-temporal-parietal language areas are functionally more developed in males.[47] In this regard it is interesting to note that Broca's area is associated not just with talking, but also with temporal–sequential motor control. When Broca's area becomes activated, so to does the right hand.

In contrast, it is well established based on a variety of independent studies that, in general, males have far more elaborately developed and thus superior visual–spatial and spatial perceptual skills than females. This includes a male superiority in the recall and detection of geometric shapes, detecting figures that are hidden and embedded in an array of other stimuli, constructing three-dimensional figures from two-dimensional patterns, visually rotating or recognizing the number of objects in a three-dimensional array, playing and winning at chess (which requires superior spatial abilities), directional sense and geographic knowledge, the solving of tactual and visual mazes, aiming and tracking such as in coordinating one's movements in relationship to a moving target, coordination in aiming and throwing, and comprehending geometrical concepts.[48] Indeed, only about 25% of females in general exceed the average performance of males on tests of such abilities. Moreover, some of these differences are present during childhood.

These are also skills associated with the functional integrity of the right half of the brain.[49] Many of these abilities are also directly related to skills that would enable an ancient hunter to track, throw a spear, and dispatch various prey without getting lost and while maintaining a keen awareness of all else occurring within the environment.

Although there are exceptions, ontogeny often replicates phylogeny, and in this regard we see that not only do modern human females demonstrate an earlier onset of the ability to use language and are less likely to suffer language disturbances, but our female ancestors may have more fully developed the temporal–sequential and social–emotional aspects of language first as well.

When we consider the many selective pressures which acted on males to inhibit speech during the hunt, whereas females were allowed to talk quite freely while gathering (as there would be no fear of scaring off game), it certainly appears that these factors strongly and differentially promoted female linguistic development and a greater capacity to discuss topics unrelated to events associated with the hunt. Language was now being used for social bonding and only later became an instrument designed more and more, at least insofar as males are concerned, for the purposes of exchanging information related to business and sports—the hunt.

Hunting, Visual–Spatial Skills, and the Strong Silent Type

Searching for large game animals requires that the hunting band roam *quietly* over a huge expanse of varied terrain which may extend up to 500 or more miles. One does not need to speak in order to be a good hunter as is evident by the cooperative success demonstrated by dogs, wolves, lions, and those two-legged dinosaurs who hunted in packs. Hence, hunting does not promote language development.

For a human to be a successful hunter requires good directional sense, so that one could wonder and also find his way home without the aid of a street sign. He would need good visual–spatial skills such as depth and distance perception, so that he could walk and run as well as anticipate the trajectory of movement of a running prey, without tripping and falling. Also of importance is an excellent capacity to recognize, comprehend, and mimic animal, environmental, and nonspeech sounds, and an increased capacity to communicate nonverbally via gesture, body language, and particularly via animals' mimicry so that game would not be scared off by the sounds of speech and so that communication with other hunters could occur. These are all skills associated with the right brain of a modern-day human being and are skills at which males excel.

When males banded together for the purposes of the big game hunt, they had to walk silently for long time periods, and their only form of communication was probably concerned only with the expression of facts and information about the hunting they were engaged in. If they were to begin talking about anything else such as their family, they possibly would have been shunned as their discussion would have been irrelevant and would have scared off the game. A premium was thus placed on men who could hold their tongue. Also decreasing the need for talk was the fact that as hunters stalk their prey they may walk at anywhere from 25 to 100 yards apart. Two or more men standing together are more likely to warn a prey of their presence, whereas men standing apart have the advantage of chasing a beast into the waiting arms of his comrade.

Finally, it is likely that once the hunt was complete and they returned to their base camp, that the men did not then begin to discuss their wives, children, feelings, or personal problems, but their own prowess as hunters, incidents related to the hunt, and whatever adventures they may have had that would enhance their status among their fellows. They became storytellers and commentators.

Gathering and Why Women Like to Talk and Share Their Feelings

In contrast to males who remained generally silent during the course of the hunt, women were able to freely chatter among themselves. Indeed, gathering fostered the development of language, and like hunting for men, also served as a social activity but in a more physically close and socially intimate manner.

Our ancient female ancestors probably gathered in large groups of 7 or more individuals, as the average size of most bands was about 25 individuals, though members of a Cro-Magnon tribe may well have numbered 500 or more. Some women were pregnant or carried infants who might be set on the ground here and there, or accompanied by young adolescents who would frolic about and play. Such gathering groups must have commonly been loud, noisy, and very gay affairs filled with the talk of the women and the sounds of games and shrieks of the children. Hence, unlike the men who remained quiet for long time periods in order to not scare off game, the women were free to chatter and talk to their heart's delight. Talking also served as a means of maintaining the location of the group so that if a gatherer or a child chanced to walk away, she (or her child) could always relocate the others by their hodgepodge of speech.

Talking thus became part of the social and gathering glue and served the purpose of keeping the group together and thus of bonding them as a collective. The women likely talked about their children, their husbands, each other, and were thus exposed and allowed to expose their own feelings and thoughts to those who valued talking as much as they. For women, to socialize and be together meant to talk, and to talk is an essential social bonding element for women even today.

When considering our evolutionary heritage and the fact that women have spent a good 100,000 years or more in female-dominated gathering groups, where socializing and freely expressing oneself about social and family matters were the norm, it should come as no surprise that modern-day females continue to respond similarly. In this regard, it perhaps should not be surprising that after almost half a million years of holding his tongue and jockeying for status and position among his fellows, that modern-day man continues to respond similarly and have the same concerns, albeit translated and modified to some degree—sex, sports, politics, business.

Multiple and Emotional Means of Communication

Languages of the Body and the Brain

The ability to communicate using complex speech patterns has been around for at least 50,000 and maybe even 130,000 years, the time at which our ancestors, the Cro-Magnon people, first appeared upon the scene. In contrast, the ability to vocally express and convey one's emotions and other motivationally significant feeling states via sound may have been well established some 250 million years ago, for it was at this time that the therapsids (reptomammals) first began to possess refined organs of hearing and to develop a transitional limbic neocortex, the four-layered cingulate gyrus. Moreover, the ability to transmit and receive emotional and social signals through olfactory chemicals and pheromones has been a dominant mode of communication for at least a billion years almost regardless of species.

Because of their greater antiquity, emotional messages not only color what we say and hear but can sometimes take control over and even determine every facet of what is being communicated, even when the triggering stimulus is based on little more than the memory of a smell. There are thus a myriad of forces acting on the human psyche and affecting human communication, some of which are as subtle as a whiff of perfume.

Sometimes a person may lose control over what he or she says and what he or she does and express information which is best left unsaid. This is because the emotional aspects of communication are more firmly embedded in the fabric of our psyche and simply overwhelm the neocortex and the more recently and incompletely developed linguistic modes of conscious and civilized means of interacting. Since vocabulary and syntax have come to overlie these more powerful and ancient forms of expression, they are more easily overthrown.

This is a function of having more than a right and left cerebral hemisphere, however, or our possession of an ancient limbic system where these ancient emotions continue to be represented and maintained. It is also a function of the way we were raised and what we observed while growing up.[50]

Yet, there is so much more to communication, all of which can also disrupt or enhance the effectiveness and outcome of what is being transmitted. Indeed, there are so many facets that have in fact nothing to do with the production of speech, that multiple messages are often being expressed and discerned simultaneously and sometimes in a contradictory manner. This is because communication can be through facial expression, posture, movement, dress, and even hairstyle, and it may be transmitted through speech, gesture, touch, and even smell as well as via the right and left halves of the brain.

In some cases these multimessages all reinforce one another and enhance what has been said, whereas in others, contradictory signals are being conveyed which may confuse the receiver or which may reflect ambivalence on the part of the sender. Sometimes the messages are meant to be confusing and ambiguous such as when a prey is attempting to elude a predator, or when a man and a woman cautiously flirt after meeting for the first time. What eventually transpires depends on their success at interweaving and separating out these competing messages. Even animals and insects which live a solitary life are dependent on the ability to correctly perceive and respond to visual, auditory, tactual, and olfactory signals. For if they fail to do so, they die.

Moreover, communication is also affected by context, cultural and personal experience and expectations, hopes, desires, delusions, and self-deceptions, as well as by memory and the manner in which our very brain acts to perceive, process, store, and express information. The fact that we have two brains and a limbic system as well as our own individualized personal experiences based on the manner in which we were raised as children complicates the picture all that much more.

Many human beings, however, consciously pay little attention to the multiple messages being transmitted by their fellow humans and often fail to realize that they are transmitting a panoply of meaningful signals as well. Instead, many human beings tend consciously to focus on vocabulary and grammar so that in consequence, love, life, business, and family relationships sometimes suffer.

Given this multiplicity of interacting variables, perhaps it is little wonder that points are sometimes missed, misunderstandings occur, fights result, sexual messages and insults are inadvertently transmitted or received, and effective communication breaks down. When communication breaks down, relationships break down. In the history of our human race, devastating wars have resulted as a consequence of miscommunication and the reception and transmission of unintended messages.

Sometimes these wars go on inside the solitude of a single individual's head only to spill forth and disrupt relationships at work, at home, and even at play. When this occurs, all of life can become a battleground. Nevertheless, these wars and battles were first fought long before human beings walked on the forest floor or their ancestors hung from the redwood bough.

Fortunately, the fusion of these divergent neurological resources, our limbic system and the right and left halves of the brain, has also given rise to love, honor, morality, creativity, the arts, music, literature, the ability to speak and think in words, and the ability to write them down and read or remember them later. Through the interactions of these three brains, we are even impelled to question and examine our own existence and to develop electronically enhanced "neurological" sensors, computerized eyes and ears which scan the Earth, the sea, the cosmos, and even our brain and mind; listening and searching for the voice and hand of God in an attempt to understand the unconscious origins of all things.

. . . and the spiraling Universe coiled back and swirled round on planets' knees and shooting starts to ponder its own depths in the Temple of Human Consciousness.
. . . in the mirror of the sea of human consciousness.
. . . to peer and reflect upon its own soul as mirrored in the rising tides of human consciousness. . . .

A Romance of Planets and Stars

The process of communication has been over a billion years in the making and has been derived from multiple sources and species, both extant and extinct. The modern mind is a veritable museum and the manner in which we communicate reflects these multiple forces and ancestral influences.

Although humans are able to walk in space, the mind is still adapted to a life spent living among the elements and as such remains profoundly aware and subject to environmental influences be they transmitted through sunlight, smell, touch, movement, gesture, sights, sounds, dreams, images, voices, or visions. The human body and brain remain geared to sense and respond to the known and the unknown, and to emergencies which threatened existence for millions of years.

The brain and its many constituent elements also retain and are sensitive to all those means of communicating, perceiving, knowing, lying, and remembering which enabled the first forms of life and now human beings to evolve, flourish, and survive. Indeed, the art and science of communication, and the many languages of the body and the brain share the same unconscious origins and a common root firmly enmeshed in the eternal, interminable antiquity of all things.

. . . for when dust becomes dust and ashes ash,
we will remain what we are and what we were,
star dust and star light . . .
A romance of Planets and Stars . . .

Notes

In the Beginning . . .

1. J. E. Lovelock, *Gaia: A New Look at Life on Earth* (New York: Oxford University Press, 1979); L. Margulis, *Origin of Eukaryotic Cells* (New Haven: Yale University Press, 1970); B. K. Napier and V. Clube, "A theory of terrestrial catastrophism," *Nature*, 282 (1979), 455–458. G. C. Reid et al., "Influence of ancient solar proton events on the evolution of life," *Nature*, 259 (1976), 177–179; J. W. Schopf, "The evolution of the earliest cells," *Scientific American*, 239 (1978), 84–102.

2. J. D. Watson, *The Double Helix* (New York: Penguin, 1979); R. Dawkins, *The Selfish Gene* (New York: Oxford University Press, 1976).

3. L. H. Hyman, "The transition from the unicellular to the multicellular individual," *Biological Symposiums*, 8 (1942), 27–42; Margulis, 1970; Schopf, 1978.

4. S. Weinberg, *The First Three Minutes* (London: Deutsch, 1977).

5. F. Hoyle, *The Black Cloud* (London: Heinemann, 1957); W. M. Irvine et al., "Thermal history, chemical composition, and the relationship of comets to the origin of life," *Nature*, 283 (1980), 748; J. Gribbin, *Genesis* (New York: Delacorte Press, 1981).

6. Margulis, 1970; Schopf, 1978; Hyman, 1942.

7. Margulis, 1970; Hyman, 1942.

8. D. E. Clayton, "A comparative study of the non-nervous elements in the nervous system of invertebrates," *Journal of Entomology and Zoology*, 24 (1932), 3–22; Lovelock, 1979.

9. J. W. Papez, *Comparative Neurology* (New York: Hafner, 1967); C. U. Ariens

Kappers, *The Evolution of the Nervous System* (Bohn: Harlem, 1929); T. L. Lentz, *Primitive Nervous Systems* (New Haven: Yale University Press, 1968).

10. R. Joseph, "The right cerebral hemisphere. Language, music, emotion, visual–spatial skills, body image, dreams, and awareness." *Journal of Clinical Psychology*, 44 (1988), 630–673; R. Joseph, "The neuropsychology of development. Hemispheric laterality, limbic language, and the origin of thought," *Journal of Clinical Psychology*, 36 (1962), 4–33; R. Joseph, *Neuropsychology, Neuropsychiatry, and Behavioral Neurology* (New York: Plenum Press, 1990).

11. K. Von Frisch and M. Lindauer, "The language and orientation of the honey bee," *Annual Review of Entomology*, 1 (1956), 45–58; K. Von Frisch, *The Dancing Bees* (New York: Harcourt, 1953); S. S. Schneider et al., "Vibration dance of the honey bee," *Animal Behavior*, 34 (1986), 377–385.

12. Joseph, 1982, 1990.

13. R. Joseph, *The Right Brain and the Unconscious* (New York: Plenum Press, 1992a); P. Maclean, *The Evolution of the Triune Brain* (New York: Plenum Press, 1990); Joseph, 1982, 1988, 1990.

14. R. Joseph, "Confabulation and delusional denial: Frontal lobe and lateralized influences," *Journal of Clinical Psychology*, 42 (1986), 507–518. R. Joseph, R. E. Gallagher, W. Holloway, and J. Kahn. "Two brains, one child: Interhemispheric transfer deficits and confabulation in children aged 3, 7, 10," *Cortex*, 20 (1988), 317–331; R. Joseph, "The limbic system: Emotion, id, unconscious mind," *The Psychoanalytic Review*, 1992b; Joseph, 1982, 1988, 1990.

15. MacLean, 1990; Joseph, 1982, 1988, 1990, 1992a,b; L. Miller, *Inner Natures* (New York: St. Martin's Press, 1990); L. Miller, *Freud's Brain* (New York: Guilford Press, 1991); R. Ornstein, *The Psychology of Consciousness* (San Francisco: W. H. Freeman, 1972).

16. Joseph, 1982, 1988, 1990, 1992a.

17. Joseph, 1982, 1988, 1990, 1992a; Miller, 1990, 1991; Ornstein, 1972.

18. Watson, 1970.

19. G. H. Bishop, "Natural history of nerve impulses," *Physiological Review*, 36 (1956) 376–399; M. C. Behrens, "The electrical response of the planarian photoreceptor," *Comparative Biochemistry & Physiology*, 5 (1961), 129–138; Papez, 1967; Ariens Kappers, 1929; Lentz, 1968; Clayton, 1932.

20. J. H. Orton, "An experiment effect of light on the sponge," *Nature*, 113 (1924), 924–925; R. H. Emson, "The reactions of the sponge to applied stimuli, *Comparative Biochemistry and Physiology*," 18 (1966), 805–827; Lentz, 1968; Bishop, 1956.

21. Papez, 1967; Ariens Kappers, 1929; Lentz, 1968; Orton, 1924; Emson, 1966.

22. Emson, 1966.

23. Lentz, 1968.

24. P. B. Applewhite, "The micrometazoa as a model systems for studying the physiology of memory," *Yale Journal of Biology and Medicine*, 39 (1966), 90–105; Papez, 1967; Ariens Kappers, 1929; Lentz, 1968.

25. A. L. Burnett and N. A. Diehl, "The nervous system of hydra," *Journal of Experimental Zoology*, 157 (1964), 237–250; Orton, 1924; Emson, 1966.

26. J. V. McConnel et al., "The effects of regeneration upon retention of a conditioned response in the planarian," *Journal of Comparative and Physiological Psychology*, 52 (1959) 1–5; J. V. McConnel et al. "The effects of ingestion of conditioned planaria on the response level of naive planaria," *Worm Runners Digest*, 3 (1961), 41–47; A. L. Jacobson, "Learning in flatworms and annelids," *Psychological Bulletin*, 60 (1963), 74–94; Behrens, 1961; Burnett and Diehl, 1964.

27. McConnel et al., 1959, 1961.

28. Papez, 1967; Ariens Kappers, 1929; Lentz, 1968; Bishop, 1956; Burnett and Diehl, 1964.

29. Papez, 1967; Ariens Kappers, 1929; Bishop, 1956.

30. Joseph, 1982, 1990, 1992a,b.

31. Papez, 1967; Ariens Kappers, 1929.

32. MacLean, 1990; Papez, 1967; Ariens Kappers, 1929.

33. R. Broom, *The Mammal-like Reptiles of South Africa and the Origin of Mammals* (London: Witherby, 1932); A. S. Brink, "Speculation on some advanced mammalian characteristics in the higher mammal-like reptiles," *Palaeontology*, 4 (1956), 77–95; G. S. Paul, *Predatory Dinosaurs* (New York: Simon & Schuster, 1988); R. T. Bakker, "Dinosaur physiology and the origin of mammals," *Evolution*, 25 (1971), 636–658; D. K. Thomas and E. D. Olson, *A Cold Look at the Warm Blooded Dinosaurs* (Washington, DC: AAAS); W. E. Scheele, *Prehistoric Animals* (New York: World Publishing Co., 1954); V. J. Maglio, *Evolution of African Mammals* (Cambridge: Harvard University Press, 1978); A. W. Crompton and F. A. Jenkins, "Mammals from reptiles," *Annual Review of Earth and Planetary Sciences*, 1 (1973), 131–155; A. W. Crompton and F. A. Jenkins, *Origin of Mammals*. In *Mesozoic Mammals*, J. A. Lillegraven et al. (Eds.) (Berkeley: University of California Press, 1979); A. S. Romer, *Vertebrate Paleontology* (Chicago, University of Chicago Press, 1966); J. C. Quiroga, "The brain of two mammal-like reptiles," *J. Hirnforsch*, 20 (1979), 341–350; J. C. Quiroga, "The brain of the mammal-like reptile," *J. Hirnforsch*, 21 (1980), 299–336.

34. W. Alvarez, "Toward a theory of impact crisis," *Eos*, 131 (1986), 248–250; M. Davis et al., "Extinction by periodic comet showers," *Nature*, 308 (1984), 715–717.

35. R. T. Bakker, "Tetrapod mass extinctions" (pp. 339–468) In *Patterns of Evolution*, A. Hallem (Ed.) (Amsterdam: Elsevier, 1977); B. K. Napier and

V. A. Clube, "Theory of terrestrial catastrophism," *Nature*, 282 (1979), 455–457; Davis et al., 1984.

36. D. Johanson and J. Shreeve, *Lucy's Child* (New York: Morrow, 1989); D. Johanson and M. Edey, *Lucy: The Beginnings of Humankind* (New York: Simon & Schuster, 1981); J. Pfeiffer, *The Emergence of Humankind* (New York: Harper & Row, 1985); L. Binford, *Bones: Ancient Men and Modern Myths* (New York: Academic Press, 1981); F. brown et al., "Early Homo erectus skeleton from west of Lake Turkana, Kenya," *Nature*, 316 (1985), 788–792; R. L. Hay and M. D. Leakey, "The fossil footprints of Laetoli," *Scientific American*, 2 (1982), 47–87; W. H. Kimbel and T. D. White, "A reconstruction of the adult cranium of Australopithecus afarensis," *American Journal of Physical Anthropology*, 52 (1980), 244; R. E. Leakey, "Hominids of Africa," *American Scientist*, 64 (1976), 174–178; C. O. Lovejoy, "Evolution of human walking," *Scientific American*, 11 (1988), 118–125.

37. A. Leroi-Gourhan, *Treasures of Prehistoric Art* (New York: Harry N. Abrams); H. Vallada et al., "Thermoluminescence dating of Mousterian 'Proto-Cro-Magnon' remains from Israel and the origin of modern man," *Nature*, 331 (1988), 614–616; H. G. Bandi, *Art of the Stone Age* (New York: Crown Publishers, 1961); Pfeiffer, 1985; R. Prideaux, *Cro-Magnon Man* (New York: Time-Life Books, 1973).

38. P. Lieberman and E. S. Crelin, "On the speech of Neanderthal man," *Linguistic Inquiry*, 2 (1971), 203–222; P. Lieberman, *The Biology and Evolution of Language* (Cambridge: Harvard University Press, 1984); G. Constable, *The Neanderthals* (New York: Time-Life Books, 1973).

39. R. Joseph, Awareness and the role of conscious self-deception in resistance and repression, *Psychological Reports*, 46 (1980), 767–781; Joseph, 1982, 1988, 1990, 1992a,b.

Chapter 1

1. H. Hecaen and M. L. Albert, *Human Neuropsychology* (New York: John Wiley, 1978); H. Goodglass and E. Kaplan, *Boston Diagnostic Aphasia Examination* (Philadelphia: Lea & Febiger, 1982); D. N. Levine and E. Sweet, *Localization of lesions in Broca's motor aphasia*. In *Localization in Neuropsychology*, A. Kertesz (Ed.) (pp. 185–207) (New York: Academic Press, 1983). R. Joseph, *Neuropsychology, Neuropsychiatry, and Behavioral Neurology* (New York: Plenum Publishing, 1990).

2. Hecaen and Albert, 1978; Goodglass and Kaplan, 1982; Joseph, 1990.

3. R. Joseph, "The neuropsychology of development: Hemispheric laterality, limbic language and the origin of thought," *Journal of Clinical Psychology*, 38

(1982), 4–33; R. Joseph, "The right cerebral hemisphere," *Journal of Clinical Psychology*, 44 (1988), 630–673; Joseph, 1990.

4. Joseph, 1982, 1988, 1990.

5. P. B. Gorelick and E. D. Ross, "The aprosodias," *Journal of Neurology, Neurosurgery, and Psychiatry*, *37* (1987) (727–737); E. Ross, "The aprosodias: Functional-anatomic organization of the affective components of language in the right hemisphere," *Archives of Neurology*, 38 (1981), 561–589; Joseph, 1982, 1988, 1990; D. Breitling et al., "Auditory perception of music measured by brain electrical activity mapping," *Neuropsychologia*, 25 (1987), 765–774; M. P. Bryden et al., "A left-ear advantage for identifying the emotional quality of tonal sequences," *Neuropsychologia*, 20 (1982), 83–87; A. Smith, "Speech and other functions after left hemisperectomy," *Journal of Neurology, Neurosurgery, and Psychiatry*, 29 (1966), 467–471; A. Yamadori et al., "Perseveration of singing in Broca's aphasia," *Journal of Neurology, Neurosurgery, Psychiatry*, 40 (1977), 221–224; A. Carmon and I. Nachshon, "Ear asymmetry in perception of emotional non-verbal stimuli," *Acta Psychologica*, 37 (1973), 351–357; H. W. Gordon, "Hemispheric asymmetries in the perception of musical chords," *Cortex*, 6 (1970), 387–398; H. R. McFarland and D. Fortin, "Amusia due to right temporal-parietal infarct," *Archives of Neurology*, 39 (1982), 725–727; K. M. Heilman et al., "Comprehension of affective and nonaffective prosody," *Neurology*, 34 (1984), 917–921; K. Heilman et al., "Auditory affective agnosia," *Journal of Neurology, Neurosurgery, and Psychiatry*, 38 (1975), 69–72; S. Weintraub et al., "Disturbances in prosody: A right hemisphere contribution to language," *Archives of Neurology*, 38 (1981), 742–744; S. Blumstein and W. E. Cooper, "Hemispheric processing of intonational contours," *Cortex*, 10 (1974), 146–158; Joseph, 1982, 1988, 1990.

6. Joseph, 1982, 1988, 1990.

7. J. L. Bradshaw et al., "Braille reading and left and right hemispace," *Neuropsychologia*, 20 (1982), 493–500; A Carmon and H. P. Bechtoldt, "Dominance of the right cerebral hemisphere for stereopsis," *Neuropsychologia*, 7 (1969), 29–39; A Carmon and A. L. Benton, "Tactile perception of direction and number in patients with unilateral cerebral disease," *Neurology*, 19 (1969), 525–532; H. D. Cohen et al., "Electroencephalographic laterality changes during human sexual orgasm," *Archives of Sexual Behavior*, 5 (1976), 189–200; S. Corkin et al., "Somotosensory thresholds: Contrasting effects of post-central gyrus and posterior parietal-lobe excisions," *Archives of Neurology*, 23 (1970), 41–58; E. DeRenzi and G. Scotti, "The influence of spatial disorders in impairing tactual discrimination of shapes," *Cortex*, 5 (1969), 53–62; J. E. Desmedt, "Active touch exploration of extrapersonal space elicits specific electrogenesis in the right cerebral hemisphere of intact right-handed man." *Proceedings of the National Academy of Sciences*, 74 (1977), 4037–4040; R. G. Ley and M. P.

Bryden, "Hemispheric differences in processing emotions and faces," *Brain and Language*, 7 (1979), 127–138; A. G. Dodds, "Hemispheric differences in tactuo-spatial processing," *Neuropsychologia*, 16 (1978), 247–254; J. Hom and R. Reitan, "Effects of lateralized cerebral damage on contalster and ipsilateral sensorimotor performance," *Journal of Clinical Neuropsychology*, 3 (1982), 47–53; E. DeRenzi, "Prosopagnosia in two patients with CT-scan evidence of damage confined to the right hemisphere," *Neuropsychologia*, 24 (1986), 385–389; R. Landis et al., "Are unilateral right posterior cerebral lesions sufficient to cause prospagnosia? Clinical and radiological findings in six additional patients," *Cortex*, 22 (1986), 243–252; D. N. Levine, "Prosopagnosia and visual object agnosia; A behavioral study," *Brain and Language*, 5 (1978), 341–365; Joseph, 1988, 1990.

8. H. Gardner et al., "Missing the point: The role of the right hemisphere in the processing of complex linguistic materials." In *Cognitive Processing in the Right Hemisphere*, E. Perceman (Ed.) (New York: Academic Press, 1983); H. H. Brownell et al., "Inference deficits in right brain-damaged patients," *Brain and Language*, 27 (1986), 310–321; M. Cicone et al., "Sensitivity to emotional expressions and situations in organic patients," *Cortex*, 16 (1980), 145–158; F. Boller et al., 1979; J. W. Dwyer and W. E. Rinn, "The role of the right hemisphere in contextual inference," *Neuropsychologia*, 19 (1981), 479–482; Joseph, 1982, 1988, 1990.

9. Joseph, 1982, 1988, 1990.

10. Joseph, 1988, 1990.

11. A. Gates and J. L. Bradshaw, "The role of the cerebral hemispheres in music," *Brain and Language*, 3 (1977), 451–460; Heilman et al., 1984, 1975; Weintraub et al., 1981; Gorelick and Ross, 1987; Ross, 1981; Joseph, 1982, 1988, 1990.

12. Y. Kim et al., "Visuoperceptual and visuomotor abilities and locus of lesion," *Neuropsychologia*, 2 (1984), 177–185; D. Benson and M. Barton, "Disturbances in constructional ability," *Cortex*, 6 (1970), 19–46; A. Benton, "Visuoperceptive, visuospatial and visuoconstructive disorders." In *Clinical Neuropsychology*, K. M. Heilman and E. Valenstein (Eds.) (pp. 186–232) (Oxford: Oxford University Press, 1979); F. W. Black and B. A.Bernard, "Constructional apraxia as a function of lesion locus and size in patients with focal brain damage," *Cortex*, 20 (1984), 111–120; R. Calvanio et al., "Left visual spatial neglect is both environment-centered and body-centered," *Neurology*, 37 (1987), 1179–1183; E. DeRenzi, *Disorder of Space Exploration and Cognition* (New York: Wiley, 1982); Hecaen and Albert, 1978; DeRenzi, 1986; Landis et al., 1986; Levine, 1978.

13. J. E. Bogen, The other side of the brain. *Bulletin of the Los Angeles Neurological Society*, 34 (1969); J. Levy, "Language, cognition and the right hemisphere,"

American Psychologist 38 (1983), 538–541; R. Sperry, "Lateral specialization in the surgically separated hemispheres," In *The Neurosciences Third Study Program*, F. O. Schmidt and F. G. Worlden (Eds.) (pp. 1–12) (Cambridge: MIT Press, 1974); R. Sperry, "Brain bisection and the neurology of consciousness," In *Brain and Conscious Experience*, J. C. Eccles (Ed.) (pp. 298–313) (New York: Springer Verlag, 1978); R. Joseph, "Dual mental functioning in a split-brain patient," *Journal of Clinical Psychology*, 44 (1988b), 770–779; R. Joseph, *The Right Brain and the Unconscious* (New York: Plenum Press, 1992); Joseph, 1982, 1988, 1990.

14. M. S. Gazzaniga and J. E. LeDoux, *The Integrated Mind* (New York: Plenum Press, 1978); J. Levy and C. Trevarthen, "Metacontrol of hemispheric function in human split-brain patients," *Journal of Experimental Psychology: Human Perception and Performance*, 2 (1976), 299–312; Sperry, 1974, 1978; Joseph, 1982, 1988a,b, 1990, 1992.

15. Heilman et al., 1975, 1984; Gardner et al., 1983; Brownell et al., 1986; Cicone et al., 1980; Gorelick and Ross, 1987; Ross, 1981; Joseph, 1982, 1988, 1990.

16. Smith, 1966; Joseph, 1988, 1990.

Chapter 2

1. M. S. Mayer and R. W. Mankin, *Neurobiology of Pheromone Perception*, 95–144. In *Comprehensive Insect Physiology, Biochemistry, and Pharmacology* G. A. Kerkut and L. I. Gilbert (Eds.) (New York: Pergamon Press, 1985); R. Bedichek, *The Sense of Smell* (New York: Doubleday, 1960); R. Burton, *The Language of Smell* (London: Routledge & Kegan Paul, 1976); R. W. Moncrieff, *Odours* (London: Heinemann, 1970); H. M. Bruce, "Pheromones," *British Medical Bulletin*, 26 (1970), 10–22; E. O. Wilson, *Sociobiology* (Cambridge: Harvard University Press, 1980); V. Tamaki, "Sex pheromones" (145–192). In *Comprehensive Insect Physiology, Biochemistry, and Pharmacology*. G. A. Kerkut and L. I. Gilbert (Eds.) (New York: Pergamon Press, 1985); M. S. Blum, "Alarm pheromones" (193–224). In *Comprehensive Insect Physiology, Biochemistry, and Pharmacology*. G. A. Kerkut and L. I. Gilbert (Eds.) (New York: Pergamon Press, 1985); K. F. Haynes and M. C. Birch, "The role of other pheromones in the behavioral responses of Insects" (225–256). In *Comprehensive Insect Physiology, Biochemistry, and Pharmacology*. G. A. Kerkut and L. I. Gilbert (Eds.) (New York: Pergamon Press, 1985); F. H. Bronson, "Pheromonal influences on mammalian reproduction." In *Perspective in Reproduction and Sexual Behavior*. M. N. Diamond (Ed.) (pp. 341–460) (Bloomington: Indiana University Press, 1968); F. H. Bronson et al., "Strange male pregnancy block in deer mice." *Biology and Reproduction*, 1 (1969), 302–310; W. H. Cade, "Insect

mating and courtship behavior." In *Comprehensive Insect Physiology, Biochemistry, and Pharmacology*. G. A. Kerkut and L. I. Gilbert (Eds.) (New York: Pergamon Press, 1985, 591–620);

2. J. W. Papez, *Comparative Neurology* (New York: Hafner, 1967); C. U. Ariens Kappers, *The Evolution of the Nervous System* (Bohn: Haarlem, 1929); T. L. Lentz, *Primitive Nervous Systems* (New Haven: Yale University Press, 1968).

3. R. Joseph, "The Limbic System," *Psychoanalytic Review*, 1992; R. Joseph, *Neuropsychology, Neuropsychiatry, Behavioral Neurology* (New York: Plenum Press, 1990); D. Ackerman, *A Natural History of the Senses* (New York: Vintage, 1990); W. Wickler, *The Sexual Code* (Garden City: Anchor, 1973); I. Eibl-Eibesfeldt, *Ethology* (New York: Holt, 1975); Bedichek, 1960; Burton, 1976; Moncrieff, 1970.

4. Wickler, 1973; Eibl-Eibesfeldt, 1975.

5. H. Keller, *My Life* (New York: Macmillan, 1960).

6. Ackerman, 1990; Burton, 1976.

7. Bedichek, 1960; Burton, 1976; Moncrieff, 1970.

8. Joseph, 1990, 1992.

9. Papez, 1967; Ariens Kappers, 1929; Lentz, 1968.

10. Papez, 1967; Ariens Kappers, 1929; Lentz, 1968.

11. Lentz, 1968.

12. Joseph, 1990, 1992.

13. Papez, 1967; Ariens Kappers, 1929; Lentz, 1968.

14. Joseph, 1990, 1992.

15. Mayer and Mankin, 1985; Blum, 1985; Bruce, 1970; Wilson, 1980.

16. Tamaki, 1985; Blum, 1985; Haynes and Birch, 1985; Cade, 1985; Bruce, 1970; Wilson, 1980.

17. Joseph, 1990, 1992.

18. Joseph, 1992.

19. E. T. Morris, *Fragrance* (New York; Scribner's, 1986); P. Suskind, *Perfume* (New York: Knopf, 1987); Burton, 1976; Moncrieff, 1970; Ackerman, 1990.

20. Bedichek, 1960.

21. Mayer and Mankin, 1985; Tamaki, 1985; Blum, 1985; Haynes and Birch, 1985; Cade, 1985.

22. Mayer and Mankin, 1985; Cade, 1985.

23. F. H. Bronson and W. K. Whitten, "Oestrus-accelerating pheromone in mice," *Journal of Reproduction and Fertility*, 15 (1968), 131–140; Bronson et al., 1969; Wickler, 1973; Eibl-Eibesfeldt, 1975.

24. Bronson, F. H., 1968; Bronson and Whitten, 1968.

25. Bronson et al., 1969.

26. Wickler, 1973; Ackerman, 1990.

27. E. O. Wilson, *The Insect Societies* (Boston: Harvard University Press, 1971); L.

Krames et al., "A pheromone associated with social dominance among male rats," *Psychonomic Science*, 16 (1969), 11–12; Tamaki, 1985; Blum, 1985; Haynes and Birch, 1985; Bronson et al., 1969; Wilson, 1980; Cade, 1985.

28. Cade, 1985; Bronson et al., 1969; Haynes and Birch, 1985.

29. E. O. Wilson and F. E. Regnier, "The evolution of the alarm-defense system in formicine ants," *American Naturalist*, 105 (1971), 279–289; Blum, 1985.

30. Blum, 1985; Haynes and Birch, 1985.

31. Tamaki, 1985; Blum, 1985; Haynes and Birch, 1985.

32. Blum, 1985; Haynes and Birch, 1985; Mayer and Mankin, 1985; Wilson, 1980, 1971.

33. Morris, 1986; Suskind, 1987.

34. R. D. Nadler, "Sexual behavior of captive lowland gorillas," *Archives of Sexual Behavior*, 5 (1976); R. D. Nadler, "Sexual behavior of captive orang-utans," *Archives of Sexual Behavior*, 6 (1977); R. D. Nadler, "Sexual behavior of the chimpanzees in relation to the gorilla and orang-utan." In *Progress in Ape Research*, B. H. Bourne (Ed.) (New York: Academic Press, 1977); J. Goodall, *Through a Window* (Boston: Houghton Mifflin Co., 1990).

35. Mayer and Mankin, 1985; Tamaki, 1985; Haynes and Birch, 1985.

Chapter 3

1. R. N. Emde and K.L. Koenig, "Neonatal smiling and rapid eye movement states," *American Academy of Child Psychiatry*, 8 (1969), 57–87; R. A. Spitz and K. M. Wolf, "The smiling response," *Genetic Psychology Monographs*, 34 (1946), 57–125; E. Milner, *Human Neural and Behavior Development* (Springfield, IL: Thomas, 1967).

2. Emde and Koenig, 1969; Spitz and Wolf, 1946; Milner, 1967.

3. R. Joseph, *Neuropsychology, Neuropsychiatry, and Behavioral Neurology* (New York: Plenum Press, 1990); R. Joseph, "The limbic system," *Psychoanalytic Review*, 1992a.

4. Joseph, 1990, 1992a.

5. R. Joseph, *Right Brain and the Unconscious* (New York: Plenum Press, 1992b).

6. P. Gloor, "Amygdala." In *Handbook of Physiology*, J. Field (Ed.) (Washington, DC: American Physiological Society, pp. 300–370, 1960); A. Kling, "Effects of amygdalectomy on social-affective behavior in non-human primates." In *The Neurobiology of the Amygdala* (pp. 127–170) (New York: Plenum Press, 1972); H. Ursin and B. R. Kaada, "Functional localization within the amygdaloid complex," *EEG and Clinical Neurophysiology*, 12 (1960), 1–20; J. O'Keefe and H. Bouma, "Complex sensory properties of certain amygdala units in the freely

moving cat," *Experimental Neurology*, 23 (1969), 384–398; M. Fukuda et al., "Functional relation among inferotemporal cortex, amygdala and lateral hypothalamus," *Journal of Neurophysiology*, 57, 1060–1077; Joseph, 1990, 1992a.

7. I. Eibl-Eibesfeldt, *Ethology* (New York: Holt, 1975); W. Wickler, *The Sexual Code* (Garden City: Anchor, 1973).

8. Eibl-Eibesfeldt, 1975; Wickler, 1973; S. C. Strum, "Life with the pumphouse gang," *National Geographic*, 5, 1975; J. Goodall, *Through a Window* (Boston: Houghton Mifflin Co., 1990); H. Kummer, "Social organization of Hamadryas baboons." In *The Baboon in Medical Research*. H. Vogtberg (Ed.) (pp. 1–16) (Austin: University of Texas Press, 1965).

9. J. Itani and A. Suzuki, "The social unit of chimpanzees," *Primates*, 6 (1967); R. D. Nadler, "Sexual behavior of captive lowland gorillas," *Archives of Sexual Behavior*, 5, (1976); R. D. Nadler, "Sexual behavior of captive orang-utans," *Archives of Sexual Behavior*, 6, 1077; R. D. Nadler, "Sexual behavior of the chimpanzees in relation to the gorilla and organ-utan," In *Progress in Ape Research*, B. H. Bourne (Ed.) (New York: Academic Press, 1977); Strum, 1975; Goodall, 1990; Kummer, 1965; Eibl-Eibesfeldt, 1975; Wickler, 1973.

10. R. Joseph, "Competition between women," *Psychology* (1986); W. Gaylin, *The Male Ego* (New York: Viking, 1992); J. M. Ross, *The Male Paradox* (New York: Simon & Schuster, 1992); C. Gilligan, *In a Different Voice* (Boston: Harvard University Press, 1982); J. Lever, "Sex differences and the games children play," *Social Problems*, 23 (1976), 478–487; H. Bord, (Ed.), *The Making of New Masculinities* (Boston: Allen & Unwin, 1987); J. Wallerstein and Blakeslee, *Second Chances: Men, Women, and Children a Decade after Divorce* (New York: Ticknor & Field, 1989); L. Glass, *He Says, She Says* (New York: G. P. Putnam's Sons, 1992); N. Henley, *Language and Sex* (Massachusetts: Newbury House Publishers, 1975); D. Tanner, *You Just Don't Understand* (New York: Ballantine, 1990).

11. Joseph, 1985.

12. Gaylin, 1992; Ross, 1992; Bord, 1987; Glass, 1992; Henley, 1975; Tanner, 1990.

13. Maclean, P. D. "New findings of brain function and sociosexual behavior." In *Contemporary Sexual Behavior*, J. Zubin and J. Money (Eds.) (pp. 90–117) (Baltimore: Johns Hopkins Press, 1973); Emde and Koenig,1969; Joseph, 1990, 1992a.

14. W. G. Lisk, "Neural localization for androgen activation of copulatory behavior," *Endocrinology*, 80 (1967), 754–780; R. Bleier et al., "Cytoarchitectonic sexual dimorphisms of the medial preoptic and anterior hypothalamic area," *Journal of Comparative Neurology*, 66 (1982), 603–605; T. C. Rainbow et al., "Sex differences in brain receptors," *Nature*, 300 (1982), 648–649; G. Raisman

and P. Field, "Sexual dimorphism in the preoptic area of the rat," *Science*, 173 (1971), 731–733.

15. Joseph, 1988, 1990, 1992a,b.
16. R. A. Spitz, "Hospitalism: An inquiry into the genesis of psychiatric conditions in early childhood," *Psychoanalytical Study of the Child*, 1 (1945), 53–74.
17. R. Joseph and V. A. Casangrade, "Visual field defects and recovery following lid closure in a prosimian primate," *Behavioral Brain Research*, 1 (1980), 150–178; R. Joseph and V. A. Casagrande, "Visual field defects and morphological changes resulting from monocular deprivation in primates," *Proceedings of the Society for Neuroscience*, 4 (1978), 2021; V. A. Casagrande and R. Joseph, Effects of monocular deprivation on geniculostriate connections in primates, *Anatomical Records*, 14 (1978), 2001–2021; R. Joseph, Effects of rearing and sex on learning and competitive exploration, *Journal of Psychology*, 101, (1979), 37–43; V. A. Casagrande and R. Joseph, "Morphological effects of monocular deprivation and recovery on the dorsal lateral geniculate nucleus in Galago." *Journal of Comparative Neurology*, 194 (1980), 413–426; R. Joseph, "The neuropsychology of development," *Journal of Clinical Psychology*, 44 (1982), 4–33; R. Joseph and R. E. Gallagher, "Gender and early environmental influences on activity, arousal, overresponsiveness, and exploration," *Developmental Psychobiology*, 13 (1980), 527–544; J. Langmeier and Z. Matejcek, *Psychological Deprivation in Childhood* (New York: Wiley, 1975); J. Bowlby, "The influence of early environment in the development of neurosis and neurotic character," *International Journal of Psycho-Analysis*, 21 (1940), 154–178; J. Bowlby, *Maternal Care and Mental Health* (Geneva: WHO, 1951); J. Bowlby, "Separation anxiety," *International Journal of Psychoanalysis*, 412 (1960), 1–25; W. Greenough, Enduring effects of differential experience and training in *Neural Mechanisms of Learning and Memory*, M. R. Rosenzweig and E. L. Bennet (Eds.) (Cambridge: MIT Press, 1976); M. Rosenzweig et al., "Chemical and anatomical plasticity of the brain." In *Macromolecules and Behavior*. J. Gaito (Ed.) (New York: Appleton, 1972); W. Dennis, "Causes of retardation among institutionalized children," *Journal of Genetic Psychology*, 96 (1975), 47–59.
18. Langmeier and Matejcek, 1975; Bowlby, 1940, 1951; Langmeier and Matejcek, 1975; Dennis, 1975; Joseph, 1982.
19. Spitz and Wolf, 1946; Bowlby, 1960.
20. G. Morgenson, "Septal-hypothalamic relationships." In *The Septal Nuclei* (J. F. DeFrance, Ed.) (New York: Plenum Press, 1976); D. Dicks et al., "Uncus and amygdaloid lesions on social behavior in the free ranging monkey," *Science*, 160 (1969), 69–71; K. R. Johanson and L. J. Enloe, "Alterations in social behavior following septal and amygdaloid lesions in the rat," *Journal of Comparative and*

Physiological Psychology, 75 (1972), 280–301; Joseph, R., 1990, 1992a; Maclean, 1990.
21. Maclean, 1990; Joseph, 1990, 1992a,b.
22. Joseph, 1990, 1992a,b.
23. Joseph, 1990, 1992a,b.
24. Langmeier and Matejcek, 1975; Joseph, R., 1982, 1988, 1990, 1992a,b; Joseph and Casagrande, 1980; Casagrande and Joseph, 1980; Joseph and Gallagher, 1980.
25. R. Melzack and T. H. Scott, "The effects of early experience on the response to pain," *Journal of Comparative and Physiological Psychology*, 50 (1956), 155–161.
26. Spitz, 1945; Langmeier and Matejcek, 1975; Dennis, 1975; Bowlby, 1940.
27. Langmeier and Matejcek, 1975.
28. H. F. Harlow, The heterosexual affectional system in monkey, *American Psychologist*, 17 (1962), 1–9.
29. Spitz, 1945.
30. Bowlby, 1940, 1951, 1960.
31. Spitz, 1945; Langmeier and Matejcek, 1975; Bowlby, 1940, 1951, 1960.
32. Gaylin, 1992; Ross, 1992; Bord, 1987.
33. Gaylin, 1992; Ross, 1992; Bord, 1987; Wallerstein and Blakeslee, 1989; Glass, 1992.
34. Joseph, 1992b; Gaylin, 1992; Ross, 1992; Bord, 1987.
35. Gaylin, 1992; Ross, 1992; Wallerstein and Blakeslee, 1989.
36. Wallerstein and Blakeslee, 1989; Gaylin, 1992.
37. B. W. Robinson, "Vocalizations evoked from forebrain," *Physiology and Behavior*, 2 (1967), 345–353; Joseph, 1982, 1988, 1990, 1992a,b; Maclean, 1990.
38. Joseph, 1982, 1988, 1990, 1992a,b.
39. Joseph, 1992b.
40. Eibl-Eibesfeldt, 1975; Wickler, 1973.
41. Eibl-Eibesfeldt, 1975.
42. Kummer, 1965; Eibl-Eibesfeldt, 1975; Wickler, 1973.
43. Eibl-Eibesfeldt, I., 1975; Wickler, 1973.
44. R. D. Lawrence, *In Praise of Wolves* (New York: Holt, 1986); L. D. Mech, *The Wolf* (Garden City: Natural History Press, 1970).
45. G. F. Oster and E. O. Wilson, *Caste and Ecology in the Social Insects* (Princeton: Princeton University Press, 1978); Eibl-Eibesfeldt, 1975; Wickler, 1973.
46. Eibl-Eibesfeldt, 1975; Wickler, 1973.
47. Eibl-Eibesfeldt, 1975; Wickler, 1973; A. L. Zihlman et al., "Pygmy chimpanzee as a possible prototype for the common ancestor of humans, chimpanzees and

gorillas," *Nature*, 275 (1978), 744–746; V. Sarich, Primate systematics. In *Old World Monkeys*, J. R. Napier and P. H. Napier (Eds.) (175–226) (New York: Academic Press, 1970).

48. S. Chevalier-Skolnikoff, "Male-female, female-female, and male-male sexual behavior in Stumptail Monkey," *Archives of Sexual Behavior*, 3 (1974); M. Galdikas and M. F. Birute, "Living with the great orange apes," *National Geographic*, 157 (1980); T. L. Maple and M. P. Hoff, *Gorilla Behavior* (New York: Van Nostrand, 1982); G. B. Schaller, *The Mountain Gorilla* (Chicago: University of Chicago Press, 1963); Zihlman et al., 1978; Sarich, 1970; Telecki, 1973; Strum, 1975; Goodall, 1990; Lawrence, 1986; Mech, 1970; Maple and Hoff, 1982; Schaller, 1963; Goodall, 1990.

49. Lawrence, 1986; Mech, 1970; Eibl-Eibesfeldt, 1975; Wickler, 1973.

50. Eibl-Eibesfeldt, 1975; Wickler, 1973; Goodall, 1990; Zihlman et al., 1978.

51. Lawrence, 1986.

52. Lawrence, 1986; Mech, 1970; Eibl-Eibesfeldt, 1975; Wickler, 1973.

53. R. Ardrey, *The Hunting Hypothesis* (New York: Bantam, 1977); G. Isaac and D. C. Crader, "To what extent were early hominids carnivorous." In *Omnivorous Primates*. R. S. O. Harding and G. Telecki (Eds.) (pp. 37–103) (New York: Columbia University Press, 1980); R. Klein, "The ecology of early man in southern Africa," *Science*, 197 (1981), 115–126; T. Gibson, "Meat sharing as a political ritual." In *Hunters and Gatherers*, T. Ingold et al. (Eds.) (pp. 165–180) (New York: Berg, 1988); H. Kaplan and K. Hill, "Hunting ability and reproductive success among male ache foragers," *Current Anthropology*, 26 (1985), 131–133; G. P. Murdock and C. Provost, "Factors in the division of labor by sex," *Ethnology*, 12, 203–235; Joseph, 1992a; Nadler, 1977; Telecki, 1973; Strum, 1975; Goodall, 1990.

54. Eibl-Eibesfeldt, 1975; Wickler, 1973.

55. Goodall, 1990.

56. Nadler, 1977; Telecki, 1973; Strum, 1975; Goodall, 1990; Eibl-Eibesfeldt, 1975; Wickler, 1973.

57. Eibl-Eibesfeldt, 1975; Wickler, 1973.

Chapter 4

1. F. Nietzsche, *The Birth of Tragedy* (New York: Anchor, 1956); C. Sachs, *World History of Dance* (New York: Norton, 1933); I. Duncan, *The Art of the Dance* (New York: Theatre Arts Books, 1969); R. Lange, *The Nature of Dance* (New York: International Publications Service, 1976); A. A. Johnstone, "Languages and non-languages of dance." In *Illuminating Dance: Philosophical Explorations*. M. Sheets Johnstone (Ed.), 167–186. (London: Associated Universities

Press, 1984); D. L. O'Keefe, *Stolen Lighting: The Social Theory of Magic* (New York: Continuum, 1982); B. Malinowski, *Magic, Science and Religion*. In *Science, Religion, and Reality*. J. Needham (Ed.) (London: Allen & Unwin, 1935); L. Spence, *Myth and Ritual in Dance Game and Rhyme* (London: Watts & Co., 1947); Frazier, *The Golden Bough* (New York: Macmillan, 1922).

2. Nietzsche, 1956; Sachs, 1933; Duncan, 1969; Spence, 1947; Frazier, 1922.

3. K. Von Frisch and M. Lindauer, "The language and orientation of the honey bee," *Annual Review of Entomology*, 1 (1956), 45–58; K. Von Frisch, *The Dancing Bees* (New York: Harcourt, 1953); S. S. Schneider et al., "Vibration dance of the honey bee," *Animal Behavior*, 34 (1986), 377–385.

4. G. F. Oster and E. O. Wilson, *Caste and Ecology in the Social Insects* (Princeton: Princeton University Press, 1978); I. Eibl-Eibesfeldt, *Ethology: The Biology of Behavior* (New York: Holt, 1974); W. Wickler, *The Sexual Code* (New York: Anchor Books, 1973).

5. Von Frisch and Lindauer, 1956; Von Frisch, 1953; Schneider et al., 1986.

6. Von Frisch and Lindauer, 1956; Von Frisch, 1953; Schneider et al., 1986.

7. Gould, J. L. and Towne, W. F., "Evolution of the dance language," *The American Naturalist*, 130 (1987), 317–338; Malinowski, 1935; O'Keefe, 1982; Sachs, 1933.

8. Duncan, 1969; Lange, 1976; Johnstone, 1984.

9. Duncan, 1969.

10. J. Harrison, *Ancient Art and Ritual* (Oxford: Oxford University Press, 1913); O'Keefe, 1982; Spence, 1947; Frazier, 1922.

11. Duncan, 1969; Lange, 1976; Johnstone, 1984.

12. R. Laban, *The Mastery of Movement* (London: Macdonald & Evan, 1960); Duncan, 1969; Lange, 1976; Johnstone, 1984; Gould and Towne, 1987.

13. Von Frisch and Lindauer, 1956; Von Frisch, 1953; Schneider et al., 1986.

14. Eibl-Eibesfeldt, 1974; Wickler, 1973; Sachs, 1933.

15. L. E. Bachman, *Religious Dances in the Christian Church and in Popular Medicine* (London: Allen & Unwin, 1952); J. Meerloo, *Dance Craze and Sacred Dance* (London: Owen & Co., 1961); Sachs, 1933; Malinowski, 1935.

16. Schneider et al., 1986.

17. Schneider et al., 1986.

18. Plato, *Laws* (London: Heinemann, 1952).

19. L. B. Lawler, *The Dance in Ancient Greece* (London: Adams & Black, 1964); Sachs, 1933.

20. Plato, 1952; Nietzsche, 1956.

21. Nietzsche, 1956.

22. Duncan, 1969.

23. Nietzsche, 1956; Spence, 1947; O'Keefe, 1982; Gould and Towne, 1987; Sachs, 1933; Malinowski, 1935; Harrison, 1913.

24. Sachs, 1933; Malinowski, 1935; Spence, 1947; Frazier, 1922.
25. Sachs, 1933.
26. O'Keefe, 1982; Sachs, 1933; Malinowski, 1935; Spence, 1947; Frazier, 1922; Harrison, 1913.
27. Lange, 1976; Johnstone, 1984; Gould and Towne, 1987; Laban, 1960.
28. Duncan, 1969; Laban, 1960.
29. R. Joseph, *Neuropsychology, Neuropsychiatry, and Behavioral Neurology* (New York: Plenum Press, 1990); R. Joseph, "The right cerebral hemisphere: Language, music, emotion, visual-spatial skills, body image, dreams, and awareness," *Journal of Clinical Psychology*, 44 (1988), 630–673; D. Benson and M. Barton, "Disturbances in constructional ability," *Cortex*, 6 (1970), 19–46; A. Benton, "Visuoperceptive, visuospatial and visuoconstructive disorders." In *Clinical Neuropsychology*, K. M. Heilman and E. Valenstein (Eds.) (pp. 186–232) (Oxford: Oxford University Press, 1979); F. W. Black and B. A. Bernard, "Constructional apraxia as a function of lesion locus and size in patients with focal brain damage," *Cortex*, 20 (1984), 111–120; R. Calvanio et al., "Left visual spatial neglect is both environment-centered and body-centered," *Neurology*, 37 (1987), 1179–1183; E. DeRenzi, *Disorder of Space Exploration and Cognition* (New York: Wiley, 1982); H. Hecaen and M. L. Albert, *Human Neuropsychology* (New York: Wiley, 1978); Y. Kim et al., "Visuoperceptual and visuomotor abilities and locus of lesion." *Neuropsychologia*, 2 (1984), 177–185.
30. Joseph, 1982, 1988, 1990.

Chapter 5

1. I. Eibl-Eibesfeldt, *Ethology* (New York: Holt, 1975); W. Wickler, *The Sexual Code* (Garden City: Anchor, 1973); J. Fast, *The Body Language of Sex, Power, Aggression* (New York: Jove Books, 1977); R. Birdwhistell, *Masculinity and Femininity as Display*. In *Kinesics and Context* (Philadelphia: University of Pennsylvania Press, 1970, 39–46); W. Dyer, *Your Erroneous Zones* (New York: Avon, 1976).
2. Birdwhistell, 1970; Fast, 1977.
3. J. Lever, "Sex differences in the games children play." *Social Problems*, 23 (1976), 478–487; S. De Beauvoir, *The Second Sex* (New York: Bantam, 1952); W. Gaylin, *The Male Ego* (New York: Viking, 1992); J. M. Ross, *The Male Paradox* (New York: Simon & Schuster, 1992); L. Glass, *He Said, She Said* (New York, 1992); Birdwhistell, 1970; Henley, 1975; Fast, 1977.
4. D. Kimura, "Manual activity during speaking." *Neuropsychologia*, 11 (1973), 45–50; D. Kimura, "Neuromotor mechanisms in the evolution of human

communication." In H. D. Steklis and M. J. Raleigh (Eds.), *Neurobiology of Social Communication in Primates* (New York: Academic Press, 1979).

5. Gaylin, 1992; Ross, 1992; Eibl-Eibesfeldt, 1975; Wickler, 1973; Henley, 1975; Glass, 1992; Birdwhistell, 1970; Fast, 1977; Dyer, 1976.

6. R. A. Spitz, "The smiling response: A contribution to the ontogenesis of social relations," *Genetic Psychological Monographs*, 34, 57–125; Eibl-Eibesfeldt, 1975; Wickler, 1973.

7. A. Jolly, *The Evolution of Primate Behavior* (New York: Macmillan, 1972); Eibl-Eibesfeldt, 1975; Wickler, 1973.

8. Eibl-Eibesfeldt, 1975; Wickler, 1973; Henley, 1975; Fast, 1977.

9. Gaylin, 1992; Ross, 1992; Lever, 1976; Bord, 1987.

10. W. G. Lisk, "Neural localization for androgen activation of copulatory behavior," *Endocrinology*, 80 (1967), 754–780, R. Bleier et al., "Cytoarchitectonic sexual dimorphisms of the medial preoptic and anterior hypothalamic area," *Journal of Comparative Neurology*, 66 (1982), 603–605; T. C. Rainbow et al., "Sex differences in brain receptors," *Nature*, 300 (1982), 648–649; G. Raisman and P. Field, "Sexual dimorphism in the preoptic area of the rat," *Science*, 173 (1971), 731–733; R. Joseph, S. Hess, and E. Birecree, "Effects of sex hormone manipulation and exploration on sex differences in learning," *Behavioral Biology*, 24 (1978), 364–377.

11. Eibl-Eibesfeldt, 1975; Wickler, 1973; Glass, 1992; Henley, 1975.

12. E. S. Kima and U. Bellugi, *The Signs of Language* (Boston: Harvard University Press, 1979); M. Critchley, *The Language of Gesture* (New York: Haskell House, 1971); N. Freedman, "Hands, words and mind." In *Communicative Structures and Psychic Structures*. N. Freedman and S. Grand (Eds.) (pp. 109–132) (New York: Plenum Press, 1971); A. Kendon, "Current issues in the study of gesture." In *The Biological Foundations of Gestures*. J. Nespoulous et al. (Eds.) (pp. 23–48) (Hillsdale, NJ: Lawrence Erlbaum & Associates, 1986); L. Spence, *Myth and Ritual Dance, Game and Rhyme* (London: Watts, 1947); Kimura, 1973, 1979.

13. A. Zahava, "Mate selection—A selection for a handicap," *Journal of Theoretical Biology*, 53 (1975), 205–214; Eibl-Eibesfeldt, 1975; Wickler, 1973.

14. R. Joseph, "Competition between women," *Psychology*, 22 4 (1985), 1–11.

15. Eibl-Eibesfeldt, 1975; Wickler, 1973; Critchley, 1971; Freedman, 1971.

16. Critchley, 1971.

17. Eibl-Eibesfeldt, 1975; Wickler, 1973.

18. J. Goodall, *Through a Window* (Boston: Houghton Mifflin, 1990); A. L. Zihlman et al. "Pygmy chimpanzee as a possible prototype for the common ancestor of humans, chimpanzees and gorillas," *Nature*, 275 (1978), 744–746; V. Sarich, "Primate systematics." In *Old World Monkeys*, J. R. Napier and P. H. Napier (Eds.) (175–226) (New York: Academic Press, 1970); S. C. Strum, "Life

with the pumphouse gang." *National Geographic*, 5, 1975; Eibl-Eibesfeldt, 1975; Wickler, 1973; Critchley, 1971; Jolly, 1975; P. Maclean, *Evolution of the Triune Brain* (New York: Plenum Press, 1990); Joseph, 1990, 1992a,b.

19. Kima and Bellugi, 1979; Critchley, 1971; Freedman, 1971; Kendon, 1986.

20. Eibl-Eibesfeldt, 1975; Wicker, 1973; Critchley, 1971; Freedman, 1971; Kendon, 1986.

21. Critchley, 1971.

22. H. Werner and B. Kaplan, *Symbol Formation* (New York: Wiley, 1963); Critchley, 1971; Kima and Bellugi, 1979.

23. J. Piaget, *The Origins of Intelligence in Children* (New York: Norton, 1952); J. Piaget, *Play, Dreams, and Imitations in Childhood* (New York: Norton, 1962); J. Piaget, *The Child and Reality* (New York: Viking Press, 1974); E. Milner, *Human Neural and Behavioral Development* (Springfield, IL: Thomas, 1967); Critchley, 1971; Werner and Kaplan, 1963; Joseph, 1982.

24. Piaget, 1952, 1962, 1974; Milner, 1967; Critchley, 1971; Werner and Kaplan, 1963.

25. Joseph, 1982, 1990; Piaget, 1952, 1962, 1974; Milner, 1967; Critchley, 1971; Werner and Kaplan, 1963.

26. Werner and Kaplan, 1963; Critchley, 1971; Piaget, 1952, 1962, 1974.

27. Werner and Kaplan, 1963; Piaget, 1952, 1962, 1974.

28. Werner and Kaplan, 1963.

29. E. Chiera, *They Wrote on Clay* (Chicago: University of Chicago Press, 1966); Joseph, 1992a.

30. K. Heilman et al., "Two forms of ideomotor apraxia," *Neurology*, 32 (1982), 342–346; L. Rothi et al., "Pantomime agnosia," *Journal of Neurology, Neurosurgery, and Psychiatry*, 49 (1986), 451–454; Joseph, 1990.

31. Werner and Kaplan, 1963.

32. M. Kinsbourne, "Brain organization underlying orientation and gestures." In *The Biological Foundations of Gestures*, J. Nespoulous et al. (Eds.) (pp. 65–76) (Hillsdale, NJ: Lawrence Erlbaum & Associates, 1986); D. Kimura, "Left-hemisphere control of oral and brachial movement and their relation to communication," *Philosophical Transactions of the Royal Society of London*, 298 (1982), 135–149; D. Kimura and Y. Archibald, "Motor functions of the left hemisphere," *Brain*, 97 (1974), 337–350; Kimura, 1973, 1979.

33. Joseph, 1982.

34. J. C. Lynch, The functional organization of posterior parietal association cortex, *Behavioral Brain Sciences*, 3 (1980), 485–499; J. C. Lynch et al., "Parietal lobe mechanisms for directed visual attention." *Journal of Neurophysiology*, 40 (1977), 362–389; B. Motter and V. B. Mountcastle, "The functional properties of the light sensitivity neurons of the posterior parietal cortex studies in waking monkey," *Journal of Neuroscience*, 1 (1981), 3–26; V. B. Mountcastle, "Modal-

ities and topographic properties of single neurons of cat's sensory cortex," *Journal of Neurophysiology*, 20 (1957), 408–434; V. B. Mountcastle, "The world around us: Neural command functions for selective attention," *Neurosciences Research Progress Bulletin*, 14 (1976), 1–47; V. B. Mountcastle et al., "Posterior parietal association cortex of the monkey," *Journal of Neurophysiology*, 38 (1975), 871–908; V. B. Mountcastle et al., "Some further observations on the functional properties of neurons in the parietal lobe of the waking monkey," *Brain and Behavioral Sciences*, 3, 520–529; V. B. Mountcastle and T. P. S. Powell, "Central nervous mechanisms subserving position sense and kinesthesis," *Bulletin of the Johns Hopkins Hospital*, 105, 173–200; Joseph, 1990.

35. Lynch, 1980; Lynch et al., 1977; Motter and Mountcastle, 1981; Mountcastle, 1957, 1976; Mountcastle et al., 1975, 1980; Mountcastle and Powell, 1959; Joseph, 1990.

36. Joseph, 1990.

37. Kimura, 1982; Kimura and Archibald, 1974.

38. Keller, H. *My Life* (New York: Macmillan, 1960).

39. R. A. Gardner and B. T. Gardner, "Teaching sign language to a chimpanzee," *Science*, 165 (1969), 664–670; T. L. Maple and M. P. Hoff, *Gorilla Behavior* (New York: Van Nostrand, 1982); F. Patterson and E. Linden, *The Education of Koko* (New York: Holt, Rinehart & Winston, 1981); G. Galdikas and M. F. Birute, "Living with the Great Orange Apes," *National Geographic*, 157 (1980); G. B. Schaller, *The Mountain Gorilla* (Chicago: University of Chicago Press, 1963); R. D. Lawrence, *In Praise of Wolves* (New York: Holt, 1986); L. D. Mech, *The Wolf* (Garden City: Natural History Press, 1970); M. Temerlin, *Lucy, Growing Up Human* (New York, 1977); Jolly, 1975; Sarich, 1970; Goodall, 1990; Eibl-Eibesfeldt, 1975; Wickler, 1973.

40. Temerlin, 1977; Goodall, 1990; Patterson and Linden, 1981; Gardner and Gardner, 1969.

41. E. S. Kima and U. Bellugi, *The Signs of Language* (Boston: Harvard University Press, 1979).

42. R. Joseph, "Confabulation and delusional denial: Frontal lobe and lateralized influences," *Journal of Clinical Psychology*, 42 (1986), 507–518; N. Geschwind, "Disconnection syndromes in animals and man," *Brain*, 88 (1965), 585–644; Joseph, 1982, 1988, 1990.

43. Joseph, 1982, 1986, 1988, 1990; Geschwind, 1965.

44. Gardner and Gardner, 1969.

45. Temerlin, 1977.

46. Goodall, 1990.

47. Goodall, 1990; Eibl-Eibesfeldt, 1975; Wickler, 1973; Telecki, 1973; Jolly, 1975; F. B. M. Waal, *Chimpanzee Politics: Power and Sex among Apes* (New York: Academic, 1989).

48. Goodall, 1990; Eibl-Eibesfeldt, 1975; Wickler, 1973; Jolly, 1975; Waal, 1989.
49. Waal, 1989.
50. Goodall, 1990.
51. Waal, 1989.
52. J. P. Scott and J. L. Fuller, *Dog Behavior* (Chicago: University of Chicago Press, 1965); Lawrence, 1986; Mech, 1970.
53. Lawrence, 1986; Mech, 1970.
54. Lawrence, 1986; Mech, 1970; Scott and Fuller, 1965.
55. Lawrence, 1986; Mech, 1970.
56. Maple and Hoff, 1982; Schaller, 1963.
57. Patterson and Linden, 1981.
58. Patterson and Linden, 1981.
59. H. Hecaen and M. L. Albert, *Human Neuropsychology* (New York: John Wiley, 1978); Joseph, 1990; Heilman et al., 1982.
60. Joseph, 1990.

Chapter 6

1. R. Joseph, *Neuropsychology, Neuropsychiatry, and Behavioral Neurology* (New York: Plenum Press, 1990); J. Cutting, "Two left hemisphere mechanisms in speech perception," *Perception and Psychophysics*, 16 (1974), 601–612; D. Kimura, "Cerebral dominance and the perception of verbal stimuli," *Canadian Journal of Psychology*, 15 (1961), 156–171. D. Kimura and S. Folb, "Neural processing of backward speech sounds," *Science*, 161 (1968), 395–396; J. Levy, "Psychological implications of bilateral asymmetry." In *Hemisphere Function in the Human Brain*, S. Diamond and J. G. Beaumont (Eds.) (London: Paul Elek, Ltd., 1974); L. Mills and G. B. Rollman, "Hemispheric asymmetry for auditory perception of temporal order," *Neuropsychologia*, 18 (1980), 41–47, G. Papcun et al., "Is the left hemisphere specialized for speech, language and-or something else?," *Journal of the Acoustical Society of America*, 55, 319–327; D. Shankweiler and M. Studdert-Kennedy, "Lateral differences in perception of dichotically presented synthetic consonant-vowel syllables and steady-state vowels," *Journal of the Acoustic Society of America*, 39 (1966), 12–56; D. Shankweiler and M. Studdert-Kennedy, "Identification of consonants and vowels presented to left and right ears," *Quarterly Journal of Experimental Psychology*, 19 (1967), 59–63.
2. A. Carmazza and E. B. Zurif, "Dissociation of algorithmic and heuristic process in language comprehension: Evidence from aphasia," *Brain and Language*, 3 (1976) 572–582; M. Critchley, *The Parietal Lobes* (New York: Hafner, 1953); E. DeRenzi et al., "The pattern of neuropsychological impairment associated with left posterior cerebral artery infarcts," *Brain*, 110 (1987), 1099–

1116; R. Efron, "The effect of handedness on the perception of simultaneity and temporal order," *Brain*, 86 (1963), 261–284; H. Goodglass and E. Kaplan, *Boston Diagnostic Aphasia Examination* (Philadelphia: Lea & Febiger, 1982); H. Hecaen and M. L. Albert, *Human Neuropsychology* (New York: Wiley, 1978); K. Heilman and R. J. Scholes, "The nature of comprehension errors in Broca's conduction, and Wernicke's aphasia," *Cortex*, 12 (1976), 258–265; A. Kertesz, "Localization of lesions in Wernicke's aphasia." In *Localization in Neuropsychology*, A. Kertesz (Ed.) (New York: Academic Press, 1983); D. N. Levine and E. Sweet, "Localization of lesions in Broca's motor aphasia." In A. Kertesz (Ed.), *Localization in Neuropsychology* (New York: Academic Press, 1983); B. Milner, "Memory and the medial temporal regions of the brain," In *Biology of Memory*, K. Pribram and D. E. Broadbent (Eds.) (New York: Academic Press, 1970); L. Vignolo, "Modality-specific disorders of written language," In A. Kertesz (Ed.), *Localization in Neuropsychology* (New York: Academic Press, 1983); E. G. Zurif and G. Carson, "Dyslexia in relation to cerebral dominance and temporal analysis," *Neuropsychologia*, 8 (1970), 239–244; Cutting, 1974; Kimura, 1961; Mills and Rollman, 1980; Shankweiler and Studdert-Kennedy, 1966, 1967; Joseph, 1990.

3. A. Luria, *Higher Cortical Functions in Man* (New York: Basic Books, 1980); Shankweiler and Studdert-Kennedy, 1966, 1967; Joseph, 1990.

4. R. Joseph, "The neuropsychology of development," *Journal of Clinical Psychology*, 38 (1982), 4–33; D. Kimura, "Manual activity during speaking," *Neuropsychologia*, 11 (1973), 45–50; D. Kimura, "Neuromotor mechanisms in the evolution of human communication," In *Neurobiology of Social Communication in Primates*, H. D. Steklis and M. J. Raleigh (Eds.) (New York: Academic Press, 1979); D. Kimura, "Left-hemisphere control of oral and brachial movement and their relation to communication," *Philosophical Transactions of the Royal Society of London*, 298 (1982), 135–149; D. Kimura and Y. Archibald, "Motor functions of the left hemisphere," *Brain*, 97 (1974), 337–350; M. Kinsbourne, "Brain organization underlying orientation and gestures," pp. 65–76. In *The Biological Foundations of Gestures*, J. Nespoulous et al. (Eds.) (Hillsdale: Lawrence Erlbaum & Associates, 1986); Levy, 1974; Joseph, 1990.

5. Joseph, 1988, 1990.

6. E. Zaidel, "Unilateral auditory language comprehension on the token test following cerebral commissurotomy and hemispherectomy," *Neuropsychologia*, 10 (1977), 405–417; Joseph, 1988, 1990.

7. R. Joseph, *The Right Brain and the Unconscious* (New York: Plenum Press, 1992); D. Kimura, "Neuromotor mechanisms in the evolution of human communication," In *Neurobiology of Communication in Primates*, H. D. Steklais and M. J. Raleigh (Eds.) (New York: Academic Press, 1979).

8. Kimura, 1973, 1979, 1982; Kimura and Archibald, 1974; Kinsbourne, 1986; Levy, 1974; Joseph, 1982, 1988, 1990.

9. Kimura, 1973, 1982; Kimura and Archibald, 1974; Kinsbourne, 1986; Levy, 1974.
10. Joseph, 1990.
11. Joseph, 1990.
12. J. Kaas et al., "Multiple representations of the body in the post central somatosensory cortex of primates." In *Cortical Sensory Organization (1)*, C. N. Woolsey (Ed.) (New Jersey: Humana Press, 1981); W. Penfield and H. Jasper, *Epilepsy and the Functional Anatomy of the Human Brain* (Boston: Little, Brown, & Co., 1954); W. Penfield and T. Rasmussen, *The Cerebral Cortex of Man: A Clinical Study of Localization of Function* (New York: Macmillan, 1950).
13. J. Gerstmann, "Syndrome of finger agnosia, disorientation for right and left, agraphia and acalculia," *Archives of Neurology and Psychiatry*, 44 (1930), 398–408; J. Gerstmann, "Problem of imperception of disease and of impaired body territories with organic lesions," *Archives of Neurology and Psychiatry*, 48 (1942), 890–913; M. Critchley, *The Language of Gesture* (New York: Haskell House, 1971); M. Critchley, *The Parietal Lobes* (New York: Hafner, 1953).
14. Y. Iwamura and M. Tanaka, "Postcentral neurons in hand region of area 2," *Brain Research*, 150 (1978), 662–666; J. Lynch, "The functional organization of posterior parietal association cortex," *Behavioral Brain Sciences*, 3 (1980), 485–499; V. B. Mountcastle, "The world around us: Neural command functions for selective attention," *Neurosciences Research Progress Bulletin*, 14 (1976) 1–47; V. B. Mountcastle et al., Posterior parietal association cortex of the monkey. *Journal of Neurophysiology*, 38 (1975), 871–908; V. B. Mountcastle et al., "Some further observations on the functional properties of neurons in the parietal lobe of the waking monkey," *Brain Behavioral Sciences*, 3 (1980), 520–529; R. H. Lamotte and C. Acuna, "Defects in accuracy of reaching after removal of posterior parietal cortex in monkeys," *Brain Research*, 139 (1978), 309–326; L. Leinonen, "Functional properties of neurons in the lateral part of associative area seven of awake monkeys," *Experimental Brain Research*, 34 (1979), 299 320.
15. V. Hrbek, "Pathophysiologic interpretation of Gerstmann's syndrome," *Neuropsychologia*, 11 (1977), 377–388; R. L. Strub and N. Geschwind, "Localization in Gerstman's syndrome." In *Localization in Neuropsychology*, A. Kertesz (Ed.) (pp. 173–190) (New York: Academic Press, 1983); J. S. Blum et al., "A behavioural analysis of the organization of the parieto-temporo-preoccipital cortex," *Journal of Comparative Physiology*, 93 (1950), 53–199; N. Geschwind, "Disconnection syndromes in animals and man," *Brain*, 88 (1965), 585–644; H. S. Levin, "The acalculias." In *Clinical Neuropsychology*, K. M. Heilman and E. Valenstein (Eds.) (New York: Oxford University Press, 1979).
16. Blum et al., 1950; Gerstmann, 1930, 1942; Geschwind, 1965; Strub and Geschwind, 1983; Hrbek, 1977; Luria, 1980.
17. Gerstmann, 1930, 1942; Hrbek, 1977; Strub and Geschwind, 1983.

18. F. Boller and J. Grafman, "Acalculia: Historical development and current significance," *Brain and Cognition*, 2 (1983), 205–223; D. Benson and W. Weir, "Acalculia: Acquired anarithmetria," *Cortex*, 8 (1972), 465–472; A. L. Benton et al., "Arithmetic ability, finger-localization capacity and right-left discrimination in normal and defective children," *American Journal of Orthopsychiatry*, 21 (1975), 756–766; Luria, 1980; Hrbek, 1977; Gerstmann, 1930.

19. Luria, 1980; Joseph, 1990.

20. Joseph, 1982, 1988, 1990; Blum et al., 1950; Geschwind, 1965.

21. Joseph, 1982, 1988, 1990; Strub and Geschwind, 1983; Geschwind, 1965.

22. S. M. Blinkov and I. I. Glezer, *The Human Brain in Figures and Tables* (New York: Basic Books, 1968); Joseph, 1982.

23. K. Kawano and M. Sasaki, "Response properties of neurons in posterior parietal cortex of monkey during visual-vestibular stimulation," *Journal of Neurophysiology*, 51 (1984), 352–360; K. Kawano et al., "Response properties of neurons in posterior parietal cortex of monkey during visual-vestibular stimulation," *Journal of Neurophysiology*, 51 (1984), 340–351; J. Hyvarinen, *The Parietal Cortex of Monkey and Man* (Berlin: Springer-Verlag, 1982); J. Hyvarinen and V. Shelepin, "Distribution of visual and somatic functions in the parietal association area of the monkey," *Brain Research*, 169 (1979), 561–564; R. H. Lamotte and C. Acuna, "Defects in accuracy of reaching after removal of posterior parietal cortex in monkeys," *Brain Research*, 139 (1978), 309–326; L. Leinonen et al., "Functional properties of neurons in the lateral part of associative area 7 of awake monkeys." *Experimental Brain Research*, 34 (1979), 299–320; J. C. Lynch et al., "Parietal lobe mechanisms for directed visual attention," *Journal of Neurophysiology*, 40 (1977), 362–389; M-M Mesulam et al., "Limbic and sensory connections of the inferior parietal lobule in the rhesus monkey," *Brain Research*, 136 (1977), 393–414; V. B. Mountcastle, "The world around us: Neural command functions for selective attention," *Neurosciences Research Progress Bulletin*, 14 (1976), 1–47; E. T. Rolls et al., "Responses of neurons in area 7 of the parietal cortex to objects of different significance," *Brain Research*, 169 (1979), 194–198; Lynch, 1980; Mountcastle et al., 1975, 1980.

24. H. Goodlass and E. Kaplan, *The Assessment of Aphasia and Related Disorders* (Philadelphia: Lea & Febiger, 1982). Hecaen and Albert, 1979; Geschwind, 1965; Gerstmann, 1930, 1942; Joseph, 1990.

25. D. F. Benson and N. Geschwind, "The alexias." In *Handbook of Clinical Neurology*, P. J. Vinken and G. W. Bruyn (Eds.) (Amsterdam: North-Holland, 1969); H. Hecaen and H. Kremin, "Neurolinguistic research on reading disorders from left hemisphere lesions," In *Studies in Neurolinguistics*, H. A. Whitkaer and H. Whitaker (Eds.) (New York: Academic Press, 1976); K. M. Heilman, "Reading and writing disorders caused by central nervous system

defects," *Geriatrics*, 30 (1975), 115–118; Geschwind, 1965; Goodlass and Kaplan, 1982; Joseph, 1990.

26. D. F. Benson et al., "Conduction aphasia," *Archives of Neurology*, 28 (1973), 339–346; H. Goodlass and Kaplan, 1982; Geschwind, 1965; Hecaen and Kremin, 1976; Joseph, 1990.

27. Joseph, 1992.

28. A. Leroi-Gourhan, *Treasure of Prehistoric Art* (New York: Harry N. Abrams); H. Bandi, *Art of the Stone Age* (New York: Crown Publishers, 1961); T. Prideaux, *Cro-Magnon Man* (New York: Time-Life Books, 1973).

29. A. Breasted, *History of Egypt*; S. N. Kramer, *History Begins at Sumer* (Philadelphia: University of Pennsylvania Press, 1981); C. L. Wooley, *Ur of the Chaldees* (New York: Norton, 1965); E. Chiera, *They Wrote on Clay* (Chicago: University of Chicago Press, 1966).

30. Leroi-Gourhan, A.

31. Kramer, 1981; Wooley, 1965; Chiera, 1966.

32. Kramer, 1981; Wooley, 1965; Chiera, 1966.

33. Chiera, 1966.

34. Geschwind, 1965; Goodlass and Kaplan, 1982; Joseph, 1982, 1988, 1990; Hecaen and Albert, 1979.

35. H. S. Levin, "The acalculias." In *Clinical Neuropsychology*. K. M. Heilman and E. Valenstein (Eds.) (New York: Oxford University Press, 1979); Strub and Geschwind, 1983; Luria, 1980.

36. Geschwind, 1965; Hrbek, 1977; Gerstmann, 1930; Goodlass and Kaplan, 1982; Heilman, 1975; Joseph, 1990.

37. Joseph, 1990.

38. J. Pfeiffer, *The Emergence of Humankind* (New York: Harper & Row, 1985).

39. Pfeiffer, 1985.

40. Leroi-Gourhan, A.; Bandi, 1961; Prideaux, 1973.

41. Kramer, 1981; Wooley, 1965; Chiera, 1966.

Chapter 7

1. J. W. Papez, "Comparative Neurology," (New York: Hafner, 1967); C. U. Ariens Kappers, *The Evolution of the Nervous System* (Bohn: Harlem, 1929); T. L. Lentz, *Primitive Nervous Systems* (New Haven: Yale University Press, 1968).

2. A. Brodal, *Neurological Anatomy* (New York: Oxford University Press, 1981).

3. R. Joseph, *Neuropsychology, Neuropsychiatry, and Behavioral Neurology* (New York: Plenum Press, 1990); R. Joseph, "The limbic system," *Psychoanalytic Review*, 1992a.

4. A. Brink, "Speculation on some advanced mammalian characteristics in the higher mammal-like reptiles," *Palaeontology*, 4 (1956), 77–95; G. S. Paul, *Predatory Dinosaurs* (New York: Simon & Schuster,1988); R. T. Bakker, "Dinosaur physiology and the origin of mammals," *Evolution*, 25 (1971), 636–658; Maglio, V. J. (Ed.), *Evolution of African Mammals* (Cambridge: Harvard University Press, 1978); A. W. Crompton and F. A. Jenkins, "Mammals from reptiles," *Annual Review of Earth and Planetary Sciences*, 1 (1973), 131–155; A. W. Crompton et al., "Origin of mammals." In *Mesozoic Mammals*, J. A. Lillegraven et al. (Eds.) (Berkeley: University of California Press, 1979); A. S. Romer, *Vertebrate Paleontology* (Chicago: University of Chicago Press, 1966); J. C. Quiroga, "The brain of two mammal-like reptiles," *J. Hirnforsch*, 20 (1979), 341–350; J. C. Quiroga, "The brain of the mammal-like reptile. Probainognatheus jenseni. A correlative paleo-neoneurological approach to the neocortex at the reptile-mammal transition," *J. Hirnforsch*, 21 (1980), 299–336; P. Maclean, *The Evolution of the Triune Brain* (New York: Plenum Press, 1990).
5. Maclean, 1990.
6. Brodal, 1981.
7. A. Luria, *The Working Brain* (New York: Basic Books, 1973); A. Luria, *Higher Cortical Functions in Man* (New York: Basic Books, 1980); W. Penfield and H. Jasper, *Epilepsy and the Functional Anatomy of the Human Brain* (Boston: Little, Brown & Co., 1954); W. Penfield and P. Perot, "The brain's record of auditory and visual experience," *Brain*, 86 (1963), 595–696; W. Penfield and T. Rasmussen, *The Cerebral Cortex of Man: A Clinical Study of Localization of Function* (New York: Macmillan, 1950); N. Geschwind, "Disconnection syndromes in man and animal," *Brain*, 1966; Joseph, 1990; H. Hecaen and M. L. Albert, *Human Neuropsychology* (New York: John Wiley, 1978); H. Goodglass and E. Kaplan, *Boston Diagnostic Aphasia Examination* (Philadelphia: Lea & Febiger, 1982); J. W. Brown, *Aphasia, Apraxia and Agnosia* (Springfield, IL: C. C. Thomas, 1972); D. F. Benson, *Aphasia, Alexia, Agraphia* (New York: Churchill-Livingstone, 1972).
8. Joseph, 1990; Brodal, 1981.
9. Papez, 1967; Ariens Kappers, 1929; Lentz, 1968.
10. Luria, 1973, 1980; Joseph, 1990.
11. D. J. Felleman and J. H. Kaas, "Receptive-field properties of neurons in middle temporal visual area (MT) of owl monkeys," *Journal of Neurophysiology*, 52 (1974), 488–513; D. H. Hubel et al., 1959; J. H. R. Maunsell and D. C. Van Essen, "Functional properties of neurons in middle temporal visual area of the macaque," *Journal of Neurophysiology*, 49 (1984), 1127–1165; L. Mills and G. B. Rollman, "Hemispheric asymmetry for auditory perception of temporal order," *Neuropsychologia*, 18 (1980), 41–47.
12. J. E. Cutting, "Two left hemisphere mechanisms in speech perception,"

Perception and Psychophysics, 16 (1974), 601–606; Z. Wollberg and V. D. Newman, "Auditory cortex of squirrel monkey," *Science*, 175 (1972), 212–214; Luria, 1973, 1980; Joseph, 1990; D. Shankweiler, "Effects of temporal lobe damage on the perception of dichotically presented melodies," *Journal of Comparative and Physiological Psychology*, 62 (1966), 115–122; D. Shankweiler and M. Studdert-Kennedy, "Lateral differences in perception of dichotically presented synthetic consonant-vowel syllables and steady-state vowels," *Journal of the Acoustic Society of America*, 39 (1966), 1256; D. Shankweiler and M. Studdert-Kennedy, "Identification of consonants and vowels presented to left and right ears," *Quarterly Journal of Experimental Psychology*, 19 (1967), 59–63; M. Studdert-Kennedy and D. Shankweiler, "Hemispheric specialization for speech perception," *Journal of the Acoustical Society of America*, 48 (1970), 579–594; I. G. Mattingly and M. Studdert-Kennedy, *Modularity and the Motor Theory of Speech Perception* (Hillsdale, NJ: Lawrence Erlbaum, 1991); H. J. Neville, "Whence the specialization of the Language Hemisphere." In *Modularity and the Motor Theory of Speech Perception*, I. G. Mattingly and M. Studdert-Kennedy (Eds.) (pp. 269–294) (Hillsdale, NJ: Lawrence Erlbaum, 1991), A. M. Liberman and I. G. Mattingly, "The motor theory of speech perception revised," *Cognition*, 21 (1985), 1–36; E. H. Lenneberg, *Biological Foundations of Language* (New York: Wiley, 1967); Mills and Rollman, 1980; Joseph, 1990.

13. Luria, 1973, 1980; Joseph, 1990.

14. H. Poizner et al., "Brain Function for Language." In I. G. Mattingly and M. Studdert-Kennedy (Eds.), *Modularity and the Motor Theory of Speech Perception* (Hillsdale, NJ: Lawrence Erlbaum, 1991); J. L. Lackner and H-L Teuber, "Alterations in auditory fusion thresholds after cerebral injury in man," *Neuropsychologia*, 11 (1973), 409–415. Lenneberg, 1967; Neville, 1991; Studdert-Kennedy, 1991; Liberman and Mattingly, 1985.

15. Joseph, 1990.

16. Mattingly and Studdert-Kennedy, 1985; Poizner et al., 1991; Lenneberg, 1967; Neville, 1991; Studdert-Kennedy, 1991; Liberman and Mattingly, 1985.

17. D. Slobin, *Psycholinguistics* (Scott Foresman, 1971); J. Fodor and J. Katz, *The Structure of Language* (Englewood Cliffs, NJ: Prentice-Hall, 1964); S. Liebersen, *Explorations in Sociolinguistics* (Bloomington: Indiana University Press, 1967); W. E. Lamberg, *Language, Psychology, and Culture* (Palo Alto: Stanford University Press).

18. Slobin, 1971; Fodor and Katz, 1964; Liebersen, 1967.

19. Z. Woolberg and V. D. Newman, "Auditory cortex of squirrel monkey," *Science*, 1175 (1972), 212–214; M. M. Merzenich and J. F. Brugge, "Representation of the cochlear partition of the superior temporal plane," *Brain Research*, 50 (1973), 275–296; Brodal, 1981.

20. D. Kimura and S. Folb, "Neural processing of backward speech sounds," *Science*, 161 (1968), 395–396; D. Kimura, "Cerebral dominance and the perception of verbal stimuli," *Canadian Journal of Psychology*, 15 (1961), 156–171; Levy, 1974; Shankweiler, 1966; Shankweiler and Studdert-Kennedy, 1967; Studdert-Kennedy and Shankweiler, 1966, 1970; Joseph, 1990.

21. Kimura and Folb, 1968; Kimura, 1961; Shankweiler, 1966; Shankweiler and Studdert-Kennedy, 1967; Studdert-Kennedy and Shankweiler, 1966, 1970; Joseph, 1990.

22. Poizner et al., 1991; Neville, 1991; Studdert-Kennedy, 1991; Liberman and Mattingly, 1985.

23. Joseph, 1990.

24. Brodal, 1981; Felleman and Kaas, 1974; Hubel et al., 1959; Maunsell and Van Essen, 1983; Mills and Rollman, 1980; Wollberg and Newman, 1972.

25. Joseph, 1990.

26. Penfield and Jasper, 1954; Penfield and Perot, 1963; Penfield and Rasmussen, 1950.

27. M. L. Albert and D. Bear, "Time to understand. A case study of word deafness with reference to the role of time in auditory comprehension." *Brain*, 97 (1974), 383–394. M. L. Albert et al., "A case of auditory agnosia. Linguistic and nonlinguistic processing," *Cortex*, 8 (1972), 427–443; Hecaen and Albert, 1978; Auerbach et al., "Pure word deafness," *Brain*, 105 (1982), 271–300; M. P. Earnest et al., "Cortical deafness," *Neurology*, 27 (1977), 1172–1175; Goodglass and Kaplan, 1982; Brown, 1972; Joseph, 1990.

28. Hecaen and Albert, 1978; Goodglass and Kaplan, 1982; Albert and Bear, 1974; Albert et al., 1972; Auerbach et al., 1982; Brown, 1972; Earnest et al., 1977; Joseph, 1990.

29. Joseph, 1990; Luria, 1980.

30. Goodglass and Kaplan, 1982; Hecaen and Albert, 1978; Brown, 1972; Luria, 1973, 1980; Geschwind, 1966; Joseph, 1990.

31. Joseph, 1982, 1988, 1990.

32. R. Abrams, M. Taylor, A. Kasprisin, et al., "Comparison of schizophrenic patients with formal thought disorder and neurologically impaired patients with aphasia," *American Journal of Psychiatry*, 14 (1983), 1348–1351; Joseph, 1990.

33. D. F. Benson, *Aphasia, Alexia, Agraphia* (New York: Churchill, 1979).

34. Faber et al., 1983.

35. Joseph, 1982, 1988, 1990.

36. K. M. Heilman et al., "Comprehension of affective and nonaffective prosody," *Neurology*, 34 (1984), 917–921; K. Heilman et al., "Auditory affective agnosia," *Journal of Neurology, Neurosurgery and Psychiatry*, 38 (1975), 69–72; Breitling et al., 1987; J. W. Dwyer and W. E. Rinn, "The role of the right hemisphere in contextual inference," *Neuropsychologia*, 19 (1981), 479–482;

D. Breitling et al., "Auditory perception of music measured by brain electrical activity mapping," *Neuropsychologia*, 25 (1987), 765–774; M. P. Bryden et al., "A left-ear advantage for identifying the emotional quality of tonal sequences," *Neuropsychologia*, 20 (1982), 83–87; C. Knox and D. Kimura, "Cerebral processing of nonverbal sounds in boys and girls," *Neuropsychologia*, 8 (1970), 227–237; S. J. Segalowitz and P. Plantery, "Music draws attention to the left and speech draws attention to the right," *Brain and Cognition*, 4 (1985), 1–6; Cutting, 1974.

37. H. Gardner et al., "Missing the point: The role of the right hemisphere in the processing of complex linguistic materials," In *Cognitive Processing in the Right Hemisphere*, E. Perceman (Ed.) (New York: Academic Press, 1983); H. H. Brownell et al., "Inference deficits in right brain-damaged patients," *Brain and Language*, 27 (1986), 310–321; M. Cicone et al., "Sensitivity to emotional expressions and situations in organic patients,"*Cortex*, 16 (1980), 145–158; H. Gardner et al., "Comprehension and appreciation of humorous material following brain damage," *Brain*, 98 (1975), 399–412; P. B. Gorelick and E. D. Ross, *Journal of Neurology, Neurosurgery, and Psychiatry*, 37 (1987), 727–737; E. Ross, "The aprosodias: Functional-anatomic organization of the affective components of language in the right hemisphere," *Archives of Neurology*, 38 (1981), 561–589; S. Weintraub et al., "Disturbances in prosody: A right hemisphere contribution to language," *Archives of Neurology*, 38 (1981), 742–744; Bryden et al., 1982; Knox and Kimura, 1970; Segalowitz and Plantery, 1985; Smith, 1966.

38. K. Heilman et al., "Comprehension of affective and nonaffective prosody," *Neurology*, 34, 917–921; K. Heilman et al., "Auditory affective agnosia," *Journal of Neurology, Neurosurgery and Psychiatry*, 38 (1975), 69–72; D. Tucker et al., "Affective discrimination and evocation of affectively toned speech in patients with right parietal disease," *Neurology*, 27, 947–950; H. R. McFarland and D. Fortin, "Amusia due to right temporal-parietal infarct," *Archives of Neurology*, 39 (1982), 725–727; D. Kimura, "Right temporal lobe damage: Perception of unfamiliar stimuli after damage," *Archives of Neurology*, 18 (1963), 264–271; Ross, 1981; Gorelick and Ross, 1987; Bryden et al., 1982; Knox and Kimura, 1970; Segalowitz and Plantery, 1985.

39. Joseph, 1982, 1988, 1990.

40. Maclean, 1990.

41. G. S. Paul, *Predatory Dinosaurs* (New York: Simon & Schuster, 1988); D. K. Thomas and E. D. Olson (Eds.), *A Cold Look at the Warm Blooded Dinosaurs* (Washington, DC: AAAS, pp. 287–310); Brink, 1956; Bakker, 1971; Maglio, 1978; Crompton and Jenkins, 1979; Romer, 1966; Quiroga, 1979, 1980.

42. Paul, 1988; Brink, 1956; Bakker, 1971; Maglio, 1978; Crompton and Jenkins, 1979; Quiroga, 1979, 1980.

43. Maclean, 1990.
44. Maclean, 1990.
45. B. W. Robinson, "Vocalizations evoked from forebrain in Macaca mulatta," *Physiology and Behavior*, 2 (1967), 345–352; Joseph, 1990, 1992; Maclean, 1990.
46. Joseph, 1982, 1988, 1990.
47. Joseph, 1982, 1988, 1990; Robinson, 1967.
48. Joseph, 1982, 1988, 1990; Robinson, 1967.
49. Joseph, 1990.
50. Maclean, 1990.
51. P. Farb, *Word Play* (New York: Bantam, 1975).
52. N. Chomsky, *Language and Mind* (New York: Harcourt, 1972); N. Chomsky, *Syntactic Structures* (The Hague: Mouton, 1957).
53. Chomsky, 1972; E. Lennenberg, *Biological Foundations of Language* (New York: Wiley, 1967).
54. Chomsky, 1957.
55. E. Sapir, *Culture, Language, and Personality* (University of California Press, 1966).
56. B. Whorf, *Language, Thought, and Reality* (New York: John Wiley & Sons, 1956).
57. A. F. Chamberlain, "Some points in linguistic psychology," *American Journal of Psychology*, 5 (1983), 116–119.
58. Farb, 1975.

Chapter 8

1. R. Joseph, *Neuropsychology, Neuropsychiatry, and Behavioral Neurology* (New York: Plenum Press, 1990); R. Joseph, "The right cerebral hemisphere: Language, music, emotion, visual-spatial skills, body image, dreams, and awareness," *Journal of Clinical Psychology*, 44 (1988), 630–673; R. Joseph, "Confabulation and delusional denial: Frontal lobe and lateralized influences," *Journal of Clinical Psychology*, 42 (1986), 507–518; R. Joseph, "The neuropsychology of development: Hemispheric laterality, limbic language, and the origin of thought," *Journal of Clinical Psychology*, 38 (1982), 4–33; H. Hacaen and M. L. Albert, *Human Neuropsychology* (New York: John Wiley, 1978); H. Goodglass and E. Kaplan, *Boston Diagnostic Aphasia Examination* (Philadelphia: Lea & Febiger, 1982); D. N. Levine and E. Sweet, "Localization of lesions in Broca's motor aphasia." In *Localization in Neuropsychology*, A. Kertesz (Ed.) (pp. 185–207) (New York: Academic Press, 1983); J. W. Brown, *Aphasia, Apraxia and Agnosia* (Springfield, IL: C. C. Thomas, 1972); D. F. Benson, *Aphasia, Alexia, Agraphia* (New York: Churchill-Livingstone, 1979).
2. Joseph, 1982, 1986, 1988, 1990.

3. Joseph, 1990.
4. Hecaen and Albert, 1978; Goodglass and Kaplan, 1982; Brown, 1972; Benson, 1979; Joseph, 1990.
5. N. Geschwind, "Disconnection syndromes in animals and man," *Brain*, 88 (1965), 585–644; Joseph, 1982, 1986, 1988, 1990.
6. Hecaen and Albert, 1978; Goodglass and Kaplan, 1982; Brown, 1972; Benson, 1979; Joseph, 1990.
7. Joseph, 1990; Hecaen and Albert, 1978; Goodglass and Kaplan, 1982; Brown, 1972; Benson, 1979.
8. Geschwind, 1965; Joseph, 1982, 1986, 1988, 1990.
9. Joseph, 1982, 1986, 1988, 1990.
10. Joseph, 1982, 1986, 1988, 1990.
11. Geschwind, 1965; Joseph, 1982, 1986, 1988, 1990.
12. Joseph, 1982, 1986, 1988, 1990; R. Joseph, "Dual mental functioning in a split-brain patient," *Journal of Clinical Psychology*, 44 (1988b), 770–779; R. Joseph, "Reversal of cerebral dominance for language and emotion in a corpus callosotomy patient," *Journal of Neurology, Neurosurgery and Psychiatry*, 49 (1986c), 628–634; R. Joseph, R. E. Gallagher, W. Holloway, J. Kahn, "Two brains, one child: Interhemispheric information transfer deficits and confabulatory responding in children ages 4, 7, 10," *Cortex*, 20 (1984), 317–331; M. Nathanson et al., "Denial of illness," *Archives of Neurology and Psychiatry*, 68 (1952), 380–387; M. Roth, "Disorders of the body image caused by lesions of the right parietal lobe," *Brain*, 72 (1949), 89–111; N. Roth, "Unusual types of anosognosia and their relation to the body image," *Journal of Nervous and Mental Disease*, 100 (1944), 35–43; E. A. Weinstein and R. L. Kahn, "The syndrome of anosognosia," *Archives of Neurology and Psychiatry*, 64 (1950), 772–791; E. A. Weinstein and R. L. Kahn, "Non-aphasic misnaming (paraphasia) in organic brain disease," *Archives of Neurology and Psychiatry*, 67 (1952), 72–78.
13. Joseph, 1986; Geschwind, 1965.
14. J-P. Sartre, *Being and Nothingness* (New York: Philosophical Library, 1956).
15. R. Joseph, "Awareness, the origin of thought and the role of conscious self-deception in resistance and repression," *Psychological Reports*, 46 (1980), 767–781; Joseph, 1982, 1988a, 1990, 1992.
16. Sartre, 1956.
17. F. C. Redlich and J. E. Dorsey, "Denial of blindness by patients with cerebral disease," *Archives of Neurology and Psychiatry*, 53 (1945), 407–417; Joseph, 1982, 1986, 1988, 1990.
18. Joseph, 1982, 1986, 1988, 1988b, 1986c, 1990; Joseph et al., 1984; Geschwind, 1965.
19. Joseph, 1986; Nathanson et al., 1952; Roth, 1949; Roth, 1944; Weinstein and Kahn, 1950.
20. J. Gerstmann, "Syndrome of finger agnosia, disorientation for right and left,

agraphia, and acalculia," *Archives of Neurology and Psychiatry*, 44 (1930), 398–408.

21. E. Bisiach and A. Berti, "Dyschiria." In *Neurophysiological and Neuropsychological Aspects of Spatial Neglect*, M. Jeannerod (Ed.) (Amsterdam: North-Holland, 1987).

22. Joseph, 1988, 1990.

23. J. Bures and O. Buresova, "The use of Leao's spreading depression in the study of interhemispheric transfer of memory traces," *Journal of Comparative and Physiological Psychology*, 59 (1960), 211–214; G. L. Risse and M. S. Gazzaniga, "Well-kept secrets of the right hemisphere: A carotid amytal study of restricted memory transfer," *Neurology*, 28 (1979), 950–953; R. W. Doty and W. H. Overman, "Mnemonic role of forebrain commissures in macaques," In *Lateralization in the Nervous System*, S. Harnad, R. W. Doty, L. Goldstein, J. Jaynes, and G. Krauthamer (Eds.) (pp. 75–88) (New York: Academic Press, 1977); M. S. Gazzaniga and J. E. LeDoux, *The Integrated Mind* (New York: Plenum Press, 1978).

24. Joseph, 1982, 1986, 1988, 1988b, 1986c, 1990.

25. Joseph, 1986, 1990.

26. Joseph, 1986, 1990.

27. D. Blumer and D. F. Benson, "Personality changes with frontal and temporal lesions." In *Psychiatric Aspects of Neurological Disease*, D. F. Benson and D. Blumer (Eds.) (Orlando, FL: Grune & Stratton, 1975).

28. N. Geschwind, F. A. Quadfasel, and J. M. Segarra, "Isolation of the speech area," *Neuropsychologia*, 6 (1968), 327–340; Geschwind, 1965; Joseph, 1982, 1986, 1990.

29. Joseph, 1990; Goodglass and Kaplan, 1982.

30. Geschwind et al., 1968; Joseph, 1982.

31. C. J. Jung, *Experimental Researches, Collected Words*, II (Princeton, NJ: Princeton University Press, 1954); N. Ach, "Determining tendencies: awareness." In D. Rappaport (Ed.) (pp. 15–28) (New York: Columbia University Press, 1951); K. Koffka, *Principles of Gestalt Psychology* (New York: Harcourt Brace, 1935); L. S. Vygotsky, *Thought and Language* (Cambridge: MIT Press, 1962); Joseph, 1982, 1986, 1988, 1990.

32. Reitman, W. R., Grove, R. B., and R. Shoup, "Argus: An information processing model of thinking," *Behavioral Science* (1964), 270–281; Jung, 1954; Ach, 1951; Joseph, 1982, 1986, 1988, 1990.

33. S. Freud, *The Interpretation of Dreams*. Standard Edition, 5 (London: Hogarth Press, 1900); Joseph, 1980, 1982, 1986, 1988, 1990; Jung, 1954.

34. J. Piaget, *The Origins of Intelligence in Children* (New York: Norton, 1952); J. Piaget, *Play, Dreams and Imitation in Childhood* (New York: Norton, 1962); J. Piaget, *The Child and Reality* (New York: Viking Press, 1974); Joseph, 1980,

1982, 1986, 1988, 1990; Koffka, 1935; Vygotsky, 1962; Joseph, 1982, 1986, 1988, 1990; Reitman et al., 1964; Jung, 1954; Ach, 1951.
35. Joseph, 1980, 1982, 1986, 1988, 1990; Jung, 1954; Ach, 1951; Vygotsky, 1962.
36. Vygotsky, 1962; Piaget, 1952, 1962, 1974; Joseph, 1980, 1982, 1988, 1990.
37. A. R. Lecours, "Myelogenetic correlates of the development of speech and language." In *Foundations of Language Development*, E. Lenneberg and E. Lenneberg (Eds.) (New York: Academic Press, 1975); P. I. Yakovlev and A. Lecours, "The myelogenetic cycles of regional maturation of the brain." In *Regional Development of the Brain in Early Life*, A. Minkowski (Ed.) (London: Blackwell, 1967); Joseph, 1982, 1988, 1990; Joseph et al., 1984.
38. R. Joseph and R. E. Gallagher, "Interhemispheric transfer and the completion of reversible operations in non-conserving children," *Journal of Clinical Psychology*, 41 (1985), 796–800; R. E. Gallagher and R. Joseph, "Non-linguistic knowledge, hemispheric laterality, and the conservation of inequivalence," *Journal of General Psychology*, 107 (1982), 31–40; D. Galin et al., "Development of the capacity for tactile information transfer between hemispheres in normal children," *Science* 204 (1979), 1330–1332; M. Finlayson, "A behavioral manifestation of the development of interhemispheric transfer of learning in children," *Cortex* 12 (1975), 290–295; D. S. O'Leary, "A developmental study of interhemispheric transfer in children aged 5 to 10," *Child Development*, 51 (1980), 743–750; Joseph et al., 1984; Joseph, 1982, 1988, 1990, 1992.
39. Joseph et al., 1984; Joseph, 1982, 1988, 1990, 1992.
40. Vygotsky, 1962; Piaget, 1952, 1962, 1974; Joseph, 1982, 1990, 1992.
41. Joseph, 1982, 1988, 1990, 1992; Vygotsky, 1962; Piaget, 1952, 1962, 1974.
42. Joseph, 1982, 1988, 1990, 1992; Jung, 1954; Ach, 1951.
43. Joseph, 1982, 1988, 1990, 1992.

Chapter 9

1. J. Day, "Right hemisphere language processing in normal right-handers," *Journal of Experimental Psychology: Human Perception and Performance* (1977), 518–528; R. Day et al., "Perception of linguistic and non-linguistic dimensions of dichotic stimuli," *Status Report of Haskins Laboratories*, 27 (1971), 1–6; G. Deloche et al., "Right hemisphere language processing: Lateral difference with imageable and nonimageable ambiguous words," *Brain and Language*, 30 (1987), 197–205; M. S. Gazzaniga, *The Bisected Brain* (New York: Appleton, 1970); M. S. Gazzaniga and J. E. LeDoux, *The Integrated Mind* (New York: Plenum Press, 1978); D. Hines, "Recognition of verbs, abstract nouns and concrete nouns from the left and right visual half-fields," *Neuropsychologia*, 14 (1976), 211–216; T. Landis et al., "Aphasic reading and writing:

Possible evidence for right hemisphere participation," *Cortex*, 18 (1982), 105–112; J. Levy, "Psychological implications of bilateral asymmetry." In *Hemisphere Function in the Human Brain*, S. Diamond and J. G. Beaumont (Eds.) (pp. 151–183) (London: Paul Elek, 1974); J. Levy, "Language, cognition and the right hemisphere," *American Psychologist*, 38 (1983), 538–541; H. R. Mannhaupt, "Processing of abstract and concrete nouns in lateralized memory-search tasks," *Psychological Research*, 45 (1983), 19–105. R. Joseph, *Neuropsychology, Neuropsychiatry, and Behavioral Neurology* (New York: Plenum Press, 1990); R. Joseph, "The right cerebral hemisphere. Language, music, emotion, visual-spatial skills, body image, dreams, and awareness," *Journal of Clinical Psychology*, 44 (1988), 630–673; R. Joseph, "The neuropsychology of development: Hemispheric laterality, limbic language, and the origin of thought," *Journal of Clinical Psychology*, 38 (1982), 4–33.

2. E. Berne, *The Games People Play* (New York: Bantam Books, 1966).

3. S. de Beauvoir, *The Second Sex* (New York: Bantam Books, 1961); J. Croates, *Women, Men and Language* (London: Longman, 1986); W. Farrell, "The politics of vulnerability," In *The Forty-Nine Percent Majority*, D. S. David and R. Brannon (Eds.) (pp. 51–54) (Menlo Park: Addison Wesley, 1976); J. O. Balswick and C. W. Peek, "The inexpressible male." In *The Forty-Nine Percent Majority*, D. S. David and R. Brannon (Eds.) (Menlo Park: Addison Wesley, 1976); J. Brooks-Gunn and W. S. Matthews, *He and She: How Children Develop Their Sex Role Identity* (Englewood Cliffs, NJ: Prentice-Hall, 1979); R. C. Savin-Williams, "Dominance and submission among adolescent boys," In *Dominance Relations*, D. R. Omark et al. (Eds.) (New York: Garland Press, 1980); P. Ekert, "Cooperative competition in adolescent girl talk," *Discourse Processes*, 13 (1990), 1–12; W. Gaylin, *The Male Ego* (New York: Viking, 1962); J. M. Ross, *The Male Paradox* (New York: Simon & Schuster, 1992); L. Glass, *He Says, She Says* (New York: Putnam's Sons, 1992); R. Birdwhistell, "Masculinity and femininity as display." In *Kinesics and Context* (Philadelphia: University of Pennsylvania Press, 1970); R. Brend, "Male-female intonation patterns in American English." In *Language and Sex*. B. Thorne and N. Henley (Eds.) (Massachusetts: Newbury House Publishers, 1975); R. Coleman, "Male and female voice quality and its relationship to vowel formant frequencies," *Journal of Speech and Hearing Research*, 14 (1971), 123–133; C. Edelsky, "Question intonation and sex role," *Language and Sociology*, 8 (1979), 15–32; N. M. Henley, "Power, sex, and nonverbal communication," In *Language and Sex*, B. Throne and N. Henley (Eds.) (Massachusetts: Newbury House, 1975).

4. R. Joseph, "Competition between women," *Psychology*, 22 (1985), 1–11; R. Joseph, *The Right Brain and the Unconscious* (New York: Plenum Press, 1992a); D. Tannen, *You Just Don't Understand* (New York: Ballantine Books, 1990); D. Eder and M. Hallinan, "Sex differences in children's friendships," *American*

Sociological Review, 43 (1978), 237–250; W. C. McGrew, *An Ethological Study of Children's Behavior* (New York: Academic Press, 1979); Birdwhistell, 1970; Glass, 1992; Ekert, 1990; Farrell, 1976; Brend, 1975; Coleman, 1971; Edelsky, 1979; Henley, 1975; Balswick and Peek, 1976; Brooks-Gunn and Matthews, 1979; Gaylin, 1992; Ross, 1992.

5. Joseph, "The limbic system: Emotion, id, unconscious mind," *Psychoanalytic Review*, 1992b; Joseph, 1992a.

6. S. Freud, *The Psychopathology of Everyday Life*, C. G. Jung, *The Collected Works of C. G. Jung. Experimental Researches, Vol. 2* (Princeton: Princeton University Press, 1973a); C. G. Jung, *The Collected Works of C. G. Jung: The Structure and Dynamics of the Psyche* (Princeton: Princeton University Press, 1973b); C. G. Jung, *Collected Papers on Analytical Psychology* (London: Bailliere, Tidall and Cox, 1922); Joseph, 1992a.

7. S. R. Rochester, "The significance of pauses in spontaneous speech," *Journal of Psycholinguistic Research*, 2 (1973), 51–81; S. Liebersen, *Explorations in Sociolinguistics* (Bloomington: Indiana University Press, 1967); H. M. Hoenigswald, *Language Change and Linguistic Reconstruction* (Chicago: University of Chicago Press); H. Hoijer, *Language in Culture* (Chicago: University of Chicago Press, 1954); J. Fodor and J. Katz, *The Structure of Language* (Englewood Cliffs, NJ: Prentice-Hall, 1964); P. Farb, *Word Play* (New York: Bantam, 1975); Tannen, 1990.

8. S. Blumstein and W. E. Cooper, "Hemispheric processing of intonational contours," *Cortex*, 10 (1974), 146–158; J. W. Dwyer and W. E. Rinn, "The role of the right hemisphere in contextual inference," *Neuropsychologia*, 19 (1981), 479–482; C. Gilligan, *In a Different Voice* (Cambridge: Harvard University Press, 1982); Joseph, 1988, 1990; Tannen, 1990; Farb, 1975; Coleman, 1971; Edelsky, 1979; Brend, 1975; Thorne and Henley, 1975.

9. B. Emil and R. Stutman, "Sex role differences in the relational control of dyadic interaction," *Women's Studies in Communication*, 6 (1983), 96–103; Coleman, 1971; Edelsky,1979; Gilligan, 1982; Farb, 1975; Glass, 1992; Eder and Hallinan, 1978; Brooks-Gunn and Matthews, 1979; Tannen, 1990; Farrell, 1976; Balswick and Peek, 1976; Savin-Williams, 1980; Brend, 1975; Throne and Henley, 1975.

10. J. Lever, "Sex differences in the games children play," *Social Problems*, 23 (1976), 478–483; Brooks-Gunn and Matthews, 1979; McGrew, 1979; Glass, 1992; Farb, 1975.

11. Farb, 1975; Brend, 1975; Thorne and Henley, 1975; Coleman, 1971; Edelsky, 1979.

12. Brend, 1975; Edelsky, 1979; Thorne and Henley, 1975; Tannen, 1990; Glass, 1992.

13. Glass, 1992; Tannen, 1990; Gaylin, 1992; Ross, 1992.

14. R. Joseph and R. E. Gallagher, "Gender and early environmental influences on

activity, overresponsiveness, and exploration," *Developmental Psychobiology*, 13 (1980), 527–544; R. Joseph et al., "Effects of sex hormones manipulations and exploration on sex differences in learning," *Behavioral Biology*, 24 (1978), 364–377; Joseph, "Competition between women," *Psychology*, 22 (1985), 1–11; L. J. Harris, "Sex differences in spatial ability: Possible environment, genetic, and neurological factors." In *Asymmetrical Function of the Brain*, M. Kinsbourne (Ed.) (pp. 405–522) (New York: Cambridge University Press, 1978); D. M. Broverman et al., "Roles of activation and inhibition in sex differences in cognitive abilities," *Psychological Review*, 76 (1968), 328–331; J. McGlone, "Sex differences in human hemispheric laterality," *Behavioral Brain Sciences*, 4 (1982), 3–331; J. Durden-Smith and D. DeSimone, *Sex and the Brain* (New York: Warner Books, 1980); J. Levy and W. Heller, "Gender differences in human neuropsychological function." In *Handbook of Behavioral Neurobiology*, A. A. Gerall et al. (Eds.) (pp. 245–274) (New York: Plenum Press, 1992).

15. Harris, 1978; McGlone, 1982; Durden-Smith and DeSimone, 1985; Levy and Heller, 1992.

16. J. Lever, "Sex differences and the games children play," *Social Problems*, 23 (1976), 478–487; Glass, 1992; Croates, 1986; Farb, 1975; Farrell, 1976; Balswick and Peek, 1976.

17. Harris, 1978.

18. Farb, 1975; Glass, 1992.

19. Glass, 1992; Birdwhistell, 1970; Brend, 1975; Edelsky, 1979.

20. Savin-Williams, 1980; Birdwhistell, 1970; Ekert, 1990; Farb, 1975.

21. Harris, 1978; Broverman et al., 1968; Glass, 1992; Tannen, 1990; Ekert, 1990; Croates, 1986; Farrell, 1976; Brooks-Gunn and Matthews, 1979; Balswick and Peek, 1976.

22. Joseph, 1982, 1988, 1990.

23. Jung, 1973a; Freud, 1900, 1937; Joseph, 1982, 1990, 1992.

24. Farb, 1975.

25. J. V. McConnell, "Subliminal stimulation: An overview," *American Psychologist*, 13 (1958), 229–244.

26. Jung, 1973a; Freud, 1900, 1937.

27. Jung, 1973a.

28. Jung, 1973a.

29. Tannen, 1990; Farb, 1975; Rochester, 1973.

30. J. Harris, *I'm OK, You're OK* (New York: Avon, 1969); A. Bandura et al., "Imitation of film-mediated aggressive models," *Journal of Abnormal and Social Psychology*, 66 (1963), 3–11; Jung, 1973a; Berne, 1964; Joseph, 1992.

31. Jung, 1973a.

32. Jung, 1973a.

33. Joseph, 1992.

Chapter 10

1. P. Lieberman, "On the evolution of human syntactic ability," *Journal of Human Evolution*, 14 (1985), 657–668; P. Lieberman and E. Crelin, "On the speech of Neanderthal man," *Linguistic Inquiry*, 2 (1971), 203–222; P. Lieberman et al., "Phonetic ability and related anatomy of the newborn and adult human, Neanderthal man, and the chimpanzee," *American Anthropologist*, 74 (1972), 287–307.
2. R. Joseph, *Neuropsychology, Neuropsychiatry, and Behavioral Neurology* (New York: Plenum Press, 1990); R. Joseph, "The right cerebral hemisphere. Language, music, emotion, visual-spatial skills, body image, dreams, and awareness," *Journal of Clinical Psychology*, 44 (1988), 630–673; R. Joseph, "The neuropsychology of development. Hemispheric laterality, limbic language, and the origin of thought," *Journal of Clinical Psychology*, 38 (1982), 4–33; R. Joseph, "The limbic system," *Psychoanalytic Review*, 1992a; R. Joseph, *The Right Brain and the Unconscious* (New York: Plenum Press, 1992b).
3. Joseph, 1982, 1988, 1990, 1992a,b.
4. Joseph, 1982, 1988, 1990.
5. P. Maclean, *The Triune Brain in Evolution* (New York: Plenum Press, 1990); B. W. Robinson, "Vocalizations evoked from forebrain in Macaca mulatta," *Physiology and Behavior*, 2 (1967), 345–352; Joseph, 1982, 1988, 1990, 1992a,b.
6. Joseph, 1982, 1988, 1990, 1992; Maclean, 1990; Robinson, 1967.
7. K. H. Worner, *History of Music* (New York: Free Press, 1973).
8. W. H. Cade, Insect mating and courtship behavior. In *Comprehensive Insect Physiology, Biochemistry and Pharmacology*, G. A. Kerkut and L. I. Gilbert (Eds.) (pp. 591–620) (New York: Pergamon Press, 1985); W. Wickler, *The Sexual Code* (Garden City: Anchor, 1973).
9. Eibl-Eibesfeldt, *Ethology* (New York: Holt, 1975); Wickler, 1973.
10. Eibl-Eibesfeldt, 1975; Wickler, 1973; Cade, 1985.
11. Cade, 1985.
12. Eibl-Eibesfeldt, 1975; Wickler, 1973; Cade, 1985.
13. Eibl-Eibesfeldt, 1975; Wickler, 1973; Cade, 1985.
14. Eibl-Eibesfeldt, 1975.
15. M. L. Albert et al., "Melodic intonation therapy for aphasia," *Archives of Neurology*, 29 (1973), 334–339.
16. T. Alajoanine, "Aphasia and artistic realization," *Brain*, 71 (1948), 229–241; A. Luria, *The Working Brain* (New York: Basic Books, 1973); A. Luria, *Higher Cortical Functions in Man* (New York: Basic Books, 1980).
17. Alajoanine, 1948.
18. Joseph, 1982, 1988, 1990.

19. W. Penfield and P. Perot, "The brain's record of auditory and visual experience," *Brain*, 86 (1963), 595–696.

20. M. P. Bryden et al., "A left ear advantage for identifying the emotional quality of tonal sequences," *Neuropsychologia*, 20 (1982), 83–87; M. Safer and H. Leventhal, "Ear differences in evaluating emotional tones and verbal content," *Journal of Experimental Psychology, Human Perception, and Performance*, 3 (1977), 75–82; S. Blumstein and W. E. Cooper, "Hemispheric processing of intonational contours," *Cortex*, 10 (1974), 146–158; H. A. Sackheim and R. C. Gur, "Lateral asymmetry in intensity of emotional expression," *Neuropsychologia*, 16 (1978), 473–481; A. Mahoney and R. Sainsbury, "Hemispheric asymmetry in the perception of emotional sounds," *Brain and Cognition*, 6 (1987), 216–233.

21. Joseph, 1982, 1988, 1990, 1992a,b.

22. Bryden, 1982; A. Carmon and I. Nachshon, "Ear asymmetry in perception of emotional non-verbal stimuli," *Acta Psychologica*, 37 (1973), 351–357; A. Gates and J. L. Bradshaw, "The role of the cerebral hemispheres in music," *Brain and Language*, 3 (1977), 451–460; H. W. Gordon, "Hemispheric asymmetries in the perception of musical chords," *Cortex*, 6 (1970), 387–398; H. Gordon and J. E. Bogen, "Hemispheric lateralization of singing after intracarotid sodium amylobarbitone," *Journal of Neurology, Neurosurgery, and Psychiatry*, 37 (1974), 727–737; D. Bowers, et al., "Comprehension of emotional prosody following unilateral hemisphere lesions: Processing defect versus distraction defect," *Neuropsychologia*, 25 (1987), 317–328; Mahoney and Sainsbury, 1987; Joseph, 1982, 1988, 1990.

23. O. Spreen, et al., "Auditory agnosia without aphasia," *Archives of Neurology*, 13 (1965), 84–92; M. Albert et al., "A case of auditory agnosia: Linguistic and nonlinguistic processing," *Cortex*, 8 (1972), 427–443; K. M. Heilman et al., "Comprehension of affective and nonaffective prosody," *Neurology*, 34 (1984), 917–921; Joseph, 1982, 1988, 1990.

24. J. Bradshaw et al., "Braille reading and left and right hemisphere," *Neuropsychologia*, 20 (1982), 493–500; R. Campbell, "Asymmetries in interpreting and expressing a posed facial expression," *Cortex*, 15 (1978), 327–342; A. Carmon and H. P. Bechtoldt, "Dominance of the right cerebral hemisphere for stereopsis," *Neuropsychologia*, 7 (1969), 29–39; A. Carmon and A. L. Benton, "Tactile perception of direction and number in patients with unilateral cerebral diseases," *Neurology*, 19 (1969), 525–532; S. Corkin et al., "Somatosensory thresholds: Contrasting effects of post-central gyrus and posterior parietal-lobe excisions." *Archives of Neurology*, 23 (1970), 41–58; S. DeKosky et al., "Recognition and discrimination of emotional faces and pictures," *Brain and Language*, 9 (1980), 206–214; E. DeRenzi, *Disorder of Space Exploration and Cognition* (New York: Wiley, 1982); E. DeRenzi, "Prosopagnosia in two patients

with CT-scan evidence of damage confined to the right hemisphere," *Neuropsy-chologia*, 24 (1986), 385–389; E. DeRenzi et al., "The performance of patients with unilateral brain damage on face recognition tasks," *Cortex*, 4 (1986), 17–34; E. DeRenzi and G. Scotti, "The influence of spatial disorders in impairing tactual discrimination of shapes," *Cortex*, 5 (1969), 53–62; J. E. Desmedt, "Active touch exploration of extrapersonal space elicits specific electrogenesis in the right cerebral hemisphere of intact right-handed man," *Proceedings of the National Academy of Sciences*, 74 (1977), 4037–4040; Joseph, 1982, 1988, 1990.

25. Joseph, 1982, 1988, 1990.
26. G. Arrigoni and E. DeRenzi, "Constructional apraxia and hemispheric locus of lesion," *Cortex*, 1 (1964), 170–197; D. Benson and M. Barton, "Disturbances in constructional ability," *Cortex*, 6 (1970), 19–46; D. Benson et al., 1976; A. Benton, "Visuoperceptive, visuospatial and visuoconstructive disorders," In *Clinical Neuropsychology*, K. M. Heilman and E. Valenstein (Eds.) (pp. 186–232) (Oxford: Oxford University Press, 1979); E. Bisiach and C. Luzzatti, "Unilateral neglect of representational space," *Cortex*, 14 (1978), 129–133; F. Black and B. A. Bernard, "Constructional apraxia as a function of lesion locus and size in patients with focal brain damage," *Cortex*, 20 (1984), 111–120; F. W. Black and R. L. Strub, "Constructional apraxia in patients with discrete missile wounds of the brain." *Cortex*, 12 (1976), 212–220; N. Butters and M. Barton, "Effect of parietal lobe damage on the performance of reversible operations in space," *Neuropsychologia*, 8 (1970), 205–214; R. Calvanio ct al., "Left visual spatial neglect is both environment-centered and body-centered," *Neurology*, 37 (1987), 1179–1183.
27. C. Hartshorne, *Born to Sing: An Interpretation and World Survey of Bird Songs* (London: Indiana University Press, 1973); C. K. Catchpole, *Vocal Communication in Birds* (Baltimore: University Park Press, 1979).
28. Hartshorne, 1973.

Chapter 11

1. A. L. Burnett and N. A. Diehl, "The nervous system of hydra," *Journal of Experimental Zoology*, 157 (1964), 237–250; J. W. Papez, *Comparative Neurology* (New York: Hafner, 1967); C. U. Ariens Kappers, *The Evolution of the Nervous System* (Bohn: Haarlem, 1929); T. L. Lentz, *Primitive Nervous Systems* (New Haven: Yale University Press, 1968); G. H. Bishop, "Natural history of nerve impulses," *Physiological Review*, 36 (1956), 376–399; M. C. Behrens, "The electrical response of the planarian photoreceptor," *Comparative Biochemistry and Physiology*, 5 (1969), 129–138; J. V. McConnel et al., "The

effects of regeneration upon retention of a conditioned response in the pla-narian," *Journal of Comparative and Physiological Psychology*, 52 (1959), 1–5; J. V. McConnel et al., "The effects of ingestion of conditioned planaria on the response level of naive planaria," *Worm Runners Digest*, 3 (1961), 41–47; A. L. Jacobson, "Learning in flatworms and annelids," *Psychological Bulletin*, 60 (1963), 74–94.

2. Burnett, A. L. and Diehl, N. A., "The nervous system of hydra," *Journal of Experimental Zoolology*, 157 (1964), 237–250; G. H. Bishop, "Natural history of nerve impulses," *Physiological Review*, 36 (1956), 376–399; Papez, 1967; Ariens Kappers, 1929; Lentz, 1968; McConnel et al., 1959, 1961; Jacobson, 1963.

3. Papez, 1967; Ariens Kappers, 1929; Lentz, 1968.

4. R. Joseph, *Neuropsychology, Neuropsychiatry, and Behavioral Neurology* (New York: Plenum Press, 1990); R. Joseph, "The limbic system," *Psychoanalytic Review*, (1992); B. Milner, "Amnesia following operations on the temporal lobe," In *Amnesia* (pp. 75–89). R. B. Kesner and R. G. Andrus, "Amygdala stimulation disrupts the magnitude of reinforcement contribution to long-term memory," *Physiological Psychology*, 10 (1982), 55–59; J. L. McGaugh et al. (Eds.), *Brain Organization and Memory* (New York: Oxford University Press, 1990); R. J. Douglas, "The hippocampus and behavior," *Psychological Bulletin*, 67 (1967), 416–442; J. Green and A. Arduini, "Hippocampal electrical activity in arousal," *Journal of Neurophysiology*, 17 (1954), 533–557; A. Routtenberg, "The two arousal hypothesis: Reticular formation and limbic system," *Psychological Review*, 75 (1968), 51–80; E. Grastyan et al., "Hippocampal electrical activity during the development of conditioned reflexes," *EEG and Clinical Neurophysiology*, 11 (1959), 409–430; W. B. Scoville and B.Milner, "Loss of recent memory after bilateral hippocampal lesions," *Journal of Neurology, Neurosurgery, and Psychiatry*, 20 (1967), 11–21; Joseph, 1990, 1992.

5. P. Gloor, "Amygdala." In *Handbook of Physiology*. J. Field (Ed.) (pp. 300–370) (Washington, DC: American Physiological Society, 1960); H. Ursin and B. R. Kaada, "Functional localization within the Amygdaloid complex," *EEG and Clinical Neurophysiology*, 12 (1960), 1–20; J. O'Keefe and H. Bouma, "Complex sensory properties of certain amygdala units in the freely moving cat," *Experimental Neurology*, 23 (1969), 384–398; S. Feldman, "Neurophysiological mechanisms modifying afferent hypothalamus-hippocampal condition," *Experimental Neurology*, 5 (1962), 269–291; J. Green and W. R. Adey, "Electro-physiological studies of hippocampal connections and excitability," *EEG and Clinical Neurophysiology*, 8 (1956), 245–262; Routtenberg, 1968; Grastyan et al., 1959; F. K. Redding, "Modification of sensory cortical evoked potentials by hippocampal stimulation," *EEG and Clinical Neurophysiology*, 22 (1967), 74–83; Milner, 1970; Scoville and Milner, 1957; Kesner and Andrus, 1952; Douglas, 1967; Green and Arduini, 1954; Joseph, 1990, 1992.

6. Feldman, 1962; Green and Adey, 1956; Douglas, 1967; Green and A. Arduini, 1954; Routtenberg, 1968; Joseph, 1990, 1992.
7. Douglas, 1967; Green and A. Arduini, 1954; Routtenberg, 1968; Grastyan et al., 1959; Joseph, 1990, 1992.
8. R. F. Schmidt, *Fundamentals of Neurophysiology* (New York: Springer-Verlag, 1978); J. R. Cooper et al., *The Biochemical Basis of Neuropharmacology* (New York: Oxford University Press, 1984); G. Lynch, *Synapses, Circuits, and the Beginnings of Memory* (Cambridge: MIT Press, 1986); McGaugh et al., 1990.
9. G. L. Wells and E. F. Loftus, *Eyewitness Testimony* (New York: Cambridge University Press, 1984); H. D. Ellis, "Practical aspects of face memory," In *Eyewitness Testimony*, G. L. Wells and E. F. Loftus (Eds.) (pp. 12–27) (New York: Cambridge University Press, 1984); D. F. Hall, E. F. Loftus, and J. P. Tousignant, "Postevent information and changes in recollection of a natural event." In *Eyewitness Testimony*, G. L. Wells and E. F. Loftus (Eds.) (pp. 124–141) (New York: Cambridge University Press, 1984).
10. Wells and Loftus, 1984.
11. Hall, Loftus, and Tousignant, 1984.
12. B. Milner, 1970; Scoville and Milner, 1957.
13. Joseph, 1990, 1992.
14. J. A. Horel, "The neuroanatomy of amnesia," *Brain* 101 (1978), 403–445; Kesner and Andrus, 1982; Gloor, 1860; Ursin and Kaada, 1960; Green and Adey, 1956; Douglas, 1967; Green and Arduini, 1954; Routtenberg, 1968; Joseph, 1990, 1992.
15. Joseph, 1990, 1992.
16. Joseph, 1982, 1988, 1990, 1992; Joseph et al., 1984.
17. J. Bures and O. Buresova, "The use of Leao's spreading depression in the study of interhemispheric transfer of memory traces," *Journal of Comparative and Physiological Psychology*, 59 (1960), 211–214; G. L. Risse and M. S. Gazzaniga, "Well-kept secrets of the right hemisphere: A carotid amytal study of restricted memory transfer," *Neurology*, 28 (1979), 950–953; R. W. Doty and W. H. Overman, "Mnemonic role of forebrain commissures in macaques." In *Lateralization in the Nervous System*, S. Harnad, R. W. Doty, L. Goldstein, J. Jaynes, and G. Krauthamer (Eds.) (pp. 75–88) (New York: Academic Press, 1977); G. L. Risse and M. S. Gazzaniga, "Well kept secrets of the right hemisphere," *Neurology*, 28 (1979), 950–953.
18. A. R. Lecours, "Myelogenetic correlates of the development of speech and language. In *Foundations of Language Development*, E. Lenneberg and E. Lenneberg (Eds.) (New York: Academic Press, 1975); P. I. Yakovlev and A. Lecours, "The myelogenetic cycles of regional maturation of the brain," In *Regional Development of the Brain in Early Life*, A. Minkowski (Ed.) (London: Blackwell, 1967).

19. R. Joseph and R. E. Gallagher, "Interhemispheric transfer and the completion of reversible operations in non-conserving children," *Journal of Clinical Psychology*, 41 (1985), 796–800; R. E. Gallagher and R. Joseph, "Non-linguistic knowledge, hemispheric laterality, and the conservation of inequivalence," *Journal of General Psychology*, 107 (1982), 31–40; R. Joseph et al., "Two brains, one child. Interhemispheric transfer deficits and confabulation in children aged 4, 7, 10," *Cortex*, 20 (1984), 317–331; D. Galin et al., "Development of the capacity for tactile information transfer between hemispheres in normal children," *Science*, 204 (1979), 1330–1332; M. Finlayson, "A behavioral manifestation of the development of interhemispheric transfer at learning in children," *Cortex*, 12 (1975), 290–295; D. S. O'Leary, "A developmental study of interhemispheric transfer in children aged 5 to 10," *Child Development*, 51 (1980), 743–750.
20. Joseph et al., 1984.

Chapter 12

1. J. W. Papez, *Comparative Neurology* (New York: Hafner, 1967); C. U. Ariens-Kappers, *The Evolution of the Nervous System* (Bohn: Haarlem, 1929); T. L. Lentz, *Primitive Nervous Systems* (New Haven: Yale University Press, 1968).
2. R. Broom, *Mammal-like Reptiles of South Africa and the Origin of Mammals* (London: Witherby, 1932); A. Brink, "Speculation on some advanced mammalian characteristics in the higher mammal-like reptiles," *Palaeontology*, 4 (1956), 77–95; E. F. Allin, "The auditory apparatus of advanced mammal-like reptiles and early mammals." In *The Ecology and Biology of Mammal-like Reptiles*, N. Hotton et al. (Eds.) (283–294) (Washington, DC: Smithsonian Institution Press, 1986); A. W. Crompton and F. A. Jenkins, "Mammals from reptiles," *Annual Review of Earth and Planetary Sciences*, 1 (1973), 131–155; A. Crompton and F. Jenkins, "Origin of mammals." In *Mesozoic Mammals*, J. A. Lillegraven et al. (Eds.) (Berkeley: University of California Press, 1979).
3. J. H. Ostrom, "Social and unsocial behavior in dinosaurs." In *Evolution of Animal Behavior*, J. H. Ostrom (Ed.) (pp. 41–61) (Oxford: Oxford University Press, 1986); D. K. Thomas and E. D. Olson (Eds.), *A Cold Look at the Warm Blooded Dinosaurs* (Washington, DC: AAAS, 1980); R. T. Bakker, "Dinosaur physiology and the origin of mammals," *Evolution*, 25 (1971), 636–658. R. T. Bakker, "The return of the dancing dinosaurs," In *Dinosaurs Past and Present*, S. J. Czerkas and E. C. Olson (Eds.) (Los Angeles: LA County Natural History Museum, 1987); J. Hopson, "Relative brain size in dinosaurs." In *A Cold Look at the Warm Blooded Dinosaurs*, D. K. Thomas and E. D. Olson (Eds.) (Washington, DC: AAAS, 1980); G. S. Paul, *Predatory Dinosaurs* (New York:

Simon & Schuster, 1988); R. Bakker, "Dinosaur physiology and the origin of mammals," *Evolution*, 25 (1971), 636–658; R. D. K. Thomas and E. C. Olson (Eds.), *A Cold Look at the Warm Blooded Dinosaurs* (Boulder: Westview Press, 1980); J. C. Quiroga, "The brain of two mammal-like reptiles," *J. Hirnforsch*, 20 (1979), 341–350; J. C. Quiroga, "The brain of the mammal-like reptile, Probainognatheus jenseni. A correlative paleo-neurological approach to the neocortex at the reptile-mammal transition," *J. Hirnforsch*, 21 (1980), 299–336; A. Keith, "Review of the mammal-like reptiles of South Africa and the origin of mammals," *Journal of Anatomy*, 66 (1932), 669–671; Brink, 1956.

4. P. MacLean, *The Triune Brain in Evolution* (New York: Plenum Press, 1990); Brink, 1956; Allin, 1986; Quiroga, 1979, 1980.

5. R. T. Bakker, "Tetrapod mass extinctions." In *Patterns of Evolution*, A. Hallem (Ed.) (pp. 339–468) (Amsterdam: Elsevier, 1977); M. Davis et al., "Extinction by periodic comet showers," *Nature*, 308 (1984), 715–717; B. K. Napier and V. Clube, "A theory of terrestrial catastrophism," *Nature*, 282 (1979), 455–460.

6. Bakker, 1977; Davis et al., 1984; Napier and Clube, 1979.

7. Crompton et al., 1973, 1979; Bakker, 1971.

8. W. Alvarez, "Toward a theory of impact crisis," *Eos*, 131 (1986), 248–250.

9. R. Leakey and R. Lewin, *Origins* (New York: Dutton, 1977); D. Johanson and J. Shreeve, *Lucy's Child* (New York: Morrow, 1989); D. Johanson and M. Edey, *Lucy: The Beginnings of Humankind* (New York: Simon & Schuster, 1981); J. Pfeiffer, *The Emergence of Humankind* (New York: Harper & Row, 1985).

10. Leakey and Lewin, 1977; Johanson and Shreeve, 1989; Johanson and Edey, 1981; Pfeiffer, 1985.

11. R. E. Leakey, "Hominids of Africa," *American Scientist*, 64 (1976), 174–178; C. O. Lovejoy, "Evolution of human walking," *Scientific American*, 11 (1988), 118–125; Leakey and Lewin, 1977; Johanson and Shreeve, 1989; Johanson and Edey, 1981; Pfeiffer, 1985.

12. Pfeiffer, 1985; Leakey and Lewin, 1977.

13. Pfeiffer, 1985; Leakey and Lewin, 1977.

14. W. H. Kimbel and T. D. White, "A reconstruction of the adult cranium of Australopithecus afarensis," *American Journal of Physical Anthropology*, 52 (1980), 224; Johanson and Shreeve, 1989; Johanson and Edey, 1981; Pfeiffer, 1985; Leakey and Lewin, 1977.

15. R. A. Dart, "The predatory transition from ape to man," *International Anthropological and Linguistics Review* 1 (1953), 201–217.

16. L. R. Binford, *Bones: Ancient Men and Modern Myths* (New York: Academic Press, 1981).

17. J. Goodall, *The Chimpanzees of Gombe* (Cambridge: Harvard University Press, 1986); J. Goodall, *Through a Window* (Boston: Houghton Mifflin, 1990); G. P. Schaller, *The Mountain Gorilla* (Chicago: University of Chicago Press, 1963);

A. L. Zihlman et al., "Pygmy chimpanzee as a possible prototype for the common ancestor of humans, chimpanzees and gorillas," *Nature*, 275 (1978), 744–746; V. M. N. Sarich, "Primate systematics." In *Old World Monkeys*, J. R. Napier and P. H. Napier (Eds.) (pp. 175–226) (New York: Academic Press, 1970).

18. Johanson and Shreeve, 1989; Johanson and Edey, 1981.
19. Pfeiffer, 1985.
20. Johanson and Shreeve, 1989; Johanson and Edey, 1981.
21. Leakey and Lewin, 1977.
22. J. K. Wolpoff, "Modern Homo sapiens origins." In *The Origins of Modern Humans*, F. Smith and F. Spender (Eds.) (pp. 411–483). (New York: Liss, 1986); H. Vallada et al., "Thermoluminescence dating of Mousterian 'Proto-Cro-Magnon' remains from Israel and the origin of modern man," *Nature*, 331 (1988), 614–616.
23. MacLean, 1990.
24. Wolpoff, 1986; Vallada et al., 1988.
25. R. Potts, "Home bases and early hominids," *American Scientist*, 72 (1984), 338–347; Leakey and Lewin, 1977; Pfeiffer, 1985; J. D. Clark and J. W. K. Harris, "Fire and its roles in early hominid lifeways," *African Archaeology Review*, 3 (1985), 3–27.
26. G. P. Rightmire, *The Evolution of Homo Erectus* (New York: Cambridge University Press, 1990); P. V. Tobias, *The Brain in Hominid Evolution* (New York: Columbia University Press, 1971).
27. Rightmire, 1990; Joseph, 1992.
28. S. J. Washburn and C. Lancaster, "The evolution of hunting." In *Man the Hunter*, R. Lee and I. DeVore (Eds.) (pp. 293–303) (Chicago: Aldine, 1968); R. Ardrey, *The Hunting Hypothesis* (New York: Bantam, 1977); G. Isaac and D. C. Crader, "To what extent were early hominids carnivorous," In *Omnivorous Primates*, R. S. O. Harding and G. Telecki (Eds.) (pp. 37–103) (New York: Columbia University Press, 1981); H. Kaplan and K. Hill, "Hunting ability and reproductive success among male ache foragers," *Current Anthropology*, 26 (1985), 131–133; G. P. Murdock and C. Provost, "Factors in the division of labor by sex," *Ethnology*, 12 (1981), 203–235; A. L. Zilman, "Women as shapers of the human adaptation." In *Woman the Gatherer*, F. Dahlberg (Ed.) (pp. 75–120) (New Haven: Yale University Press, 1981); M. K. Martin and B. Voorhies, *Females of the Species* (New York: Columbia University Press, 1975); S. Slocum, "Women in Cross-Cultural Perspective." In *Toward an Anthropology of Women*, R. R. Reiter (Ed.) (pp. 36–50) (New York: Monthly Review Press, 1975); M. K. Roper, "A survey of evidence for intrahuman killing during the Pleistocene," *Current Anthropology*, 10 (1969), 427–459; L. Fedigan, *Primates Paradigms: Sex Roles and Social Bonds* (Montreal: Eden Press, 1982); J. K.

Brown, "A note on the division of labor by sex," *American Anthropologist*, 72 (1970), 1073–1078; W. H. Oswalt, *An Anthropological Analysis of Food Getting Technology* (New York: Wiley, 1976); F. Dahlberg (Ed.), *Woman the Gatherer* (New Haven: Yale University Press, 1981).

29. A. Leroi-Gourhan, *Treasure of Prehistoric Art* (New York: Harry N. Abrams, 1964); H. Vallada et al., "Thermoluminescence dating of Mousterian 'Proto-Cro-Magnon' remains from Israel and the origin of modern man," *Nature*, 331 (1988), 614–616; H. G. Bandi, *Art of the Stone Age* (New York: Crown Publishers, 1961); Pfeiffer, 1985; T. Prideaux, *Cro-Magnon Man* (New York: Time-Life Books, 1973).

30. G. Constable, *Neanderthal Man* (New York: Time-Life Books, 1973); E. Trinkaus, *The Shanidar Neanderthals* (New York: Academic Press, 1983); R. S. Solecki, *Shanidar: The First Flower People* (New York: Knopf, 1971).

31. Leroi-Gourhan, 1964.

32. Solecki, 1971.

33. P. Lieberman and E. S. Crelin, "On the speech of Neanderthal man," *Linguistic Inquiry*, 2 (1971), 203–222.

34. Joseph, 1992.

35. Constable, 1973; Pfeiffer, 1985.

36. Pfeiffer, 1985; Prideaux, 1973.

37. Leroi-Gourhan, 1964; Bandi, 1961; Pfeiffer, 1985; Prideaux, 1973.

38. Joseph, 1992.

39. Joseph, 1982, 1988, 1990, 1992.

40. D. Falk, "Brain lateralization in primates and its evolution in hominids," *Yearbook of Physical Anthropology*, 30 (1987a), 107–125.

41. R. Lee and I. DeVore (Eds.), *Man the Hunter* (Chicago: Aldine, 1968); R. Harding and G. Teleki (Eds.), *Omnivorous Primates* (New York: Columbia University Press, 1981); K. Hill, "Hunting and hominid evolution," *Journal of Human Evolution*, 11 (1982), 521–544; S. J. Washburn and C. S. Lancaster, "The evolution of hunting." In *Man the Hunter*, R. Lee and I. DeVore (Eds.) (pp. 293–303) (Chicago: Aldine, 1968); R. Ardrey, *The Hunting Hypothesis* (New York: Bantam, 1977); G. Isaac and D. C. Crader, "To what extent were early hominids carnivorous." In *Omnivorous Primates*, R. S. O. Harding and G. Telecki (Eds.) (pp. 37–103) (New York: Columbia University Press, 1981); T. Gibson, "Meat sharing as a political ritual." In *Hunters and Gatherers*, T. Ingold et al. (Eds.) (pp. 165–180) (New York: Berg, 1988); H. Kaplan and K. Hill, "Hunting ability and reproductive success among male ache foragers," *Current Anthropology*, 26 (1985), 131–133; G. P. Murdock and C. Provost, "Factors in the division of labor by sex," *Ethnology*, 12, 203–235; J. Goodall, *Tiwi Wives* (Seattle: University of Washington Press, 1971); G. P. Murdock, "Comparative data on the division of labor by sex," *Social Forces*, 16, 551–553;

M. K. Martin and B. Voorhies, *Females of the Species* (New York: Columbia University Press, 1975); S. Slocum, "Women in cross-cultural perspective." In *Toward an Anthropology of Women*, R. R. Reiter (Ed.) (pp. 36–50) (New York: Monthly Review Press, 1975); L. Fedigan, *Primate Paradigms: Sex Roles and Social Bonds* (Montreal: Eden Press, 1982); J. Brown, "A note on the division of labor by sex," *American Anthropologist*, 72 (1970), 1073–1078; R. A. Dart, 1953.

42. Joseph, 1992.
43. Washburn and Lancaster, 1968; Gibson, 1988; Kaplan and Hill, 1985; Martin and Voorhies, 1975; Slocum, 1975; Brown, 1970; Oswalt, 1976; Dahlberg, 1981.
44. L. J. Harris, "Sex differences in spatial ability: Possible environmental, genetic, and neurological factors." In *Asymmetrical Function of the Brain*, M. Kinsbourne (Ed.) (pp. 405–522) (New York: Cambridge University Press, 1978); D. M. Broverman et al., "Roles of activation and inhibition in sex differences incognitive abilities," *Psychological Review*, 76 (1968), 328–331.
45. Harris, 1978; Broverman et al., 1968.
46. J. Wada et al., "Cerebral hemisphere asymmetry in humans: Cortical speech zones in 100 adult and 100 infant brains," *Archives of Neurology*, 32 (1975), 239–246.
47. J. Levy and W. Heller, "Gender differences in human neuropsychological function." In *Handbook of Behavioral Neurobiology*, A. Gerall et al. (Eds.) (New York: Plenum Press, 1992); J. McGlone, "Sex differences in human hemispheric laterality," *Behavioral Brain Sciences*, 4 (1982), 3–33; J. Durden-Smith and D. DeSimone, *Sex and the Brain* (New York: Warner Books, 1980); Levy, 1990.
48. Harris, 1978; Broverman et al., 1968; McGlone, 1982; Durden-Smith and DeSimone, 1980.
49. Joseph, 1982, 1988, 1990.
50. Joseph, 1992.

Index